The Influence of Antibiotics on the Host-Parasite Relationship II

Edited by

D. Adam H. Hahn and W. Opferkuch

With 98 Figures and 60 Tables

Springer-Verlag
Berlin Heidelberg New York Tokyo

Professor Dr. Dr. med. Dieter Adam
Kinderklinik der Universität München
Lindwurmstraße 4
8000 München 2

Professor Dr. med. Helmut Hahn
Institut für Medizinische Mikrobiologie
Freie Universität Berlin
Hindenburgdamm 27
1000 Berlin 45

Professor Dr. med. Wolfgang Opferkuch
Abteilung für Theoretische und Klinische Medizin
Lehrstuhl für Medizinische Mikrobiologie und Immunologie
Ruhr-Universität Bochum
Universitätsstraße 150
4630 Bochum 1

ISBN-13: 978-3-642-70750-6 e-ISBN-13: 978-3-642-70748-3
DOI: 10.1007/978-3-642-70748-3

Softcover reprint of the hardcover 1st edition 1985

2127/3140-543210

Preface

The Second International Symposium on "The Influence of Antibiotics on the Host-Parasite Relationship" was held in Munich, F.R.G., from March 28 to 30, 1985. The topics of the meeting dealt with the aspects of changes in bacterial metabolism and structure which occur under the influence of antibiotics, and with the effects of such changes on the antibacterial host resistance. The influence on pathogenicity factors, changes in the outer membrane of bacteria, as well as the influence on the individual components of the defence system were analysed in detail. In addition, these studies showed that antibiotics proved to be an excellent tool for the examination of bacterial physiology, so that, 50 years after the introduction of antibiotics, additional important knowledge can be gained about the effect of these substances on bacteria. Considering the observations reported, it appears justifiable to postulate that new antibiotics should be routinely tested with respect to their possible effects on antiinfectious resistance. Of course, a consensus will have to be found on which to base methods and criteria employed.

The symposium documented an increasing interest of microbiologists and clinicians for this field of research. It would not have been possible to organize it without the substantial support of the Paul Ehrlich Society as well as of Squibb-Von Heyden Pharma, Inc. Particular help concerning the organization has been given by Werner Kremer of Squibb-Von Heyden Pharma. The organizing committee would like to express its gratitude and appreciation for this financial and personal help. It also thanks Bayer (Wuppertal), Boehringer (Mannheim), Grünenthal-Takeda (Aachen), Merck (Darmstadt), Merck, Sharp & Dohme (Munich), Pfizer (Karlsruhe), and Squipp-Von Heyen (Munich) for supporting the publication of these proceedings. The editors hope that this publication will help to spread the knowledge and interest about these interesting topics as widely as possible.

October 1985

D. Adam
H. Hahn
W. Opferkuch

Contents

b) Phagocytosis

IV. The Influence of Antibiotics on the Pathogenicity Factors and Virulence

a) Adherence

b) Toxin Production and Virulence

V. Miscellaneous

List of Contributors

Anwar, H., Birmingham, United Kingdom

Baker, N. R., Columbus, Ohio, USA

Beachey, E. H., Memphis, Tennessee, USA

Belsheim, J., Gävle, Sweden

Borinski, R., Jerusalem, Israel

Brown, M. R. W., Birmingham, United Kingdom

Büscher, K. H., Bochum, Federal Republic of Germany

Casey, S. W., Lubbock, Texas, USA

Ciborowski, P., Warsaw, Poland

Courtney, H. S., Memphis, Tennessee, USA

Dalhoff, A., Wuppertal, Federal Republic of Germany

Döring, G., Tübingen, Federal Republic of Germany

Friedman, H., Tampa, Florida, USA

Geers, T., Columbus, Ohio, USA

Gemmell, C. G., Glasgow, United Kingdom

Ghione, M., Milano, Italy

Giesbrecht, P., Berlin, Federal Republic of Germany

Gillissen, G., Aachen, Federal Republic of Germany

Ginsburg, I., Jerusalem, Israel

Gliozzi, T., Milano, Italy

Gnarpe, H., Gävle, Sweden

Goroncy-Bermes, P., Bochum, Federal Republic of Germany

Hahn, H., Berlin, Federal Republic of Germany

Hentges, D. J., Lubbock, Texas, USA

IACONIS, J. P., Lubbock, Texas, USA

IIDA-TANAKA, K., Kagawa, Japan

IRINO, S., Kagawa, Japan

JELJASZEWICZ, J., Warsaw, Poland

JOYCE, D., Bronx, New York, USA

KADURUGAMUWA, J. L., Birmingham, United Kingdom

KARCH, H., Bochum, Federal Republic of Germany

KLIMETZEK, V., Wuppertal, Federal Republic of Germany

KO, H. L., Cologne, Federal Republic of Germany

KROLL, H.-P., Wuppertal, Federal Republic of Germany

KUBENS, B., Bochum, Federal Republic of Germany

LAHAV, M., Jerusalem, Israel

LEYING, H., Bochum, Federal Republic of Germany

LORIAN, V., Bronx, New York, USA

MANDELL, L. A., Hamilton, Ontario, Canada

MARRE, R., Lübeck, Federal Republic of Germany

MARTIN, H. H., Darmstadt, Federal Republic of Germany

McGEE, Z. A., Salt Lake City, Utah, USA

MEIXELSPERGER, G., Milano, Italy

MELZER, B., Aachen, Federal Republic of Germany

NAGAYAMA, A., Saga, Japan

NEALON, T. J., Memphis, Tennessee, USA

NIXDORFF, K., Darmstadt, Federal Republic of Germany

O'BRIEN, A., Bethesda, Maryland, USA

OPFERKUCH, W., Bochum, Federal Republic of Germany

PAWELZIK, M., Bochum, Federal Republic of Germany

PELIZZONI, G., Milano, Italy

PULVERER, G., Cologne, Federal Republic of Germany

QUE, J. U., Lubbock, Texas, USA

ROLFE, R. D., Lubbock, Texas, USA

ROSZKOWSKI, K., Warsaw, Poland

ROSZKOWSKI, W., Warsaw, Poland

SADOVNIK, M., Jerusalem, Israel

SCHUMACHER-PERDREAU, F., Cologne, Federal Republic of Germany

SHAULI, S., Jerusalem, Israel

SIEGMUND-SCHULTZE, N., Darmstadt, Federal Republic of Germany

SIMPSON, W. A., Memphis, Tennessee, USA

SPERLING, U., Berlin, Federal Republic of Germany

STEPHENS, D. S., Atlanta, Georgia, USA

STÜBNER, G., Hannover, Federal Republic of Germany

TANAKA, T., Kagawa, Japan

TOSCH, W., Basel, Switzerland

VERHOEF, J., Utrecht, The Netherlands

VERINGA, E., Utrecht, The Netherlands

VOIGT, W. H., Wuppertal, Federal Republic of Germany

WARREN, R. L., Dayton, Ohio, USA

WECKE, J., Berlin, Federal Republic of Germany

WIEMER, C., Bochum, Federal Republic of Germany

WOS, B., Berlin, Federal Republic of Germany

ZAK, O., Basel, Switzerland

I. The Influence of Antibiotics on the Morphology and the Surface of Bacteria

Effect of Subinhibitory Concentrations of Antibiotics on Surface Proteins of *Neisseria gonorrhoeae* and *Neisseria meningitidis*: Changes that Alter Attachment to Human Cells*

D. S. STEPHENS[1], and Zell A. McGEE[2]

Veterans Administration Medical Center (Atlanta), Decatur, Georgia 30033, and Department of Medicine, Emory University School of Medicine, Atlanta, Georgia 30303[1] and the Center for Infectious Diseases, Diagnostic Microbiology and Immunology, University of Utah School of Medicine, Salt Lake City, Utah 84132[2]

Introduction

Studies of antibiotic-bacteria interactions have focused upon concentrations of antibiotics that kill or inhibit the growth of bacteria. Recent studies have shown that antibiotics in concentrations which fail to inhibit growth nonetheless may have marked effects on the structural characteristics of bacteria [1–5]. The changes resulting from exposure to subinhibitory concentrations of antibiotics may alter virulence rather than growth of the bacteria [6–10]. These results suggest that subinhibitory concentrations of antibiotics, whose modes of action are well known, can be used to examine the production, structure or function of bacterial surface components that are important in virulence. We used subinhibitory concentrations of antibiotics as probes to study the surface proteins of pathogenic *Neisseria spp.* that are important mediators of meningococcal and gonococcal attachment to human cells [11, 12].

Materials and Methods

Colony type 1 (T_1, piliated) and colony type 4 (T_4, nonpiliated) colonies of three isolates of *Neisseria gonorrhoeae* and piliated colonies of three isolates of *Neisseria meningitidis* were used in these studies [11]. The identification of the isolates as *N. gonorrhoeae* or *N. meningitidis*, the strain designation, source and additional characteristics of these isolates have been previously described [11, 13, 14]. Transparent and opaque colony variants of each strain were derived by single-colony passage. Media for cultivating meningococci and gonococci and for tube and agar dilution sensitivities were employed as previously described [11].

Penicillin, tetracycline and spectinomycin were used in these studies. The minimal inhibitory concentration (MIC) of the antibiotics for the meningococcal and gonococcal isolates was determined by two-fold dilution of the antibiotics in broth or

* This work was supported by the Medical Research Service of the Veterans Administration and by Public Health Research grant AI 19818 and AI 20265 from the National Institutes of Health

The Influence of Antibiotics on the
Host-Parasite Relationship II
Edited by D. Adam, H. Hahn, W. Opferkuch

agar. The effect of ½ MIC or less of the antibiotics on surface morphology was determined by transmission electron microscopy (TEM) using negatively-stained preparations [15]. The biochemical effects of subinhibitory antibiotics on surface proteins was determined by sodium dodecyl sulfate-polyacrylamide gel electrophoresis (SDS-PAGE) analysis of outer membrane preparations [11, 14]. Different solubilization conditions were employed, and the proteins were detected by silver staining.

Results

Surface morphology of gonococci and meningococci

The transmission electron microscopic appearance of negatively stained preparations of *N. meningitidis* grown in vitro is shown in Fig. 1. Gonococci have a similar electron microscopic appearance. The meningococcal and gonococcal surface is characterized by multiple pili. Pili appear to be present on gonococci and meningococci in vitro [16, 17] and in vivo [18, 19, 20]. Recent studies have suggested that pili are important mediators of meningococcal and gonococcal attachment to human cells [16, 21].

By electron microscopy, pili are approximately 6 nm in width and extend 2,000 to 4,000 nm from the cell surface [15, 22]. In our studies with subinhibitory antibiotics, the mean number of pili per diplococcus was 20.8 for the meningococcal strains and 26.3 for the gonococcal strains [11]. 80–100% of diplococci from primary clinical cultures are piliated.

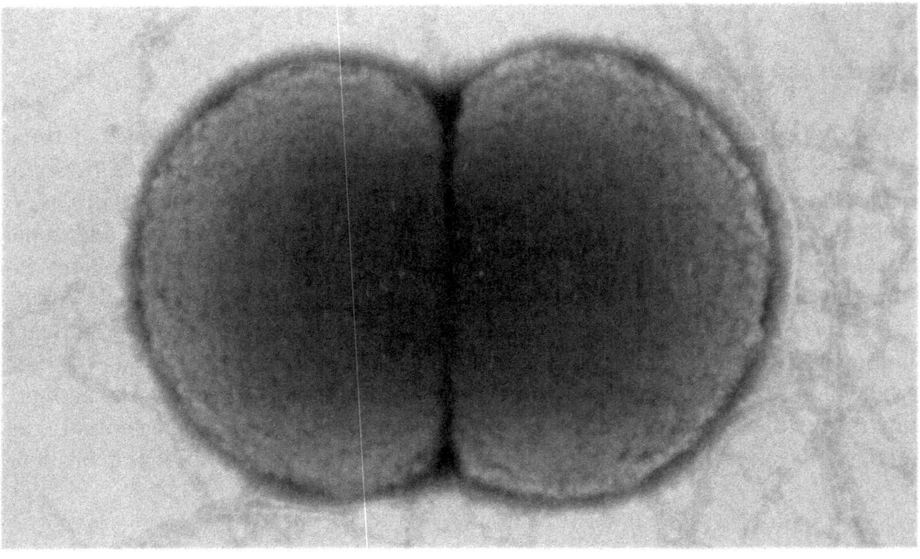

Fig. 1. Transmission electron micrograph of negatively stained preparations of *Neisseria meningitidis* strain 269B. Pili are prominent (× 68,000)

Rapidly growing gonococci and meningococci release substantial amounts of surface material in the absence of cell lysis [23]. These outer membrane blebs contain lipopolysaccharide and protein and may represent vehicles for delivery of toxic moieties to host cells [20, 24, 25].

Effect of subinhibitory concentrations of antibiotics on surface morphology of gonococci and meningococci

The effect of subinhibitory concentrations of tetracycline (½, ¼, ⅛, ¹⁄₁₆ MIC) on the electron microscopic appearance of *N. meningitidis* is shown in Fig. 2. The loss of a significant number of pili per diplococcus is evident and was typical for such preparations. Similar results were seen with subinhibitory concentrations (½, ¼, ⅛, ¹⁄₁₆, ¹⁄₃₂ MIC) of penicillin. Meningococci and gonococci lost not only a significant number of pili per diplococcus (Fig. 3a) but also there was a decrease in the percentage of organisms exhibiting pili (Fig. 3b). The susceptibility (MIC) of the strains to the antibiotics did not influence these results although highly resistant strains were not employed. Concentrations of penicillin less than ¹⁄₆₄ MIC and concentrations of tetracycline less than ¹⁄₃₂ MIC had no effect on the number or the percentage of diplococci with pili.

Subinhibitory concentrations of penicillin and tetracycline that decreased the number of pili also had a discernible effect on the electron microscopic characteristics of the outer membrane. In treated preparations the outer membrane was

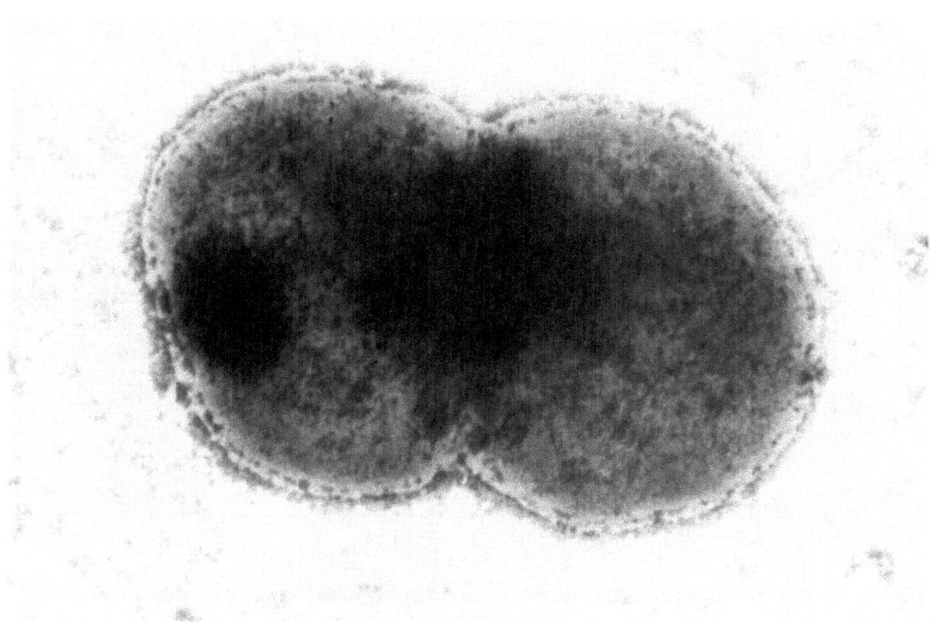

Fig. 2. Transmission electron micrograph of strain 269B grown with ¼ MIC of tetracycline (× 68,300). The loss of a significant number of pili per diplococcus is evident and was typical for such preparations. Similar results were seen with subinhibitory concentrations of penicillin

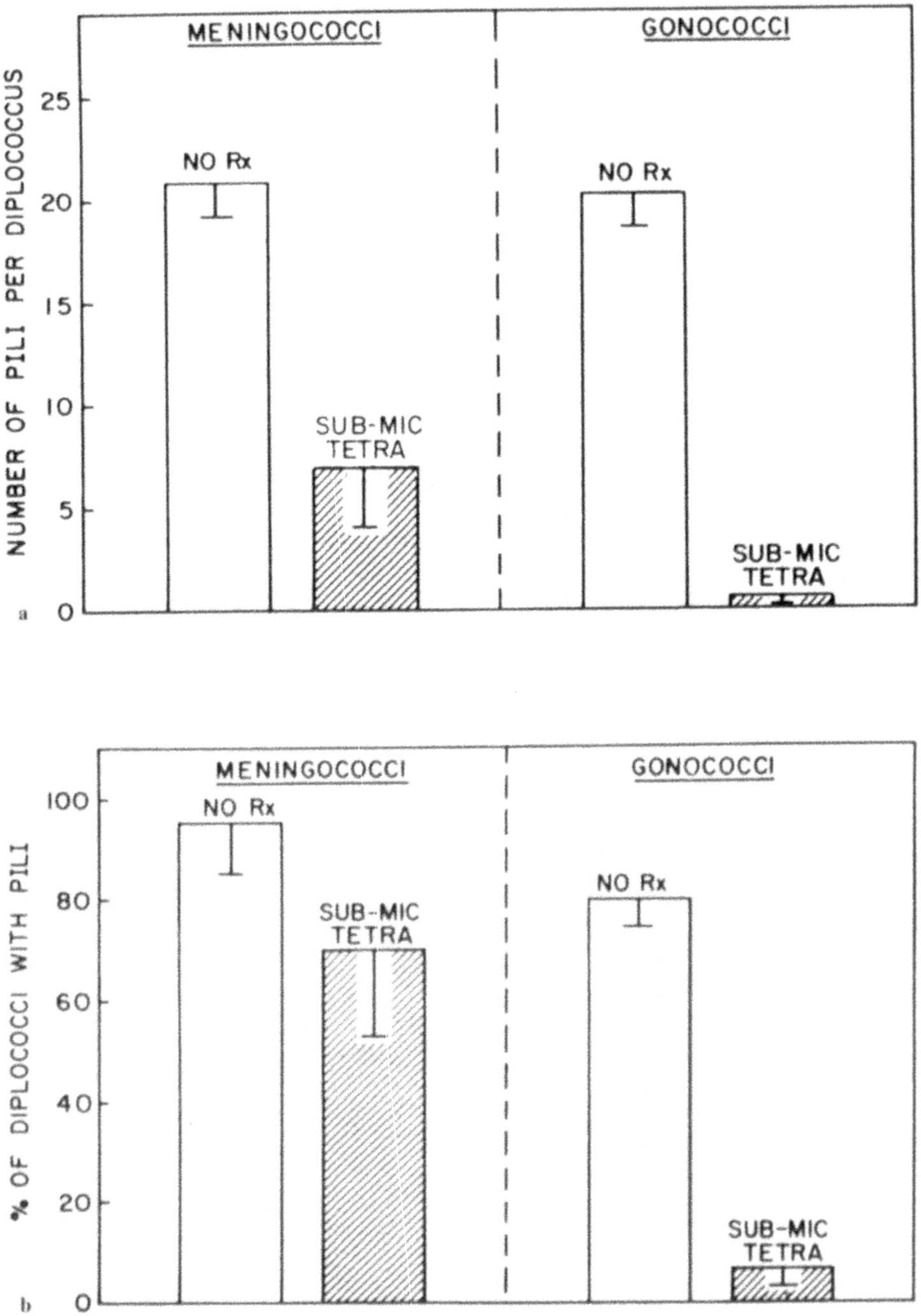

Fig. 3. Effect of subinhibitory concentrations of tetracycline on **a**) the number of pili per meningococcus or gonococcus and **b**) the percentage of meningococci or gonococci with pili. As shown by the cross-hatched columns meningococci or gonococci exposed to ½ MIC of tetracycline demonstrated a reduction in the number of pili per diplococcus and the proportion of meningococci or gonococci that were piliated

irregular in appearance and the number of blebs observed was increased [11]. Occasionally tetrads and other aberrant forms were noted in preparations exposed to subinhibitory concentrations of penicillin. Tetrads associated with increased cross-wall thickness have been previously reported by Lorian [1] after exposure of *N. gonorrhoeae* to subinhibitory concentrations of penicillin.

When gonococci or meningococci grown with subinhibitory concentrations of tetracycline or penicillin were removed to drug-free media for 18 hrs, the number of pili per diplococcus increased to levels near those of control. Likewise the percentage of piliated gonococci and meningococci returned to levels not significantly different from control [11]. These data suggest that the effects of subinhibitory concentrations of penicillin and tetracycline on pili are reversible and that mutant clones were not selected.

Biochemical characteristics of gonococcal surface proteins

Figure 4 shows the major outer membrane proteins of *N. gonorrhoeae* as they appear on SDS-PAGE and in a schematic representation to illustrate their location

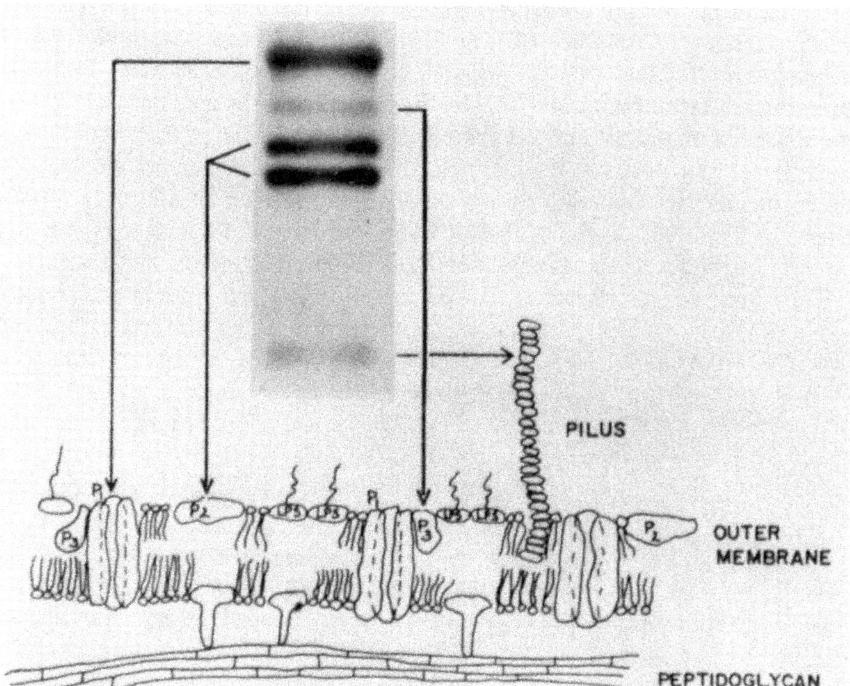

Fig. 4. Schematic representation of the outer membrane of *N. gonorrhoeae*. Major gonococcal outer membrane proteins between 45,000 and 14,000 MW are shown in the insert of an 10% SDS-polyacrylamide gel and in the schematic drawing to illustrate their location and association with other surface structures. Modified from Frasch, CD [37]

and association with other surface structures. Protein I is the major porin protein, MW 34,000 to 38,000 creating aqueous channels through the outer membrane and which may directly insert into adjacent host cell membranes [26]. Protein III, MW ~ 31,000, is closely associated with Protein I in the outer membrane of the gonococcus and may be species specific [26]. Proteins II (PII's), MW 26–34,000 are a group of surface exposed outer membrane proteins with a highly ordered tertiary structure and show heat modifiable migration in SDS-PAGE [26, 27]. PII's are found on gonococci isolated from the urethra of males with symptomatic gonococcal urethritis and on gonococci isolated from the cervix of females in the mid-menstrual cycle [28]. The presence of certain PII proteins on gonococci has been associated with opaque colony variants, with increased attachment to human cells and with increased resistance to killing by normal human sera, proteolytic enzymes and antimicrobial agents [27]. Pili are composed of identical repeating pilus subunits, ~ MW 16,000–22,000, termed pilin [22].

Effect of subinhibitory concentrations of antibiotics on gonococcal surface proteins

We studied the effects of ½ MIC of spectinomycin, tetracycline, and penicillin on the biochemical characteristics of PI, PII, PIII and pilin of gonococcal strain F62. In some outer membrane preparations from gonococci exposed to tetracycline, there was a decrease of stainable PI (Fig. 5), but the molecular weight was not altered. Exposure of opaque colony variants of gonococcal strain F62, containing PII proteins, to tetracycline and spectinomycin resulted in an alteration in the heat modifiability of at least one PII protein such that conversion from its fast migrating form (~ 24.5K) to the slow migrating form (30.5K) was incomplete even after the outer membrane preparations were boiled for 10 min at 100°C in SDS buffer (Fig. 5). Exposure of transparent colony variants of F62, which lack PII's, to antibiotics produced no change in electrophoretic migration.

These data suggest that the changes observed with tetracycline or spectinomycin may not indicate a change in the primary structure of PII but may reflect stabilization of PII in its native form. PII was not altered in OMP's isolated from gonococci of the same passage cultured on drug-free media. Penicillin did not effect the biochemical characteristics of PI or PII. PIII was not altered by any of the antibiotics tested. Lipopolysaccharide profiles of proteinase K digests of outer membrane preparations from gonococci grown with and without subinhibitory antibiotics were similar.

Pilus subunits were also affected by subinhibitory concentrations of antibiotics. Pilus subunits of ~ 21,500 molecular weight are present in OMP's from T_1 colonies (Fig. 5). When exposed to subinhibitory concentrations of tetracycline, pilus subunit production was absent or markedly decreased in some outer membrane preparations. Pilus subunits in SDS-PAGE of outer membrane preparations of gonococci exposed to penicillin had identical molecular weights and similar concentrations to that seen in outer membrane preparations from untreated organisms. Although both penicillin and tetracycline decreased the number and percentage of attached pili, only tetracycline appeared to alter production of gonococcal pilus subunits.

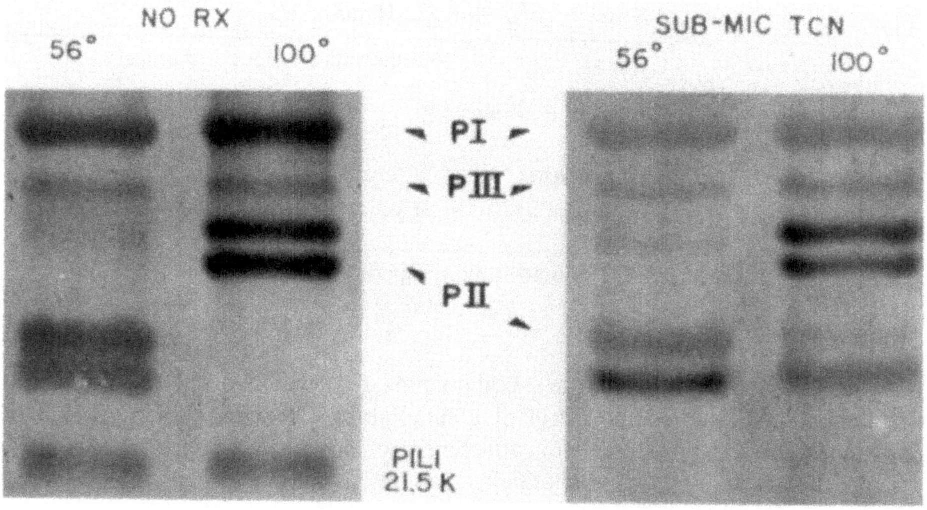

Fig. 5. SDS-PAGE profile (10% total acrylamide, silver stain) comparing the outer membrane proteins of *N. gonorrhoeae* (strain F62, opaque colony type) grown on medium containing no antibiotics (no Rx) or subinhibitory concentrations of tetracycline (SUB-MIC TCN). Samples were heated at 56°C for 30 min or boiled at 100°C for 3 min. Pilin subunits of 21,500 molecular weight were present in the outer membrane preparations of gonococci grown on medium without antibiotics. Pilus subunits were not found or were found in decreased concentrations among outer membrane preparations of gonococci grown on medium containing ½ MIC tetracycline. Tetracycline also appeared to alter the amount of protein I (PI) and the electrophoretic migration of protein II (PII). Reprinted with permission from Infection and Immunity [11]

Effect of subinhibitory concentrations of antibiotics on attachment of gonococci and meningococci to human cells

Piliated gonococci or meningococci grown in ½ MIC of tetracycline attached to buccal epithelial cells significantly less than piliated gonococci or meningococci from untreated cultures [11]. The % attachment of piliated meningococci decreased from 14.4 to 1.2 ($p < .001$) and the % attachment of piliated gonococci decreased from 9 to 1.6 ($p < .001$) when the organisms were grown on ½ MIC of tetracycline. The reduced levels of attachment of piliated organisms grown on tetracycline are similar to attachment of nonpiliated organisms (see Table 1). Similar results were obtained with ½ MIC of penicillin. Brief exposure during the attachment assay of piliated gonococci or meningococci (grown in drug-free media) to the same subinhibitory concentration of tetracycline and penicillin did not significantly affect attachment. In addition, the washing of gonococci grown in the subinhibitory concentration of tetracycline and penicillin did not restore the suppressed activity. Thus subinhibitory concentrations of antibiotics needed to be present during the period of active bacterial growth to affect binding activity. The decrease in the number and percentage of pili induced by either antibiotic suggests that the decreased attachment of meningococci and gonococci to human cells was due to loss of functional pili.

Table 1. Influence of Subinhibitory Concentrations of Tetracycline and Spectinomycin on Attachment of Nonpiliated *Neisseria gonorrhoeae* to Human Cells

Isolate, Colony Type (Rx)	% attachment*	P value
F62T_4Op (Control)	3.12 ± .71	
F62T_4Op (½ MIC Spectinomycin)	1.08 ± .22	p<.07
F62T_4Op (½ MIC Tetracycline)	1.59 ± .48	p<.13

* Data are mean values ± SEM of four or more experiments

As noted above, other surface components may be involved in gonococcal attachment. We studied the effect of a subinhibitory concentration (½ MIC) of spectinomycin and tetracycline on attachment of nonpiliated gonococci (from T_4 opaque colony variants) to human cells (Table 1). Nonpiliated gonococci adhered poorly to human buccal cells when compared to adherence of piliated gonococci of the same strain and colony phenotype (see above). Thus, the effect of antibiotics on adherence of nonpiliated gonococci was difficult to detect. The attachment of nonpiliated gonococci to buccal epithelial cells was decreased by exposure to subinhibitory antibiotics, but this decrease did not achieve statistical significance.

Discussion

Despite extensive studies into the epidemiology, immunobiology, and structure of *Neisseria meningitidis* and *Neisseria gonorrhoeae*, only partial control of the worldwide health problems caused by these organisms has been accomplished. The currently available meningococcal vaccines containing A and C or A/C/Y/W–135 capsular polysaccharides have been successful in the control of meningococcal disease among military recruits and in epidemic outbreaks [29], but are less effective for the groups most at risk, infants and young children. No vaccine is currently available for meningococcal disease due to serogroup B, the major cause of endemic meningococcal infections in the United States and many European countries. Similarly, gonococcal infections remain epidemic despite the availability of effective antimicrobial agents and public health measures designed to control the disease by epidemiologic means [30].

Efforts to develop immunologic or other approaches to disease caused by *N. gonorrhoeae* and *N. meningitidis* are most likely to be successful if they are targeted at those mechanisms which allow gonococci and meningococci to infect the host. Attachment of gonococci and meningococci to mucosal surfaces appears to be the first step in the pathogenesis of disease due to these organisms [16, 18, 21, 22, 31, 32]. There is increasing evidence that gonococci and meningococci have specific ligands (e.g. pili, outer membrane proteins) that interact either nonselectively or with selective receptor sites on epithelial cells of human mucosal surfaces. However, the production, modulation, structure and structural relationship of these ligands are not fully understood.

Subinhibitory concentrations of antibiotics with defined mechanisms of action appear to be useful probes to study the surface ligands of bacteria that are involved in attachment. For example, Eisenstein et al. [9] found that growth of a resistant *E. coli* strain in subinhibitory concentrations of streptomycin resulted in the production of structurally altered nonfunctional type 1 (mannose-sensitive) pili. They suggest that the aberrant pilus protein was caused by misreading of messenger RNA.

Although several investigators have observed that subinhibitory concentrations of antibiotics affect adherence of pathogenic *Neisseria spp* [33, 34, 35], the mechanisms that produced this effect were not fully explored. In our studies, summarized in this report, we used subinhibitory concentrations of three antibiotics (penicillin, tetracycline and spectinomycin) to study pili of meningococci and gonococci or protein II of gonococci, surface proteins which appear to mediate attachment of meningococci and gonococci to human cells.

The data from our studies provide further information about the structure, regulation and production of meningococcal and gonococcal surface proteins involved in attachment. For example, both tetracycline and penicillin in subinhibitory concentrations decreased the number of pili and the percentage of gonococci and meningococci with pili. However, only subinhibitory concentrations of tetracycline decreased the concentration of pilus subunits in the outer membrane. These findings suggest a block in synthesis of pilus subunits by subinhibitory concentrations of tetracycline. In view of the mode of action of tetracycline on the 30S ribosomal subunit [36], the block in pilus subunit synthesis probably occurs at the level of the ribosome. In contrast, pilus subunits were present in the outer membrane of gonococci exposed to subinhibitory concentrations of penicillin. This suggests that penicillin interferes with assembly or anchoring of pili in the outer membrane. Importantly, the decrease in the number and percentage of pili induced by either antibiotic was associated with decreased attachment of the treated gonococci to human cells. Immunologic or other means to block pilus function may be successful if directed not only at the intact pilus but also at steps in the synthesis of the pilus subunit, at the assembly of pilus subunits or at anchoring of the intact pilus in the outer membrane.

We conclude that concentrations of antibiotics less than those that inhibit multiplication of gonococci and meningococci may be useful probes to study the surface proteins of these organisms that are important in virulence. In vivo, alteration of surface proteins by antibiotics may lead to decreased adherence to mucosal surfaces or increased susceptibility of meningococci and gonococci to host defenses such as phagocytosis.

Summary

Surface proteins of pathogenic bacteria, which are important in attachment to human mucosal surfaces, may be altered on bacteria grown in the presence of subinhibitory concentrations of antibiotics. We studied the effects of subinhibitory concentrations of antibiotics on pili and other outer membrane proteins of *N. gonorrhoeae* and *N. meningitidis*. We focused our studies on those proteins which mediate attachment of gonococci and meningococci to human cells. Subinhibitory concentrations of tetracycline and penicillin were found to markedly reduce the number of pili per meningococcus or gonococcus and the percentage of meningococci or gonococci with pili. SDS-polyacrylamide gel electrophoresis of outer membrane preparations suggested that tetracycline decreased expression of pili by inhibiting synthesis of pilus subunits. In contrast pilus subunit synthesis was unaltered by penicillin, suggesting a defect in assembly of pilus subunits or anchoring of assembled pili. The decrease in the number of pili that occurred with subinhibitory concentrations of both tetracycline and penicillin was accompanied by a marked decrease in the ability of the organisms to attach to human cells.

Outer membrane Protein II (PII) of *Neisseria gonorrhoeae* may also influence attachment of gonococci to human cells. Exposure of transparent nonpiliated colony variants of gonococcal strain F62, which lack PII's, to subinhibitory antibiotics produced no changes in the electrophoretic migration of isolated outer membrane protein preparations (OMP's). However, exposure of opaque colony variants, containing PII proteins, to spectinomycin and tetracycline resulted in an alteration in the heat modifiability of at least one PII protein such that conversion from its fast migrating form (~24.5K) to the slow migrating form (~30.5K) was incomplete even after OMP's were boiled for 10 min. at 100°C in SDS buffer. These data suggest that spectinomycin or tetracycline may result in a change in the conformation of some PII proteins. The changes in PII associated with subinhibitory concentrations of tetracycline and spectinomycin were associated with decreased adherence of the treated gonococci to human cells, but the significance of this difference is unclear.

Subinhibitory concentrations of antibiotics are useful probes to study the structure and function of surface proteins of the pathogenic *Neisseria spp*.

References

1. Lorian V (1980) Effects of subminimum inhibitory concentrations of antibiotics on bacteria, p. 342–408. In V Lorian (ed.), Antibiotics in laboratory medicine. Williams and Wilkins. Baltimore, Md
2. Bermingham MAC, Deol BS, and Still JL (1970) Effect of streptomycin on lipid composition with particular reference to cyclic depsipeptide biosynthesis in *Serratia marcescens* and other micro-organisms. Biochem. J. 119: 861–869
3. Klainer AS, Russell RBB (1974) Effect of the inhibition of protein synthesis on the *Escherichia coli* cell envelope. Antimicrob. Agents Chemother. 6: 216–224
4. Sud IJ, and Feingold DS (1975) Detection of agents that alter the bacterial cell surface. Antimicrob. Agents Chemother. 8:34–37
5. Hirashima A, Childs F, Inouye M (1973) Differential inhibitory effects of antibiotics on the biosynthesis of envelope proteins of *Escherichia coli*. J. Mol. Biol. 79: 373–389
6. Gemmell CG, Peterson PK, Schmeling D, Kim Y, Mathews J, Wannamaker L, Quie PG (1981) Potentiation of opsonization and phagocytosis of *Streptococcus pyogenes* following growth in the presence of clindamycin. J. Clin. Invest. 67: 1249–1256
7. Michel J, Stressman P, and Stressman T (1980) Effects of subminimal inhibitory concentration of erythromycin, clindamycin, pristinamycin on the penicillinase production of *Staphylococcus aureus*. Antimicrob. Agents Chemother. 17: 13–15
8. Eisenstein BI, Beachey EH, Ofek I (1980) Influence of sublethal concentrations of antibiotics on the expression of the mannose-specific ligand of *Escherichia coli*. Infect. Immun. 28: 154–159

9. Eisenstein BI, Ofek I, and Beachey EH (1981) Loss of lectin-like activity in aberrant type 1 fimbriae of *Escherichia coli*. Infect. Immun. 31: 792–797
10. Swanborg-Eden C, Sanberg T, Stemqvist K, Ahlsted S (1978) Decrease in adhesion of *Escherichia coli* to human urinary tract epithelial cells in vitro by subinhibitory concentrations of ampicillin. Infect. 6: 5121–5124
11. Stephens DS, Krebs JW, McGee ZA (1984) Loss of pili and decreased attachment to human cells by *Neisseria meningitidis* and *Neisseria gonorrhoeae* exposed to subinhibitory concentrations of antibiotics. Infection and Immunity. 46: 507–513
12. Stephens DS, Krebs JW, Whitney AM (1983) Alteration of outer membrane protein II (PII) of *Neisseria gonorrhoeae* by subinhibitory concentrations of spectinomycin and tetracycline [Abst. no 51] in Program and Abstracts. Twenty-third Interscience Conference on Antimicrobial Agents and Chemotherapy. Washington D.C. American Society for Microbiology
13. McGee ZA, Dourmashkin DR, Gross JG, Clark JB, Taylor-Robinson D (1977) Relationship of pili to colonial morphology among pathogenic and nonpathogenic species of *Neisseria*. Infect. Immun. 143: 413–422
14. Stephens DS, McGee ZA (1983) Association of virulence of *Neisseria meningitidis* with transparent colony type and low-molecular-weight outer membrane proteins. J. Infect. Dis. 147: 282–292
15. McGee ZA, Street CH, Chappell CL, Cousar ES, Morris F, Horn RG (1979) Pili of *Neisseria meningitidis*: effect of media on maintenance of piliation, characteristics of pili, and colonial morphology. Infect. Immun. 24: 194–201
16. Stephens DS, McGee ZA (1981) Attachment of *Neisseria meningitidis* to human mucosal surfaces: influence of pili and type of receptor cell. J. Infect. Dis. 143: 525–532
17. Swanson J, Kraus J, Gotschlich EC (1971) Studies on gonococcus infection. I. Pili and zones of adhesion: their relationship to gonococcal growth patterns. J. Exp. Med. 134: 886–906
18. Craven DE, Peppler MS, Frasch CE, Mocca LF, McGrath PP, Washington G (1980) Adherence of isolates of *Neisseria meningitidis* from patients and carriers to human buccal epithelial cells. J. Infect. Dis. 142: 556–568
19. James JF, Swanson J (1978) Piliation of gonococci in vivo. J. Infect. Dis. 137: 94–96
20. Stephens DS, Edwards KM, Morris F, McGee ZA (1982) Pili and outer membrane appendages on *Neisseria meningitidis* in the cerebrospinal fluid of an infant. J. Infect Dis. 146: 568
21. Pearce WA, Buchanan TM (1978) Attachment role of gonococcal pili. Optimum conditions and quantitation of adherence of isolated pili to human cells in vitro. J. Clin. Invest. 61: 931–943
22. Schoolnik GK, Fernandez R, Tai JY, Rothbard J, Gotschlich EC (1984) Gonococcal pili primary structure and receptor binding domain. J. Exp. Med. 159: 1351–1370
23. DeVoe IW, Gilchrist JE (1973) Release of endotoxin in the form of cell wall blebs during in vitro growth of *Neisseria meningitidis*. J. Exp. Med. 138: 1156–1167
24. Gregg CR, Melly MD, Hellerqvist CG, Coniglio JG, McGee ZA (1981) Toxic activity of purified lipopolysaccharide of *Neisseria gonorrhoeae* for human fallopian tube muscosa. J. Infect. Dis. 143: 432–435
25. DeVoe IW (1981) The meningococcus and mechanisms of pathogenicity. Microbiol Rev. 46: 162–190
26. Blake MS, Gotschlich EC (1983) Gonococcal membrane proteins: speculation on their role in pathogenesis. Prog. Allergy 33: 298–313
27. Lambden PR, Heckels JE, James LT, Watt PJ (1979) Variations in surface protein composition associated with virulence properties in opacity types of *Neisseria gonorrhoeae*. J. Gen. Micro. 114: 305–312
28. James JF, Swanson J. (1979) Studies on gonococcal infection XIII. Occurrence of color/opacity colony variants in clinical cultures. Infect. Immun. 19: 332–340
29. Lepow ML, and Gold R (1983) Meningococcal A and other polysaccharide vaccines. A five-year progress report. NEJM 308: 1158
30. Center for Disease Control Annual Summary 1981. MMWR 30: 42,1982
31. McGee ZA, Johnson AP, and Taylor-Robinson D (1981) Pathogenic mechanisms of

Neisseria gonorrhoeae: Observations on damage to human fallopian tubes in organ culture by gonococci of colony type 1 or type 4. J. Infect. Dis. 143: 413–422

32. Melly MA, Gregg ZA, and McGEE ZA (1981) Studies of toxicity of *Neisseria gonorrhoeae* for human fallopian tube mucosa. J. Infect. Dis. 143: 432–439
33. Kristensen BE, Rustad L, Spanne O, Bjorvatan B (1983) Effect of subminimal inhibitory concentration of antimicrobial agents on the piliation and adherence of *Neisseria meningitidis*. Antimicrob. Agents Chemother. 24: 731–734
34. Salit IE (1983) Effect of subinhibitory concentrations of antimicrobials on meningococcal adherence. Can. J. Microbiol. 29: 369–376
35. Turgeon L, Jacques M, Mathieu LG, DeRepontigny J (1985) Influence of subminimal inhibitory concentrations of penicillin G and tetracycline on adherence of a *Neisseria gonorrhoeae* isolate to human fallopian tube tissues. [Abst No. A76]. In Program and abstracts 85th annual meeting American Society for Microbiology, Washington D.C. American Society for Microbiology
36. Pratt WB (1977) In Chemotherapy of Infection, New York, Oxford University Press. p. 128
37. Frasch CE (1979) Noncapsular surface antigens of *Neisseria meningitidis*. In L. Weinstein and BN Fields [ed.]. Seminars in infectious disease. Vol. 2 Stratton Intercontinental Medical Book Corp., NY. p. 304–337

Discussion

Brown: How did Dr. Stephens culture his bacteria? It is known that the medium, including nutrient depletion, will alter many major surface properties. Also, at high sub-MIC's the growth rate will be very slow. This also may have a big influence on structure. In vivo, iron restriction seems probable.

Stephens: Gonococci or meningococci were grown on gonococcal agar base or in gonococcal broth containing 2% (vol/vol) Isovitale X with or without subinhibitory concentration of antibiotics. It is well recognized that environmental conditions can, and do, influence surface components of bacteria and that surface components expressed in vitro culture may or may not be expressed in vivo. However, we studied the surface proteins (PI, PII, pili) because of considerable evidence that they are produced in vivo and that they are important in gonococcal or meningococcal pathogenesis. We have not examined the effects of subinhibitory concentrations of antibiotics on the iron regulated proteins of the pathogenic *Neisseria spp.* but agree these are also important surface proteins in vivo.

Vosbeck: We have observed with *E. coli* Ss142 that full recovery of piliation after exposure to sub-MIC of tetracycline or trimethoprim required growth for at least ten or more generations. For how long did you grow bacteria in the absence of antibiotic?

Stephens: We removed meningococci or gonococci to drug free media for 18 hrs before examining for recovery of piliation. This length of time provides for multiple generations of growth in the absence of antibiotics and may correlate with your findings with the *E. coli* strain.

Vosbeck: Did you do a time course?

Stephens: We did not look at earlier time points to determine how rapidly meningococci or gonococci recovered piliation.

Verhoef: As you showed, antibiotics may decrease piliation. This may have in vivo complications in the future. A lot of work is being done on the assembly of a vaccine against gonococcal disease. People are using outer membrane proteins or pilus material. When these vaccines are successful, and antibiotics are given to a patient who is immunized, antibodies may then not be able to react with the target because the target is changed due to the sub MIC of the antibiotic.

Stephens: It is possible that subinhibitory concentrations of antibiotics may alter antigenicity. For example Edlemann and Gallant (Cell 10:131–132, 1977) studied changes in flagellin, the protomeric subunit of flagella, isolated from E. coli that had been grown in subinhibitory concentrations of streptomycin. Streptomycin produced misreading of the CGU and CGC arginine codons and caused illegitimate incorporation of cysteine in the flagellin. Therefore alterations in primary or secondary structure induced by sub-MIC antibiotics might result in changes in antigenicity. We do not have data from the pathogenic *Neisseria spp.* regarding this question but agree that it is important.

Gemmell: You mentioned that tetracycline and spectinomycin at sub-MIC act similarly on PII outer membrane protein stabilization. What effect does spectinomycin or other drugs have on pilus formation?

Stephens: We have not extensively studied the effects of spectinomycin on pilus formation. Preliminary data suggest that pilus subunit production in outer membrane preparations is unaltered. This is interesting since both tetracycline and spectinomycin act on the 30S ribosomal subunit to inhibit protein synthesis. We need to do further studies with spectinomycin and other agents that bind to ribosomes.

Hahn: Penicillin G and tetracycline act by *different* mechanisms. How can they result in the same net effect, i.e. a reduction in pilus formation?

Stephens: Yes, both tetracycline and penicillin were found to markedly reduce the number of pili per meningococcus or gonococcus as determined by negative staining electron microscopy. Tetracycline appeared to decrease expression of pili by inhibiting synthesis of pilus subunits. In contrast, pilus subunit synthesis was unaltered by penicillin, suggesting a defect in assembly of pilus subunits or anchoring of assembled pili. The decrease in the number of pili that occurred with subinhibitory concentrations of *both* tetracycline and penicillin was accompanied by a marked decrease in the ability of the organisms to attach to human cells.

Hentges: Effect of subinhibitory concentrations of antibiotics on virulence of *Neisseria*. Is it possible to measure this?

Stephens: The ability of the pathogenic *Neisseria spp.* to attach to mucosal surfaces is highly correlated with the ability to cause disease. The decrease in the number of pili and the changes in pII induced by subinhibitory concentrations of antibiotics were accompanied by a decrease in the ability of the organisms to attach to human cells. Recently, Kristensen et al. (33) noted a remarkable decrease in in vitro piliation after exposure of meningococci to lincomycin. They also were able to decrease meningococcal colonization in carriers by giving subinhibitory concentrations of lincomycin. In addition, Turgeon et al. (35) reported that penicillin and tetracycline decreased attachment of gonococci to human fallopian tube organ cultures. Thus, both in vitro and in vivo experiments can be performed to examine the effect of subinhibitory concentrations of antibiotics on virulence of gonococci and meningococci.

Lorian: Did you see increases in cell size?

Stephens: Yes, occasionally tetrads and other large aberrant forms were noted in preparations exposed to subinhibitory concentrations of penicillin.

Interaction of Aminoglycosides and Ciprofloxacin with Bacterial Membranes

A. Dalhoff

Bayer Pharma Research Centre, Institute for Chemotherapy, P.O.Box 101 709, 5600 Wuppertal 1

Introduction

Alterations in outer membrane permeability caused by polycationic antibiotics (e.g. polymyxins) are caused by displacement of divalent cations by competition; on the other hand EDTA removes divalent cations by chelation and thus causes outer membrane perturbation [1]. A wide range of outer membrane permeabilisation events is mediated by polymyxin B: the uptake of hydrophobic fluorescent compounds and nitrocefin respectively is enhanced and rod like projections (i.e. formation of outer membrane blebs) is caused by polymyxin B (2–10). Furthermore polymyxins cause a release of soluble constituents from washed cells of P.aeruginosa; there was an optimum concentration of polymyxin for maximum release of cell constituents, higher concentrations inhibiting the release [11,12]. Interaction of polymyxin B with the divalent cation-binding sites on LPS and permeabilisation of the outer membranes of gram-negative bacteria is reversible upon addition of magnesium or potassium.

Aminoglycosides are to be characterized as polycationic antibiotics too; they were found to interact with divalent cation binding sites on P. aeruginosa LPS [13, 14] and thus it is tempting to speculate that aminoglycosides may affect the membrane(s) of gram-negative bacteria. In the early years of aminoglycoside research it was reported that bactericidal streptomycin concentrations caused an efflux of ribonucleotides immediately after exposure of bacteria to streptomycin [15]. As these ribonucleotides originated from de novo protein biosynthesis it must be assumed that membrane perturbation preceeded protein biosynthesis inhibition. Because of this observation Anand and Davis [16] postulated that the mode of action of streptomycin and penicillin respectively may be similar: “the action of streptomycin on membrane formation is thus analogous to that of penicillin on bacterial wall formation”. Similarly an increased efflux of aminoacids or potassium ions was considered to be the consequence of outer membrane permeabilisation events mediated by streptomycin [16, 17].

However, it was found that efflux of nucleotides from E.coli is strain specific; some strains liberated nucleotides immediately after exposure to streptomycin whereas the release of nucleotides was retarded or even not existent in others [18].

This study was presented in part at the Workshop of the Deutsche Gesellschaft für Hygiene und Mikrobiologie” Mainz, September 1982 (Zbl.Bakt.Hyg., I Abt. Orig. A (1983) 253: 423–461)

The Influence of Antibiotics on the
Host-Parasite Relationship II
Edited by D. Adam, H. Hahn, W. Opferkuch

Hancock et al proposed that aminoglycoside uptake and killing of P.aeruginosa requires interaction of aminoglycosides with outer membranes thus causing self-promoted uptake of aminoglycosides across the outer membrane and bacterial death [13, 14]. However, it was described earlier that interaction of aminoglycosides with the outer membrane was not correlated with the bactericidal action of aminoglycosides [19].

Materials and methods

Strains and media

In vitro studies were carried out with mutant strains of E.coli K12. In vitro liberation of enzymic markers was monitored in E.coli K12 pR6K; as this strain is not viable in vivo animal experiments were carried out with E.coli strain No. C 165 harbouring the TEM1 plasmic pR6K. All the other experiments described below were carried out with E. coli strain K12 (Hfr. H thi). A chemically defined minimal medium according to Davis and Mingioli [20] was used throughout this study.

Enzyme determinations

Beta-lactamase was determined iodometrically using penicillin G as substrate [21]; 1 unit was defined as the amount of enzyme hydrolysing 1 mMol penicillin G per min. β-lactamase from in vivo samples was determined as described recently [22]. *Galactosidase*-determination: β-galactosidase was determined as described by Pardee et al [23] using o-nitrophenyl-β-D-galactopyranosid (ONPG) as substrate. 1 unit of β-galactosidase is defined as the amount of enzyme hydrolysing 1×10^{-9} mol ONPG per min/ml.

Glucose-6-phosphate dehydrogenase determination: the activity of glucose-6-phosphate dehydrogenase was determined spectrophotometrically by measuring the rate of reduction of $NAD(P)^+$ at 340 nm; 1 unit of enzyme is defined as the amount which catalysed the reduction of 1 mMol $NAD(P)^+$ per min. under the conditions of the test [24].

Arrhenius plot

The temperature dependent diffusion of ONPG into E.coli was determined as an indicator for membrane fluidity. One hour after tenfold dilution of an overnight culture 50 mg chloramphenicol per ml was added and bacteria were incubated at 37°C for 30 minutes. Thereafter the aminoglycoside (1 mMol) was added and incubation at 37°C was continued for an additional 30 minutes. Bacteria were washed by centrifugation at 4°C and the sediment was resuspended in 1/10 of the original volume in physiological saline supplemented with 0.33 mMol ONPG. Aliquots of identical optical density were incubated at 0°C, 5°C, 10°C, 15°, 20°C, 25°C, 30°C, 37°C and 45°C for one hour. Therafter suspensions were centrifuged and ONPG concentrations in the supernatants were quantitated. The difference between the initial and final ONPG-concentration respectively was considered to represent ONPG uptake by E.coli; unspecific absorption of ONPG to the surface of E.coli was determined by using heat-killed bacteria. Data summarized in Fig. 3. represent the increase in ONPG uptake due to aminoglycoside exposure as compared to the drug-free controls. ONPG concentrations were determined photometrically at 420 nm following hydrolysis at 100°C for 6 minutes with 0.1 NaOH.

Free flow electrophoresis

Free flow electrophoretic separation of intact bacteria was carried out as described earlier [25]. Bacteria subjected to free flow electrophoresis (FFE) were washed two times by centrifugation in the separation buffer at 4°C. The pellet was resuspended with separation buffer to which various aminoglycosides were added; final aminoglycoside concentrations are indicated in Fig. 4. As bacteria used for FFE were not preincubated with aminoglycosides they came into contact with the antibiotics during electrophoretic separation only. 48 fractions were collected and the number of colony forming units per ml (CFU/ml) was quantitated in every second fraction.

Animal experiments were carried out by adopting the granuloma pouch model as described earlier [26, 27].

Results

Release of enzymic markers from aminoglycoside treated bacteria

The effect of aminoglycosides on the integrity of bacterial membranes was assessed by adding the antibiotics to bacterial cultures in the early logarithmic phase of growth. Generally, aminoglycoside uptake by bacteria was inhibited, firstly: by catabolite repression using glucose without cAMP supplementation as carbon source; as aminoglycoside transport is catabolite repressible aminoglycosides are not taken up under these specific growth conditions [28, 29]. Secondly, experiments were carried out at 4°C.

Following exposure of E.coli to the aminoglycosides tested for 15 minutes enzyme activities were determined in the culture supernatant following centrifugation. All data were corrected for unspecific liberation of enzymes in the drug-free controls. Data summarized in Fig. 1 clearly demonstrate that all the aminoglycosides caused a release of β-lactamase as well as glucose-6-phosphate dehydrogenase from E.coli. This release was concentration dependent in so far that there was an optimum concentration of aminoglycosides for maximum release of enzymic markers higher concentrations inhibiting the release. In addition to the periplasmic and intracellular enzyme respectively the efflux of β-galactosidase, a membrane-bound enzyme, was studied. Liberation of β-galactosidase was not determined in parallel to the above mentioned markers as β-galactosidase synthesis had to be induced specifically. Neither aminoglycoside caused a release of β-galactosidase under these test conditions (Table 1).

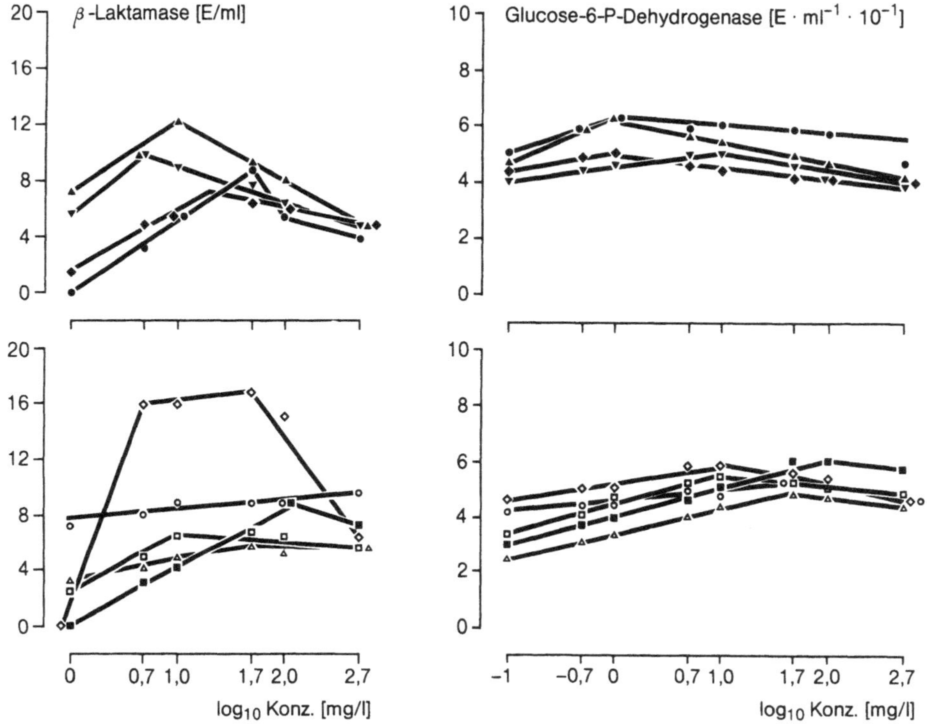

Fig. 1. Aminoglycoside mediated marker release from in vitro grown E.coli. Slopes indicate extracellular enzyme concentrations only; total intra- as well as extracellular β-lactamase concentrations amounted to 18 U/ml and glucose-6-P-dehydrogenase to 7.8 U/ml
● Gentamicin; ▲ kanamycin; ■ neomycin; ▼ streptomycin; ◆ amikacin;
△ netilmicin; □ tobramycin; ◇ dibekacin; ○ sisomicin

Table 1. Influence of aminoglycosides on β-galactosidase release from E.coli

	β-Galactosidase (U/ml)									
	extracellular aminoglycoside (mg/l)					membranebound aminoglycoside (mg/l)				
	500	50	10	1	0.1	500	50	10	1	0.1
Centamicin	0	0	0	0	0	10.8	10.3	11.0	10.9	11.2
Kanamycin	0	0	0	0	0	10.2	9.8	10.5	11.4	10.8
Neomycin	0	0	0	0	0	11.1	11.0	10.9	11.8	11.7
Streptomycin	0	0	0	0	0	10.9	10.2	10.8	11.0	11.2
Amikacin	0	0	0	0	0	10.9	11.0	11.1	10.9	11.5
Netilmicin	0.007	0.007	0	0	0	10.8	10.5	10.8	11.2	11.4
Tobramycin	0	0	0	0	0	10.6	10.8	10.5	11.0	10.9
Dibekacin	0	0	0	0	0	10.4	10.6	10.7	10.5	10.8
Sisomicin	0.014	0.010	0.010	0	0	10.2	10.1	10.5	10.4	10.9

drug-free controls: extracellular: 0.015 (U/ml); intracellular: 11.9 (U/ml)

Release of markers from aminoglycoside treated bacteria was studied in vivo as well. However, because of methodological reasons β-lactamase activity was determined as the only marker. Therapy of infected animals was commenced 24 hours after infection; at this point of time oxygen tension within the infected foci is extremely low [27] and thus aminoglycosides are not taken up by the bacteria growing in vivo (data not shown). The aminoglycosides were injected intramuscularly twice daily individual doses being 50, 25 and 10 mg/kg respectively. Data summarized in Fig. 2 clearly indicate that an individual dose of 10 mg/kg caused a more marked release of β-lactamase than therapy of animals with 50 or 25 mg/kg respectively. As in animals aminoglycoside kinetics are not linearly dose-dependent

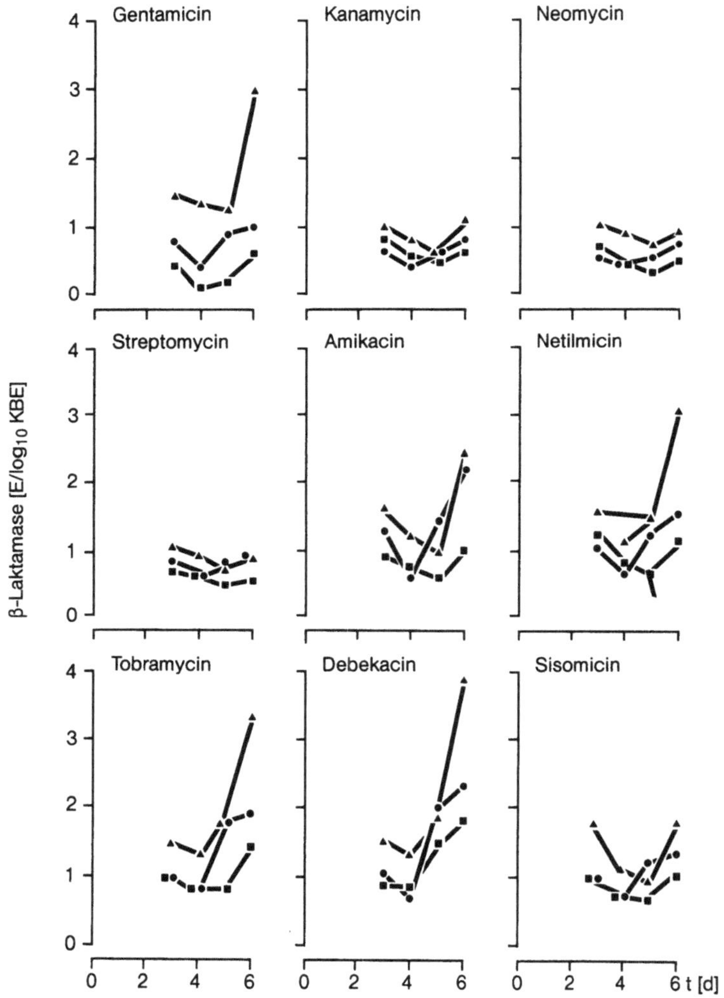

Fig. 2. Effect of intramuscular aminoglycoside therapy of animals suffering from a local E.coli infection on β-lactamase release.
• 50 mg/kg b.i.d., ■ 25 mg/kg b.i.d., ▲ 10 mg/kg b.i.d.

Table 2. β-lactamase (units/mg protein) release from E.coli K12 harbouring different plasmids incubated for 3h with subinhibitory ciprofloxacin concentrations (½ MIC)

strain	drug-free controls			ciprofloxacin		
	extra-cellular	intra-cellular	extra-cellular, % of total activity	extra-cellular	intra-cellular	extra-cellular, % of total activity
pR GK	23.2	35.8	39.3	23.0	25.7	47.2
p1822	202.4	549.2	26.9	596.3	385.6	60.7
pR GN	12.0	2.8	81.1	8.4	1.2	87.5
pR 46	16.1	8.1	66.5	14.3	4.4	76.5
p 576	19.0	5.6	77.2	17.0	2.6	86.7

(by doubling the dose, pouch-exudate levels do *not* increase correspondingly) these differences become even more significant if marker release is correlated to pharmacokinetics of the drugs within the pouch exudate i.e. to the area under the concentration versus time-curve.

Preliminary studies on the effect of subinhibitory ciprofloxacin concentrations (½ MIC) on the outer membrane of E.coli indicate, that ciprofloxacin, too, causes a release of enzymic markers from the periplasmic space. In addition to the results summarized in Table 2 the release of alkaline phosphatase was studied as well (data not shown); liberation of alkaline phosphatase differed not significantly from β-lactamase release.

Arrhenius plot

The effect of aminoglycosides on bacterial membrane fluidity was determined by monitoring the temperature dependent diffusion of ONPG into E.coli. Aminoglycoside transport as well as synthesis of inducible ONPG transport proteins was inhibited by cataboliterepression as well as addition of chloramphenicol; thus, an increase in ONPG uptake as a consequence of aminoglycoside exposition of E.coli is due to facilitated permeation as compared to the drug-free controls. Exposure of E.coli to aminoglycosides (except dibekacin and netilmicin) caused a significant increase in ONPG uptake (Fig. 3).

Morphological studies by electron microscopy carried out under the same experimental condition revealed that the aminoglycosides studied caused similar blebbing as described earlier for polymyxin [10]. Furthermore, ruptures in the outer as well as inner membrane of E.coli were to be seen (data not shown).

Free flow electrophoresis

In addition to the physiological and morphological parameters the effect of aminoglycosides on bacterial membranes was studied by physico-chemical methods.

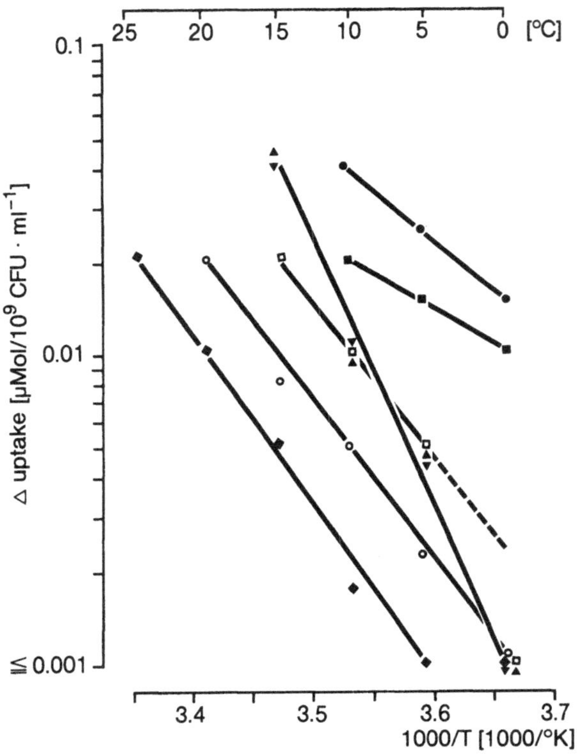

Fig. 3. Increase in ONPG uptake by E.coli due to aminoglycoside exposition as compared to the drug free controls. Symbols see Fig.1

Varying drug concentrations as indicated in Fig. 4 were used for the different antibiotics in order to avoid co-migration of bacteria with the polycationic aminoglycosides. The various concentrations used for the different aminoglycosides are indicated by the maximal number on the ordinate in Fig. 4. Although bacteria were in contact with the aminoglycosides only for a limited period in time (i.e. the electrophoretic separation) physico-chemical characteristics of drug exposed bacteria are clearly altered as compared to the drug-free controls peaking in fraction 28; in contrast, addition of aminoglycosides to the separation buffer caused a significant shift of drug exposed bacteria to the anode whereas the polycationic aminoglycosides migrated to the cathode.

Furthermore, the interaction of sisomicin with charged *phospholipid monolayers* was studied. Two synthetic phospholipids were used: Dipalmitoylphosphatidylcholin (DPPC) and dipalmitylphosphatidylglycerol (DPPG). At pH = 7 DPPC is an zwitter ion with zero nett charge wheras DPPG carries one negative charge due to its phosphate group. The films were spread onto phosphate-buffered saline (PBS) at pH = 7 and sisomicin was added to the PBS subphase prior to the addition of the phospholipids. The film area (Å) normalised per phospholipid molecule versus the resultant compressional force (π) for the phospholipids (DPPC) and (DPPG) respectively on PBS with and without sisomicin was recorded (data not shown).

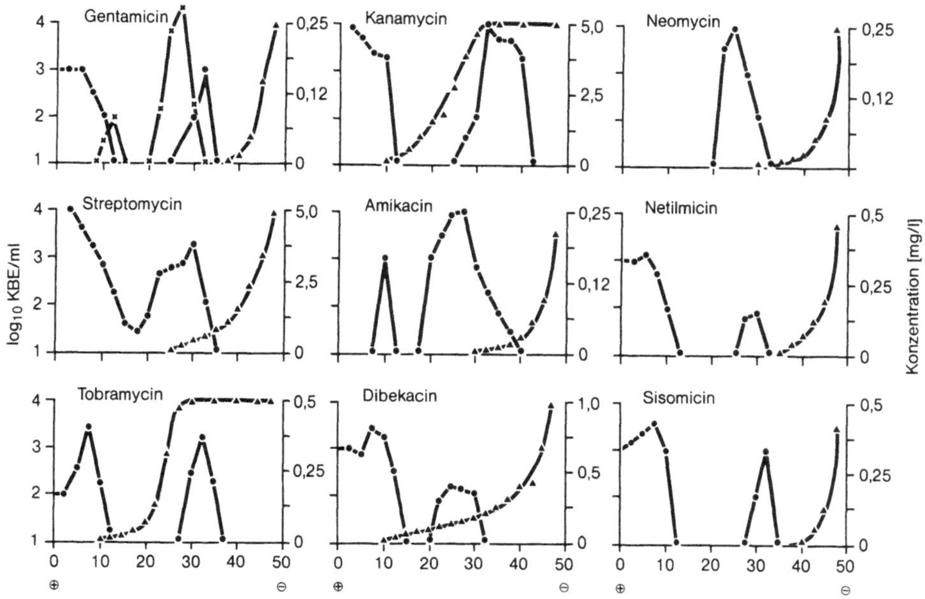

Fig. 4. Free flow electrophoretic separation of aminoglycoside exposed bacteria. × CFU drug free control; • CFU aminoglycoside exposed bacteria; ▲ drug concentration

Sisomicin has a pronounced effect on the π/Å characteristics of the DPPG film whereas the effect on the DPPC film is less pronounced. This finding indicates firstly, that sisomicin binds to head groups of DPPG probably to more than 1 DPPG molecule only. Due to 4 amine groups of sisomicin ionized at pH = 7 one sisomicin molecule is capable of binding more than 1 DPPG head group. Secondly, π/Å characteristics indicate that sisomicin is incorporated into the highly compressed DPPG-film. Thirdly, sisomicin causes a negligible effect on the π/Å characteristics of DPPC-films. The different effect of sisomicin on the DPPG and DPPC films respectively are likely due to differences in electrostatic attraction between the polycationic sisomicin on the one hand and the negatively charged phospholipid DPPG versus the zero charged phospholipid DPPC on the other hand.

Furthermore, the H-bonding capacity of sisomicin with DPPG may be different to that with DPPC.

Discussion

Data illustrated above clearly indicate that the effect of aminoglycosides on enzymic marker release from E.coli was concentration-dependent in so far that there was an optimum concentration of aminoglycosides for a maximum release of periplasmic as well as intracellular enzymic markers, higher concentrations inhibiting the release. Analogous concentration-dependencies were noted with respect to the effect of aminoglycoside on bacterial respiration as well as antibacterial efficacy in vivo

(Dalhoff data on file [45]). Analogous effects were described for various aminoglycosides with respect to their interaction with 80S ribosomes; inhibition of protein synthesis and the rate of misreading in cell free systems derived from eucaryotic organisms such as saccharomyces cereviseae and tetrahymena respectively were influenced by aminoglycosides in a similar manner higher drug concentrations being less effective than low concentrations [30–32]. This specific self-inhibiting effect of chemicals and antibiotics on bacterial growth and/or physiological markers is typical for membrane perturbing agents such as polymyxins [11, 12] and the cationic detergent cetyltrimethylammoniumbromide [11]. The efflux of periplasmatic as well as intracellular enzymic markers from E.coli as well as morphological studies by electron microscopy clearly indicate that aminoglycosides act on the inner and outer membrane of the strains tested. The membrane perturbing activity of aminoglycosides is not dependent from intracellular uptake as these data were generated under conditions of inhibited aminoglycoside transport. Thus, permeabilisation of the outer as well as inner membrane of E.coli by aminoglycosides is not a secondary effect due to inhibition of protein-biosynthesis; in contrast, interaction of aminoglycosides with bacterial membranes represents an additional primary target for their mode of action.

Similarly, exposition of E.coli to ciprofloxacin causes a release of periplasmic enzymes indicating that ciprofloxacin causes a permeabilization of the outer membrane. These preliminary experimental data have to be substantiated by additional in vitro and in vivo studies. However, it was reported recently that at high concentrations ciprofloxacin desintegrated bacterial membranes and caused lysis of gram negative bacteria [33]. In agreement with the data of this study Cohen and McConnell [34] described the release of endotoxin from E.coli exposed to ciprofloxacin; the authors concluded that ciprofloxacin may affect cell membranes. The mode of action of ciprofloxacin on bacterial membranes is unknown at the time being.

Evidence exists indicating that aminoglycosides act at the same site as do polymyxins and EDTA [5, 35, 36]; thus, the divalent cation crossbridging of lipopolysaccharide is disrupted causing localized destabilization or destortion of the membrane [36]. Due to membrane permeabilization diffusion of ONPG is facilitated (Fig. 3) and uptake of nitrocefin, lysozyme, and hydrophobic fluorescent compounds is enhanced [13, 14, 37]. Furthermore, dehydrostreptromycin decreased prolin transport being dependent from the proton motive force [38]. This finding indicates that streptomycin permeabilized the inner membrane of E.coli and P. aeruginosa thus affecting their ability to maintain a proton gradient across the cell membrane.

Binding of aminoglycosides to and insertion into model membranes corroborates the interaction of aminoglycosides with biological membranes. Aminoglycosides interact with the binding of divalent cations to artificial membranes and/or bind directly to negatively charged lipid layers because of physico-chemical interactions [39–44].

The microbiological and clinical implications of these findings are manifold; briefly first, aminoglycosides enhance outer membrane permeability of gram-negative bacteria thus facilitating transport or diffusion of nutrients, chemicals and antibiotics respectively. Second, membrane perturbation and antibacterial efficacy were directly correlated even if aminoglycosides were not taken up by the bacteria. (Dalhoff data on file [14,45]). Thus aminoglycosides act bactericidally even if transport is impaired or inhibited and consequently protein-biosynthesis is *not* affected. Third, the mechanism of aminoglycoside toxicity may be correlated to cytoplasmic and membrane effects of aminoglycosides [44].

References

1. Hancock R E W (1984) Alterations in outer membrane permeability. Ann Rev Microbiol 38: 237–264
2. Cerny G, Teuber M (1971) Differential release of periplasmic versus cytoplasmic enzymes from Escherichia coli B by Polymyxin B Arch Microbiol 78: 166–179
3. Hancock R E W, Wong P G W (1984) Compounds which increase the permeability of the Pseudomonas aeruginosa outer membrane. Antimicrob. Agents Chemother 26: 48–52
4. Newton B A (1954) Site of action of polymyxin B on Pseudomonas aeruginosa: antagonism by cations. J Gen Microbiol 10: 491–499
5. Nicas T I, Hancock R E W (1980) Outer membrane protein H1 of Pseudomonas aeruginosa: involvement in adaptive and mutational resistance to ethylenediaminetetraacetate, polymyxin B, and gentamicin. J Bacteriol 143: 872–878
6. Schindler M, Osborn M J (1979) Interaction of divalent cations and polymyxin B with lipopolysaccharide. Biochemistry 18: 4 425–4 430
7. Schindler P R G, Teuber M (1978) Ultrastructural study of Salmonella typhimurium treated with membrane active agents: specific reaction of dansylchloride with cell envelope components. J Bacteriol 135: 190–206
8. Gilleland H E, Murray R G E (1976) Ultrastructural study of polymyxin resistant isolates of Pseudomonas aeruginosa. J Bacteriol. 125: 267–281
9. Schindler P R G, Teuber M (1975) Action of polymyxin B on bacterial membranes: morphological changes in the cytoplasm and in outer membrane of Salmonella typhimurium and Escherichia coli 13. Antimicrob Agents Chemother 8: 95–104
10. Vaara M, Vaara T (1983) Polycations as outer membrane-disorganizing agents. Antimicrob Agents Chemother 24: 114–122
11. Newton B A (1953) The release of soluble constituents from washed cells of Pseudomonas aeruginosa by the action of polymyxin. J Gen Microbiol 9: 54–64
12. Newton B A (1956) The properties and mode of action of polymyxins. Bacterial Rev 20: 14–27
13. Hancock R E W, Raffle V J, Nicas T I (1981) Involvement of the outer membrane in gentamicin and streptomycin uptake and killing in Pseudomonas aeruginosa. Antimicrob Agents Chemother 19: 777–785
14. Loh B, Grant C, Hancock R E W (1984) Use of fluorescent probe 1-N-phenylnaphthylamine to study the interactions of aminoglycoside antibiotics with the outer membrane of Pseudomonas aeruginosa. Antimicrob Agents Chemother 26: 546–551
15. Roth H, Amos H, Davis B D (1960) Purine nucleotide excretion by Escherichia coli in the presence of streptomycin. Biochem Biophys. Acta 37: 398–405
16. Anand N, Davis B D (1960) Damage by streptomycin to the cell membrane of Escherichia coli. Nature Lond 185: 22–23
17. Dubin D T, Davis B D (1961) The effect of streptomycin on potassium efflux in Escherichia coli. Biochem Biophys Acta 52: 400–402
18. Gregroy H, Umbreit W W (1961) Effect of streptomycin on nucleotide leakage in various strains of E.coli. Bact Proc 1961: 189
19. Tzagoloff H, Umbreit H H (1963) Influence of streptomycin on nucleotide excretion in Escherichia coli. J Bacteriol 85: 49–52
20. Davis B D, Mingioli E S (1950) Mutants of Escherichia coli requiring methionine or vitamin B_{12}. J Bacteriol 60: 17–28
21. Perret C J (1954) Iodometric assay of penicillinase. Nature 174: 1 012 –1 013

22. Dalhoff A (1982) Interaction of mezlocillin and cefoxitin against Proteus morganii in the granuloma pouch model. Eur J Microbiol 1: 243–247
23. Pardee A B, Jacob F, Monod I (1959) The genetic control and cytoplasmic expression of "inducibility" in the synthesis of β-galactosidase by E.coli. J Mol Bol 1: 165–178
24. Langdon R G (1966) Glucose-6-phosphate dehydrogenase from erythrocytes. In: Colowick SP, Kaplan ND (eds): Methods in enzymology Vol.9: Academic Press, New York, pp 126–131
25. Dalhoff A (1983) Shift in serum sensitivity of in vivo grown Escherichia coli—a preliminary report. In: Keusch G, Waldström T (eds): Experimental bacterial and parasitic infections. Elsevier Science Publishing Co., Inc., New York, Amsterdam, Oxford, pp 317–323
26. Dalhoff, Frank G, Luckhaus G (1982) The granuloma pouch, an in vivo model for pharmacokinetic and chemotherapeutic investigations. I. Biochemical and histological characterization. Infection 10: 354–360
27. Dalhoff A, Frank G, Luckhaus G (1983) The granuloma pouch: an in vivo model for pharmacokinetic and chemotherapeutic investigations. II Microbiological characterization. Infection 11: 41–46
28. Höltje J V (1978) Streptomycin uptake via an inducible polyamine transport system in Escherichia coli. Eur J Biochem. 86: 345–351
29. Dalhoff A (1983) Transport of aminoglycosides in Escherichia coli. Zbl Bakt Hyg, I Abt Orig A 254: 379–387
30. Singh A, Ursic D, Davies J (1979) Phentoypic suppression and misreading in Saccharomyces cerevisiae. Nature (Lond) 277: 146–148
31. Palmer E, Wilhelm J M (1978) Mistranslation in a eucaryotic organism. Cell 13: 329–334
32. Palmer E, Wilhelm J M, Sherman F (1979) Phenotypic suppression of nonsense mutants in yeast by aminoglycoside antibiotics. Nature (Lond) 277: 148–150
33. Voigt W H, Zeiler H J (1984) Influence of ciprofloxacin on the ultrastructure of grampositive and gramnegative bacteria. 24th Interscience conference on antimicrobial agents and chemotherapy, Washington, October 1984, Abstract No. 982
34. Cohen J, McConnell J S (1985) Endotoxin release from bacteria exposed to ciprofloxacin and other bactericidal antibiotics. 14th International Congress of Chemotherapy, Kyoto, June 1985, Abstract No. 553–10
35. Hancock R E W (1981) Aminoglycoside uptake and mode of action with specific reference to streptomycin and gentamicin. I. Antagonists and mutants. J Antimicrob Chemother 8: 249–276
36a. Hancock R E W (1981) Aminoglycoside uptake and mode of action—with special reference to streptomycin and gentamicin. II Effects of aminoglycosides on cells. J Antimicrob Chemother 8: 429–445
36b. Brown M R W (1975) The role of cell envelope in resistance. In: Brown M R W (ed): Resistance of Pseudomonas aeruginosa. John Wiley and Sons, Inc., New York, pp 71–107
37. Bryan L E van den Elzen H M (1977) Effects of membrane energy mutations and cations on streptomycin and gentamicin accumulation by bacteria: a model for entry of streptomycin and gentamicin in susceptible and resistant bacteria. Antimicrob Agents Chemother 12: 163–177
38. Eagon R G, Hartle R J, Rake J B, Abdel-Sayed S (1982) Effect of dihydrostreptomycin on active transport in isolated bacterial membrane vesicles. Antimicrob Agents Chemother 21: 844–845
39. Corrado A P, Prado W A, Pimenta de Morais I (1975) Competitive antagonism between calcium and aminoglycoside antibiotics in skeletal and smooth muscle: In: Rocha e Silva M, Suarez-Kurtz G (eds.): Concepts of membranes in regulation and excitation. Raven Press, New York, pp. 212–226
40. Lüllmann H, Vollmer B (1982) An interaction of aminoglycoside antibiotics with Ca binding to lipid monolayers and to biomembranes. Biochem Pharmacol 31: 3769–3773
41. Brasseur R, Laurent G, Ruysschaert J M, Tulkens P (1984) Interactions with negatively charged lipid layers. Biochemical and conformational studies. Biochem Pharmacol 33: 629–637

42. Tulkens P, Brasseur R, Laurent G, Ruysschaert JM, Claes PJ, Vanderhaeghe H (1984) Conformational and biochemical studies of aminoglycoside-phospholipid interactions, a tool for the design of lesser toxic aminoglycosides. 6th International Symposium on Future Trends in Chemotherapy, Tirrenia, May 1984, Abstract Vol p. 197
43. Khanamuryan RM, Badzhinyan SA (1981) Ototoxicity of some antibiotics and permeability of model membranes. Zh Ushum Nos Oorl Bolezn 3: 59–62
44. Lipsky JJ, Lefkowith J, Lietman PS (1984) Cytoplasmic and membrane effects of aminoglycosides. A hypothesis for the mechanism of aminoglycoside toxicity. Kidney Dis 4: 261–271
45. Dalhoff A (1983) Studies on the action of aminoglycosides on bacterial membranes. Zbl Bakt Hyg I Abt Orig A 253: 427

Discussion

Ghione: Aminoglycosides are strong chelating agents chiefly towards Ca^{++} and Mg^{++}. Can you exclude that part at least of the phenomena you observed can be attributed to the chelating activity of aminoglycosides?

Dalhoff: As already mentioned, aminoglycosides act very likely at the same site as polymyxins and EDTA. Thus, a displacement of cations either by competition or by chelation is a prerequisite for the membrane directed mode of action of aminoglycosides.

Ghione: Have you determined the influence of pH changes on the course of the phenomena you described? The activity of aminoglycosidic antibiotics is in fact strongly dependent on pH.

Dalhoff: The experiments have been carried out at physiological pH values only and the animal experiments demonstrated that the interaction of aminoglycosides with bacterial membranes is relevant in vivo, too.

Vosbeck: You are postulating an effect of aminoglycosides by an alternative mechanism to their primary action or translation. Do you see these effects also in bacteria that are resistant to the primary effect of aminoglycosides?

Dalhoff: Streptomycin-resistant bacteria were included as additional controls. The aminoglycisode mediated permeabilization of bacterial membranes was also seen in these resistant strains.

Haag: Have you done your experiments also using polycationic substances instead of aminoglycosides, e.g. polyamines, and have you seen comparable effects with such chemicals?

Dalhoff: Control-experiments have been carried out with other polycationic substances (Spermidine and ribonuclease). The effects caused by the exposure of E.coli to either aminoglycosides or such chemicals were highly comparable.

Ginsburg: Can anticoagulants like heparin modulate the action of aminoglycosides in patients treated for long periods with such anticoagulants?

Dalhoff: Such interactions are not known to me but probably you may have some data pointing to an interference of heparin and aminoglycosides?

Brown: These are exciting results. Has Dr. Dalhoff considered using magnesium depleted bacteria? This may reduce the effect and indicate that the OM target is the metal cation in LPS on the outer leaflet of E.coli and of P. aeruginosa.

Dalhoff: No, I have not carried out this experiment, but published evidence exists indicating that aminoglycosides act at the same site as do polymyxins and EDTA, i.e. the membrane is destabilized because of disruption of divalent cation crossbridging of LPS.

The Influence of Ciprofloxacin on the Ultrastructure of Gram-negative and Gram-positive Bacteria

W.-H. VOIGT

BAYER AG, Institute of Immunology and Oncology, Wuppertal, F.R.G.

Introduction

Serum resistent Escherichia coli C14, Pseudomonas aeruginosa 220 and Staphylococcus aureus 25151 as test organisms were used to investigate the effect of ciprofloxacin in pooled active human serum. Ciprofloxacin was added in concentrations of 0.025–2,0 mcg/ml. 2 hours later samples were taken for electron microscopy and to determine the number of viable bacteria. Following drug concentration-dependent results were obtained: At low concentrations, elongation of gram-negative bacteria cells was observed. In the early phase of the bactericidal process in E. coli C14 distinct loosenings of the peripheral cytoplasm could be observed. In addition osmiophilic condensations in the areas of the bacterial DNA in Pseudomonas aerug. 220 could be detected. At higher concentrations cell membranes ruptured and lysis occurred. In staphylococci surface deformation and disturbances of septum formation occurred first and, in the final stage, proteolysis and empty envelopes were seen. In serum, cell damage could be detected earlier and was more severe than in broth.

Ciprofloxacin, a new quinolone derivative with high in vitro activity against gram-negative and gram-positive bacteria [1, 2, 5, 7, 11] will be developed for oral and parenteral treatment of systemic infections as well as urinary tract infections. The bacterial drug kills most bacteria in less than one generation time, at concentrations, which are 4–8 times grater than the MIC. It is proposed that ciprofloxacin like other quinolones, is an inhibitor of bacterial DNA-gyrase [9].

This paper describes results from investigations concerned with the influence of ciprofloxacin (CIP) on the ultrastructure of bacteria cultivated in human serum in correlation to the viability of different bacterial species.

Materials and Methods

The microbiological examinations were performed by Dr. H.-J. Zeiler, Institute of Chemotherapy, BAYER AG.

Serum resistant strains (Escherichia coli C14, Pseudomonas aeruginosa 220, Staphylococcus aureus 25151) from the culture collection of the Institute of Chemotherapy were used as test organisms.

Ciprofloxacin (CIP) used in this study was synthesized at BAYER AG. Minimal inhibitory concentrations (MICs) were determined by microtiter dilution technique.

The Influence of Antibiotics on the
Host-Parasite Relationship II
Edited by D. Adam, H. Hahn, W. Opferkuch

Serial delutions of 0,1 ml of ciprofloxacin were prepared in active human serum, containing 20% Isosensitest broth. To these dilution 0,1 ml were added of a 10^{-3}–10^{-4} dilution of an overnight culture of organisms ending in a final inoculum of $1–2 \times 10^{-5}$ colony forming units per ml. The microtiter plates were incubated at 37 °C for 18 hours. MICs were determined by visual inspection for turbidity.

For time-kill studies, exponentially growing bacteria in active human serum were incubated at 37 °C and various concentrations of the drug were added (0.025–2.0 mcg/ml). Dilutions were plated on Isosensitest agar and the number of colony forming units were counted after incubation for 18 hours at 37 °C.

Ciprofloxacin treated cells were harvested by centrifugation at two hours after administration of the drug. The pellet was washed twice with phosphate buffered saline. Thereafter, cells were initially fixed with 0,5% glutaraldehyde in cacodylate buffer (0,2 M, pH 7.2) containing 10 mM $CaCl_2$ by adding the fixative to the culture broth and leaving for 15 minutes at room temperature. Fixation was continued with 2% glutaraldehyde in the same buffer for 2 hours at 4 °C. The cells were washed repeatedly in cacodylate buffer and were then concentrated in 2% (wt/vol) agar. At this step small blocks were cut and postfixed in 2% OsO_4 in the same buffer for 4 hours at 4 °C. After washing and staining in 0,5% uranyl acetat overnight, the blocks were dehydrated in graded ethanol series and embedded in Epon. Thin sections were stained with lead citrate and uranyl acetate and were then observed under a Philips 400 T electron microscope.

Results

The MIC for Escherichia coli C14 was found to be ≤0,015 mcg/ml, for Pseudomonas aeruginosa 220 0,125 mcg/ml and for Staphylococcus aureus 25151 0,125 mcg/ml.

The results of the killing curves are shown in Fig. 1. Ciprofloxacin at concentrations of 0,1 mcg/ml showed a rapid bactericidal effect with E. coli. After 3 hours the initial number of viable bacteria were reduced by about 5 logs. Even 0.025 mcg/ml reduced the number ob viable bacteria by 99% within 2 hours. A strong bactericidal effect was also seen with Pseudomonas at 2,0 mcg/ml. Also 0,5 mcg/ml killed more than 99% of the initial inoculum within 2 hours. With Staphylococcus, a slower killing effect was observed even at the high concentration of 4 mcg/ml although 0,5 mcg/ml resulted in a good bactericidal effect already, which reached its maximum after 5 hours.

Figure 2 shows untreated cells of E. coli C14, having a normal cellular wall contiguous to the plasma membrane. The cytoplasm was homogenous. When E.coli C14 was treated with 0,025 mcg/ml of CIP numerous elongated forms were observed. In thin sections these filaments showed significant changes in the bacterial periphery. In the peripheral ares of the cytoplasm local loosenings of different sizes were seen. These are marked in Fig. 3 by the arrow heads. These ultrastructural changes become larger with higher concentrations of CIP. Cells of E. coli C14 which were treated for 2 hours with 0,1 mcg/ml were severely damaged and most of the cells were lysed (Fig. 4).

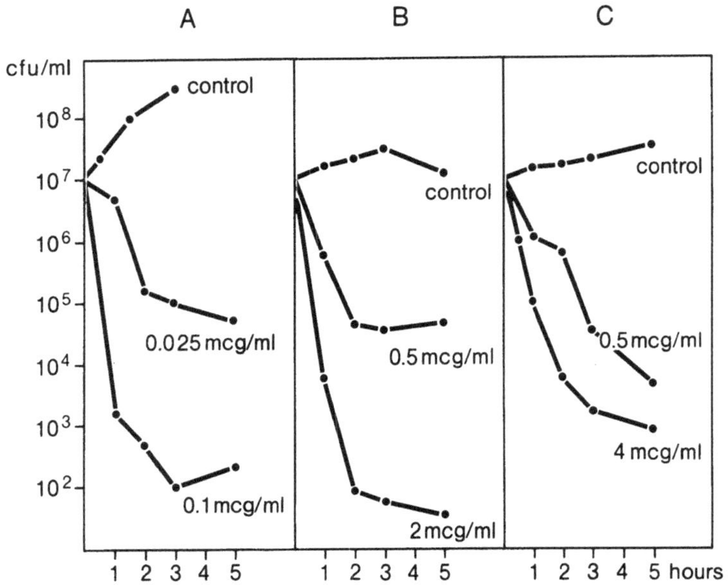

Fig. 1. Killing of Escherichia coli V14 (**a**), Pseudomonas aeruginosa 220 (**b**) and Staphylococcus aureus 25151 (**c**) cultivated in human serum by various concentrations of ciprofloxacin

The effects of ciprofloxacin on the ultrastructure of Pseudomonas aerug. 220 are shown in Fig. 5–7. Fig. 5 gives an example of the shape and structure of control cells that were not exposed to CIP. Two hours after addition 0,25 mcg/ml ciprofloxacin large areas of central loosenings in the plasm as well as zones of optically dense condensations of the cytoplasm could be observed (Fig. 6). Furthermore numerous cells being in the process of lysis were detected. Incubation of Pseudomonas aerug. 220 for two hours with 2,0 mcg/ml CIP showed that most of the cells were irreversibly damaged (Fig. 7).

Cells of Staphylococcus aureus 25151 were influenced in a different way by ciprofloxacin as compared to the results with the gram-negative species. Fig. 8 shows untreated cells of Staphylococcus aureus 25151, having a normal cell wall with septa. Treatment of these cells with 0,25 mcg/ml ciprofloxacin for two hours led to severe disturbances of septum formation and cell shape (Fig. 9). In addition, swollen cells which were frequently observed without their envelopes were as well detected as completely lysed bacteria. In the final stage (Fig. 10), only lysis and empty envelopes were found.

Discussion

There are some reports in the literature dealing with scanning electron microscopical observations on the influence of newer quinolones with regard to shape and structure of E. coli [48]. This study, performed with transmission electron microscopy clearly demonstrates that the ultrastructure of gram-negative and gram-positive bacterial cells is severely affected by a killing effect of ciprofloxacin. It could

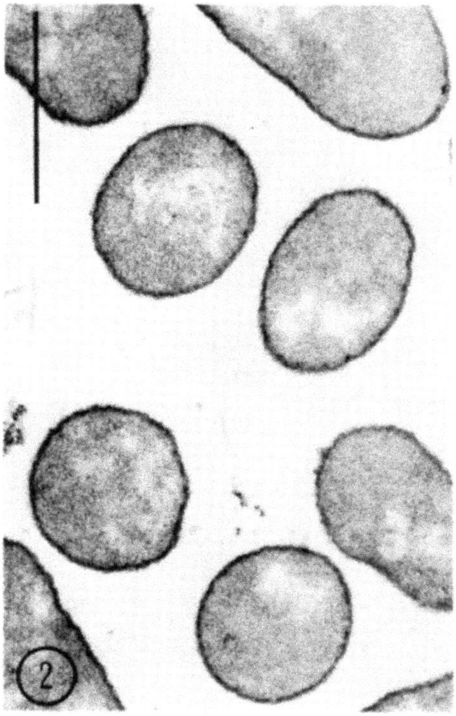

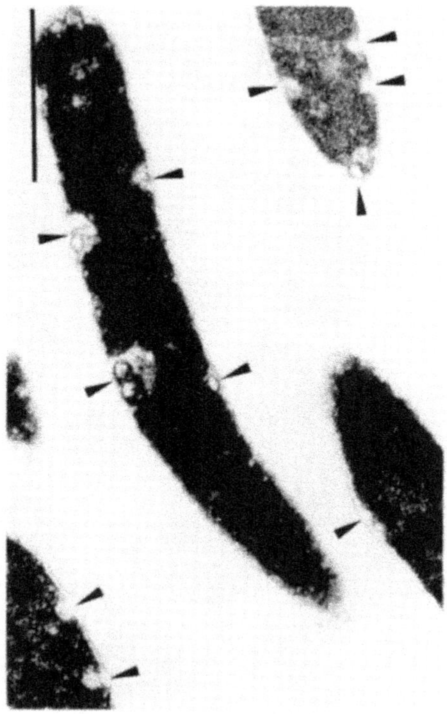

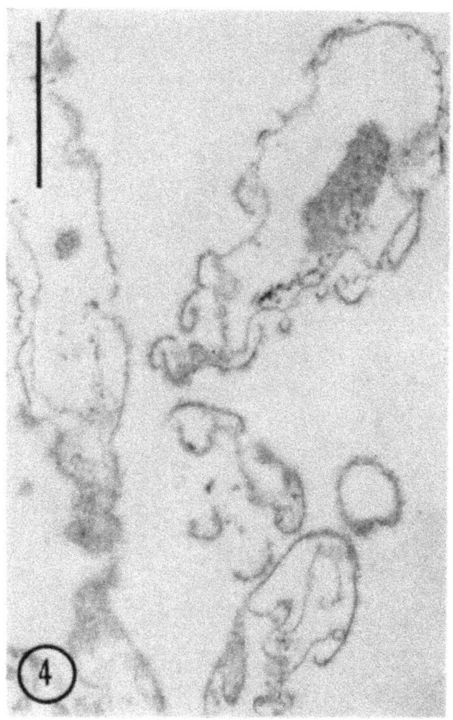

Fig. 2. Escherichia coli C14 grown in 100% human serum. Control section. Magn.: 25 000 × (all bars = 1 μm)

Fig. 3. Escherichia coli C14 grown in 100% human serum, 2 hours after treatment with 0.025 mcg/ml ciprofloxacin. Arrows show typical loosenings at the periphery of filamentous bacteria. Magn.: 21 000 ×

Fig. 4. Escherichia coli C14 grown in 100% human serum, 2 hours after treatment with 0,1 mcg/ml ciprofloxacin. At this time most of the bacteria are lysed and can be seen as empty cell remnants. Magn.: 21 000 ×

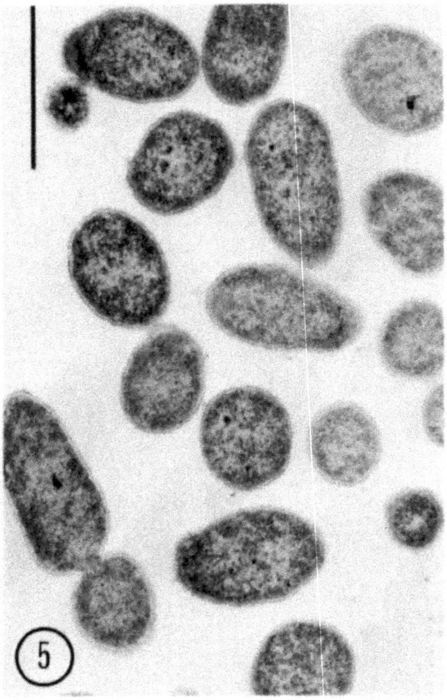

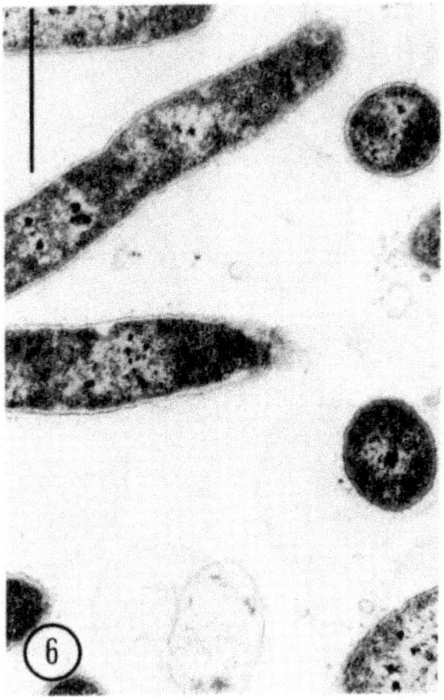

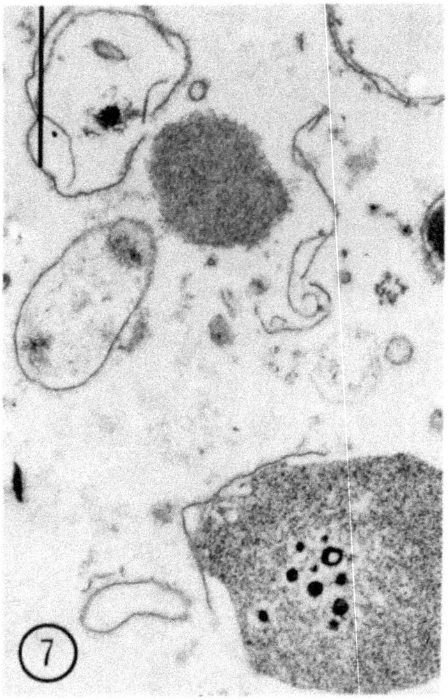

Fig. 5. Pseudomonas aeruginosa 220, cultivated in 100% human serum. Control section. Magn.: 21 000 ×

Fig. 6. Pseudomonas aeruginosa 220, cultivated in 100% human serum, 2 hours after treatment with 0,25 mcg/ml ciprofloxacin. Partial loosenings and condensations in the cytoplasm of elongated bacteria are detectable. Magn.: 21 000 ×

Fig. 7. Pseudomonas aeruginosa 220 cultivated in 100% human serum, 2 hours after treatment with 2,0 mcg/ml ciprofloxacin. At this concentration most of the cells were severely damaged or lysed. Magn.: 21 000 ×

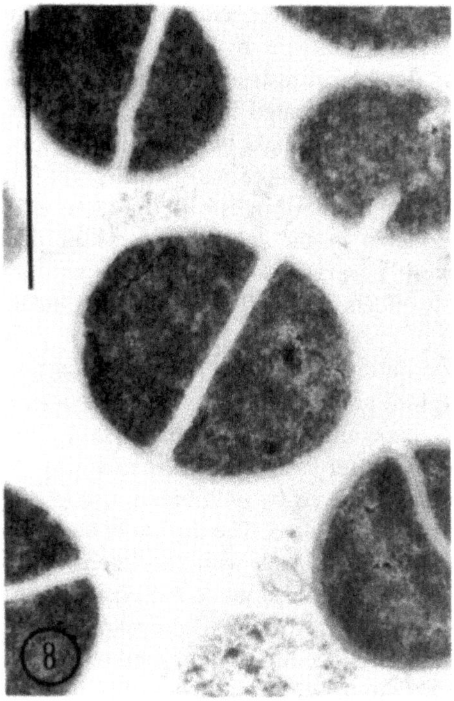

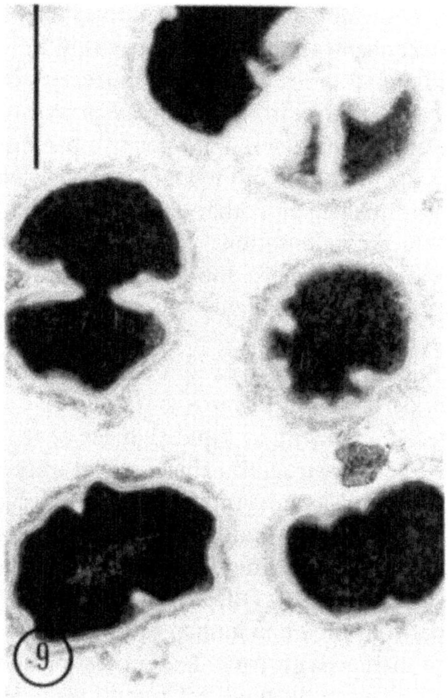

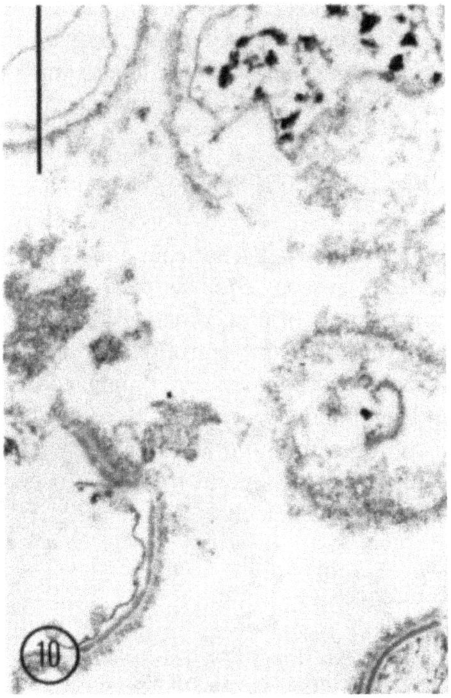

Fig. 8. Staphylococcus aureus 25151, grown in 100% human serum. Control section. Magn.: 35 700 ×

Fig. 9. Staphylococcus aureus 25151, grown in 100% human serum, 2 hours after treatment with 0,25 mcg ciprofloxacin/ml. Early changes show severe disturbances of septum formation and deformation of their shape. Magn.: 21 000 ×

Fig. 10. Staphylococcus aureus 25151, grown in 100% human serum, 2 hours after treatment with 0,5 mcg ciprofloxacin/ml. Most of bacteria are lysed or appeared as empty envelopes. Magn.: 21 000 ×

be shown by in vitro killing curves that CIP has a strong bactericidal effect in serum in concentrations which can easily be achieved in humans by oral or parenteral administrations [3, 6, 10]. The results of the present ultrastructural study demonstrates that ciprofloxacin may primarily affect areas located in the membrane of gram-negative and gram-positive bacteria. These effects might be the consequences of the inhibition of the DNA-gyrase which is supposed to be primarily the target for ciprofloxacin and other quinolones [8]. In additional experiments with Pseudomonas strong condensations of cytoplasmic components in the loosened areas of the cytoplasm (where the DNA is located) were observed soon after starting the treatment. This might be an indication of early effects of ciprofloxacin in this cellular region.

Staphylococci showed a different inhibition pattern suggesting differences in the mode of action towards gram-negative and gram-positive bacteria. This difference became evident by rapid changes of the cell shape and disturbed cell division.

The ultrastructural changes that were observed after two hours of treatment with CIP were distintegration of bacterial cytoplasm followed by membrane disruption and plasmolysis in dependence of CIP concentration applied. The degree of damage observed with different drug concentrations correlated well with the intensity of bactericidal effect measured in the time-kill-curves. We also made the observation that cells grown in human serum seemed to be more severly damaged by ciprofloxacin than cells grown in Isosentitest brothes (Voigt and Zeiler, unpublished results).

Studies performed by Crumplin et al [4] revealed that norfloxacin may also induce the formation of filamentous cells whilst also permitting anomalous cell-division events. In this case, cells of Escherichia coli K12 were treated for 4 or 5 hours with 1 mcg/ml norfloxacin. However, disintegration of cells or lysis at such high concentrations was not observed in this study by scanning electron microscopical techniques.

Acknowledgement. I would like to thank Dr. G.E. Hübner for his assistence in the preparation of this manuscript.

Summary

Serum resistant Escherichia coli C 14, Pseudomonas aeruginosa 220 and Staphylococcus aureus 25151 as test organisms were used to investigate the effect of ciprofloxacin in pooled active human serum. Visualized in electron microscopy at low concentrations ciprofloxacin elongation of gram-negative bacterial cells was observed. In addition osmiophilic condensations in the areas of the bacterial DNA in Pseudomonas aerug. 220 could be detected. At higher concentrations cell membranes ruptured and lysis occurred. In staphylococci surface deformation and disturbances of septum formation occured first and finaly, proteolysis and empty envelopes were seen.

References

1. Barry AL, Jones RN, Thornsberry C, Ayers LW, Gerlach EH, and Sommers HM (1984) Antibacterial activities of ciprofloxacin, norfloxacin, oxolinic acid, cinoxacin and nalidixic acid. Antimicrob Agents Chemother 25: 633–637

2. Chin NX and Neu HC (1983) In vitro activity of ciprofloxacin Bay 09867. Antimicrob Agents Chemother 24: 568–574
3. Crump B, Wise R, and Dent J (1983) Pharmacokinetics and tissue penetration of ciprofloxacin. Antimicrob Agents Chemother 24: 784–786
4. Crumplin GC, Kenwright M, and Hirst T (1984) Investigations into the mechanisms of action of the antibacterial agent norfloxacin. J Antimicrob Chemother 13 (Suppl. B): 9–23
5. Fass RJ (1983) In vitro activity of ciprofloxacin (Bay 09867). Antimicrob Agents Chemother 24: 568–574
6. Gonzales MA, Uribe F, Moisen SD, Fuster AP, Selen A, Welling PG, and Painter B (1984) Multiple-dose pharmakokinetics and safety of ciprofloxacin in normal volunteers. Antimicrob Agents Chemother 26: 741–744
7. Reeves DS, Bywater MJ, Holt HA, and White LO (1984) In vitro studies with ciprofloxacin, a new 4 quinolone compound. J Antimicrob Chemother 13: 333–346
8. Selwyn S, and Baktiar M (1984) Comperative in vitro studies of a new azaquinolone, enoxacin. Drugs Exptl Clin Res 10: 653–668
9. Smith JT (1984) Awakening the slumbering potential of the 4 quinolone antibacterials. The Pharmaceutical Journal 233: 299–305
10. Wingender W, Graefe HH, Gau W, Förster D, Beermann D, and Schacht P (1984) Pharmacokinetics of ciprofloxacin after oral and intraveneous administration in healthy volunteers. Eur J Clin Microbiol 3: 355–359
11. Wise R, Andrews JM, and Edwards LJ (1983) In vitro activity of Bay 09867, a new quinolone derivative, compared with those of other antimicrobial agents. Antimicrob. Agents Chemother 23: 559–564
12. Zeiler HJ and Grohe K (1984) The in vitro and in vivo activity of ciprofloxacin. Eur J Clin Microbiol 3: 339–343

Discussion

Ginsburg: Is the lytic effect noted on the cell-wall concerned with the activation of autolytic wall enzymes?

Voigt: This is not known at the time being.

Haag: How can a gyrase-inhibitor cause the ultrastructural changes you showed? Is it not necessary to look for another mode of action, apart from gyrase inhibition, in order to explain your findings?

Voigt: My colleague will comment on this topic.

Ghione: Quinolone derivatives are intercalating agents. Possibly they can intercalate in DNA and cell-membrane as well. A similar phenomenon can be observed with other intercalating agents, such as enthracyclines. Can you comment on this point?

Voigt: It has been proved that ciprofloxacin interacts with one of the subunits of the DNA-gyrase; it is not known up to now that ciprofloxacin binds directly to the DNA.

Stephens: The similar effects seen with nalidixic acid the compound from which the quinolones are derived?

Voigt: There is no published literature on this aspect.

Dalhoff: This more a comment than a question. Based on the data published by J.T. Smith (London) and the preliminary evidence presented in my paper, I would like to speculate that ciprofloxacin may not only inhibit the gyrase but may also interact with bacterial membranes. Probably the phenomena described by Dr. Voigt so impressingly may be due to a direct interaction with the bacterial membrane and may not just be a secondary effect only.

Exposure of protein antigens and reduction of siderophore production after growth of Gram-negative bacteria in Subinhibitory concentrations of cephalosporins

J. L. Kadurugamuwa[1], H. Anwar[1], M. R. W. Brown[1]* and O. Zak[2]

[1] Microbiology Research Group, Institute of Pharmaceutical Sciences, Aston University in Birmingham, Gosta Green, Birmingham B9 7ET, U.K.
[2] Ciba-Geigy Limited, Pharma Research, Infectious Diseases, CH-4402 Basel, Switzerland

Introduction

The β-lactams are among the most commonly used antibiotics in clinical practice. It is known that the molecular targets of β-lactam antibiotics are penicillin-binding proteins (PBPs) located in the cytoplasmic membrane of both Gram-negative and Gram-positive bacteria [24,32]. The PBPs are enzymes involved in the synthesis of peptidoglycan which plays an important role in confering rigidity and osmotic stability on the bacterial envelope [32]. Disruption of the synthesis of peptidoglycan by β-lactam antibiotics causes morphological changes and eventually lysis of the bacterial cells. Recently we found that growth of *K. pneumoniae* in the presence of subinhibitory concentrations of cephalosporins (not affecting growth rate) caused reduction in the production of siderophores and also in capsule formation [12]. The reduction in capsule resulted in rapid agglutination of the antibiotic treated cells of encapsulated *K. pneumoniae* by antisera raised against a non-encapsulated strain of *K. pneumoniae*[12]. This was found to be due to the exposure of outer membrane proteins antigens and possibly also lipopolysaccharide on the surface of encapsulated *K. pneumoniae* after growth in the presence of cephalosporins (submitted for publication).

In this study, we have investigated several strains of *K. pneumoniae* and also a strain of *Escherichia coli* carrying a plasmid Col V^+ encoding for the synthesis of aerobactin. The results indicated that cephalosporins reduced the production of enterochelin and aerobactin in both species and also caused reduction in capsule polysaccharide production in *K. pneumoniae.*

Materials and Methods

Bacteria: Three clinical isolates of *Klebsiella pneumoniae* (JK; GS; HA) obtained from patients with urinary tract infection were used in the study. *Escherichia coli* Col V^+ was a generous gift from Dr E. Griffiths.

Media and glassware: Cells were grown under iron-depleted (Fe-CDM) and iron sufficient (Fe+CDM) conditions, in a chemically defined medium described in a previous study [12].

Antibiotics: CGP 17520 a cephalosporin antibiotic with an α-aminoacid residue in its side chain was synthesized at Ciba-Geigy Ltd., Basel, Switzerland, and ceftriax-

The Influence of Antibiotics on the
Host-Parasite Relationship II
Edited by D. Adam, H. Hahn, W. Opferkuch

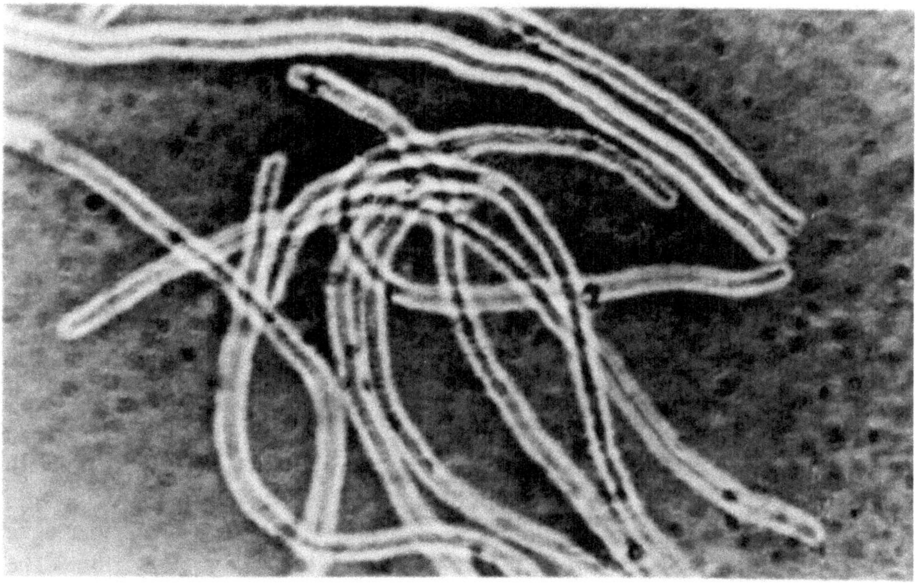

Fig. 1. A photomicrograph showing filamented *K. pneumoniae* (JK) grown in the presence of subMIC of a cephalosporin

one Hofman-La Roche, Basel, Switzerland, were used in the study. The sub-MIC for each antibiotic was determined as previously described [12]. The lowest concentration of antibiotic which had no effect on growth, yet induced filamentation (Fig. 1) on each bacterial strain used in this study was 0.3 μg/ml (1:10 MIC) *K. pneumoniae* (JK); 0.075 μg/ml (1:10 MIC) *K. pneumoniae* (GS); 0.15 μg/ml (1:5 MIC) *K. pneumoniae* (HA) for CGP-17520 and were 0.02 μg/ml (1:5 MIC) *K. pneumoniae* (JK); 0.005 μg/ml (1: 40 MIC) *K. pneumoniae* (GS); 0.015 μg/ml (1:13 MIC) *K. pneumoniae* (HA) for ceftriaxone. The sub-MIC of ceftriaxone against *E.coli* col V^+ was 0.008 μg/ml (1:10 MIC).

Assay of siderophores in culture supernatant

The method used to assay phenolate siderophores (enterochelin) was based on that described by Arnow with 3,4-dihydroxyphenylalanine as a standard and hydroxamate siderophores (aerobactin) were detected by the method of Payne [19] with benzohydroxamic acid (Aldrich chemical Co. LH, Dorsed, England) as a standard.

Preparation of antisera

The three strains of *Klebsiella* were grown in Fe-CDM in the presence or absence of CGP 17520 or ceftriaxone cells were harvested, washed twice in normal saline and resuspended in saline to a cell density of 10^9 cells/ml. The cells were killed by addition of formalin (final concentration, 4% vol/vol). Half Lop rabbits (3 kg) were immunized with 1 ml of the washed formalin treated cells intradermally at each of three sites in the neck at weekly intervals for 8 weeks. The rabbits were bled from the marginal ear vein, and then bulked serum from the rabbits were kept at −20° in 2 ml aliquots.

Surface exposed antigen preparation

The surface exposed antigens were investigated by modification of the method described by Swanson [27], in which the iodine label step is omitted and antigenic sites visualised by immunoblotting using goat anti-rabbit IgG immunoglobulin. 2-Liters of cells grown either in the absence or presence of ceftriaxone or CGP 17520 were harvested by centrifugation at 10,000 × g for 10 min and washed once with prewarmed (37°) 10 mM phosphate buffer saline (PBS), pH 7.4. The cells were treated with 6 ml of homologus antisera raised against cells grown under the same growth conditions and incubated at 37 °C in an orbital shaker for 15 min. The viability was unaffected by incubating with antisera raised against each of the *Klebsiella* strains used in the study. The cell suspension was then centrifuged at 10,000 × g for 10 min and washed three times with warmed PBS. The cell pellet was resuspended in 10 ml 1% *N*-tetradecyl-*N*, *N*-dimiethyl-3-ammonio-1-propanesulfonate (Zwittergent 3-14, Calbiochem-Behring Corp., La Jolla, Calif. USA) in 10 mM PBS, pH 7.4. The mixture was well mixed, incubated at 37 °C for 30 min and centrifuged at 6000 × g for 10 min. The supernatant was carefully removed and transferred to a cylindrical 10 ml column containing 5 ml previously swollen protein A-Sepharose 4B CL (Sigma Chemical Co., St Louis, USA). The mixture was incubated for 20 min at room temperature with regular end to end rotation, and the liquid was run off the bottom of the column which was then washed with 120 ml PBS. The antigen-antibody complexes bound to sepharose protein-A were then eluted with 100 ml 0.1 M glycine, pH 2.5 with regular end to end rotation. The eluate was dialysed against distilled water at 4 °C overnight, lyophilised and resuspended in 1 ml denaturing buffer containing 4% sodium dodecyl sulfate (specially pure, British Drug House, Poole, United Kingdom), 8% 2-mercaptoethanol, 20% glycerol, 0.125 M tris (hydroxymethyl) aminomethane (pH 6.8) and 0.01% bromophenol blue. The sample was stored at −20 °C.

Preparation of outer membrane and sodium dodecyl sulphate-polyacrylamide gel electrophoresis

Orter membranes were prepared and proteins separated on SDS-PAGE as described in our previous study [12].

Immunoblotting: OM proteins separated on polyacrylamide gels were transferred onto nitrocellulose paper and antigenic sites visualised according to Towbin et al [28].

Table 1. Effect of sub-inhibitory levels* of cephalosporins on phenolate and hydroxamate siderophore production in *K. pneumoniae*

	Siderophore µg/mg dry wt					
	Enterochelin			*Aerobactin*		
Strain	*Fe-CDM*	*Fe-CDM + subMIC antibiotic*		*Fe-CDM*	*Fe-CDM + subMIC antibiotic*	
K. pneumoniae (JK)	16.6	8.1	CGP17520	212.8	129.2	CGP17520
		8.4	ceftriaxone		156.8	ceftriaxone
K. pneumoniae (GS)	19.5	7.03	CGP17520	ND	ND	
		10.3	ceftriaxone	ND	ND	
K. pneumoniae (HA)	17.21	9.5	ceftriaxone	458.9	390.9	ceftriaxone

ND Not detected
* The highest concentration of cephalosporins not affecting the growth rate was added to culture media

Results

Table 1 shows the production of phenolate (enterochelin) and hydroxamate (aerobactin) siderophores by three strains of *K. pneumoniae* (JK, HA & GS) in the absence and presence of sub-MIC of either CGP 17520 or ceftriaxone at concentrations which had no effect on the growth rate. Production of phenolate siderophore was observed in all three strains used in the study. Phenolate siderophore production was reduced by half when these strains were grown in the presence of either CGP 17520 or ceftriaxone. Production of hydroxamate siderophore could be observed only in two strains (JK, HA) and it was barely detectable in *K. pneumoniae* (GS). The production of hydroxamate siderophore was also reduced when those two strains, which were known to produce this type of siderophore, were grown in the presence of CGP 17520 or ceftriaxone. Both types of siderophores were barely detectable when these strains of *K. pneumoniae* were grown under iron sufficient conditions.

Figure 2 illustrates the production of (a) enterochelin and (b) aerobactin by *E. coli* Col V^+ during growth in the absence or presence of ceftriaxone under iron depleted or iron sufficient conditions. The production of enterochelin (Fig. 2b) was slightly but significantly reduced when *E. coli* was grown in the presence of ceftriaxone. Simultaneous production of both phenolate and hydroxamate siderophores by Gram-negative bacteria has been reported by other workers [20]. Increased production of enterochelin was observed and its production paralleled the growth cycle [Fig. 2a]. However, only a trace amount of aerobactin was detected during the exponential phase of growth. The production of aerobactin dramatically increased when the culture entered late exponential or stationary growth phase. Similar results have been obtained by other workers [9, 26, 31].

Figure 3a illustrates the outer membrane protein (OMP) profiles of *K. pneumoniae* grown under iron-depleted conditions (lane 2) and also under iron sufficient conditions (lane 1). Six to seven high molecular weight OMPs (Fig. 3a,

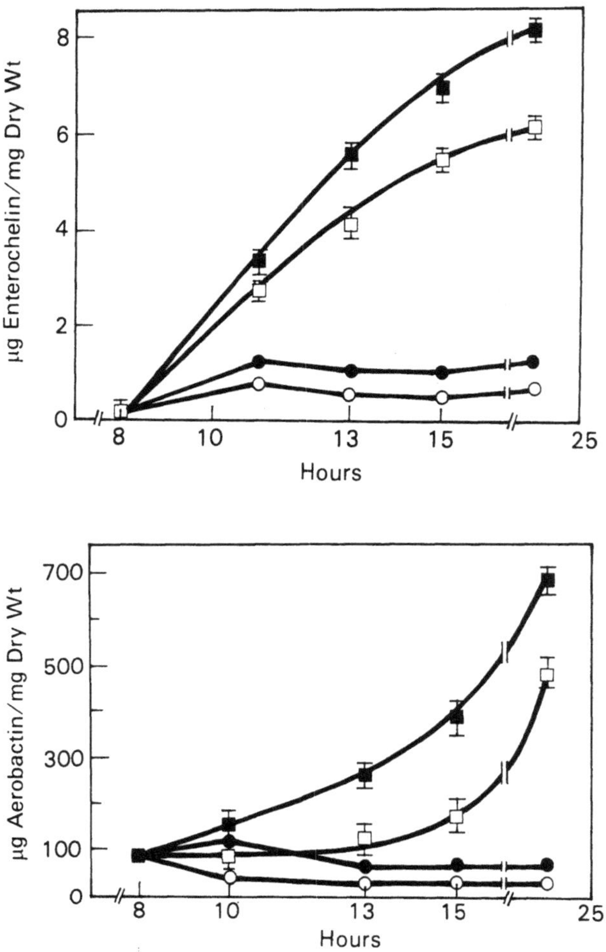

Fig. 2. Effect of subMIC of ceftriaxone on (a) enterochelin and (b) aerobactin production by *E. coli* V^+. Symbol: ■ Fe-CDM, □ Fe-CDM plus subMIC of ceftriaxone, • Fe+CDM, ○ Fe+CDM plus subMIC ceftriaxone. The highest concentration of ceftriaxone (see text) not affecting the growth rate was added to Fe+CDM and Fe-CDM. Siderophore production was monitored during batch growth

lane 2), known as iron-regulated membrane proteins (IRMP) 69-83K, were induced when the bacteria were grown under iron-depleted conditions. Growth of the bacteria under iron-sufficient conditions completely repressed the expression of the IRMPs (lane 1). The proteins with M_r of 32.5K, 35.5K and 39K have been suggested to be related to Omp A, Omp C and Omp F of *E. coli* respectively. Furthermore, the proteins with M_r of 35.5K and 39K have been found to be pore-forming proteins which allow the penetration of small hydrophilic molecules through the OM and were non-covalently associated with peptidoglycan [33]. Closely similar OMP profiles were observed in all strains used in the study (data not shown) and growth of the organisms in the presence of cephalosporins did not alter the OMP profiles.

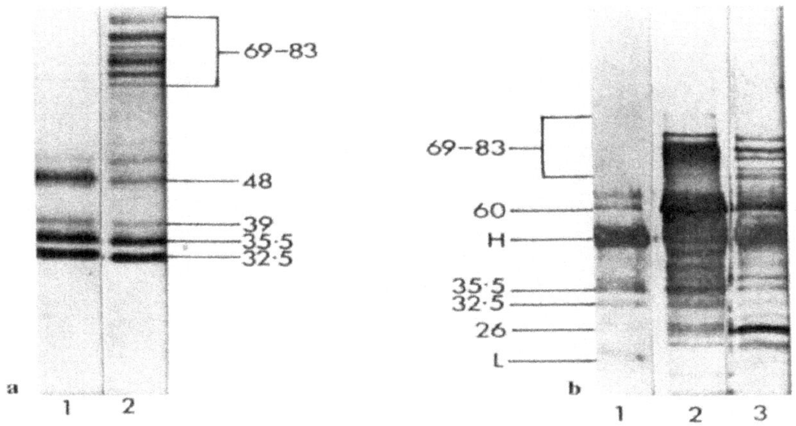

Fig. 3a. OMP profiles of *K. pneumoniae* (JK) during growth under Fe+CDM (Lane 1) and Fe-CDM (Lane 2). Seven high molecular weight proteins M_r 69–83K were observed under Fe-CDM. These are referred to as iron regulated membrane proteins (IRMPs). Similar results were obtained with other strains of *Klebsiella*. OMP profile was not affected by subMIC of cephalosporins. A protein with M_r of 48K was heavily induced in Fe + CDM; **b.** Immuno-blot of the surface-exposed antigens (SEPA) of *K. pneumoniae* (JK) electrophoretically transferred onto nitrocellulose paper and reacted with antisera raised against cells grown in Fe-CDM. Lane 1 SEPA of cells grown in Fe-CDM. Lane 2 cells grown in Fe-CDM plus CGP-17520. Lane 3 SEPA of cells grown in Fe-CDM plus ceftriaxone. *K. pneumoniae* (JK) had three protein antigens (60K, 35.5K, 32.5K) exposed on the surface (Lane 1). Growth of the same strain in Fe-CDM in the presence of sub inhibitory concentrations of cephalosporins (Lane 2 and 3) resulted in expression of a greater number of protein antigens including iron-regulated membrane proteins (M_r 69-83K). H and L the heavy and light chains of IgG molecule

However, changes in the surface properties were observed when the bacteria were grown in the presence of cephalosporins. The surface hydrophobicity of the organisms increased when they were grown in the presence of cephalosporins [12]. Furthermore, reduction of capsule was also observed and the cells became agglutinable by antisera raised against a non-encapsulated strain of *K. pneumoniae*. We used a modification of the method described by Swanson [27] to investigate the exposure of protein antigens on the surface of filamented *K. pneumoniae* as a result of growth in the presence of either CGP 17520 (Fig. 3b, lane 2) or ceftriaxone (Fig. 3b, lane 3). The results indicated that the reduction of capsular polysaccharide resulted in the exposure of a number of additional protein antigens including the IRMPs on the surface of encapsulated *K. pneumoniae* (Fig. 3b). Capsular polysaccharide is known to be located on the outermost layer of the bacterial cell envelope. It is either loosely or firmly anchored in the OM through the fatty acid residues of the molecules [11]. Capsular polysaccharide occlude the cell and prevents access of antibodies to antigenic determinants, including proteins exposed at the surface. Reduction of the capsular polysaccharide resulted in the exposure of protein antigens and they became accessible to antibodies.

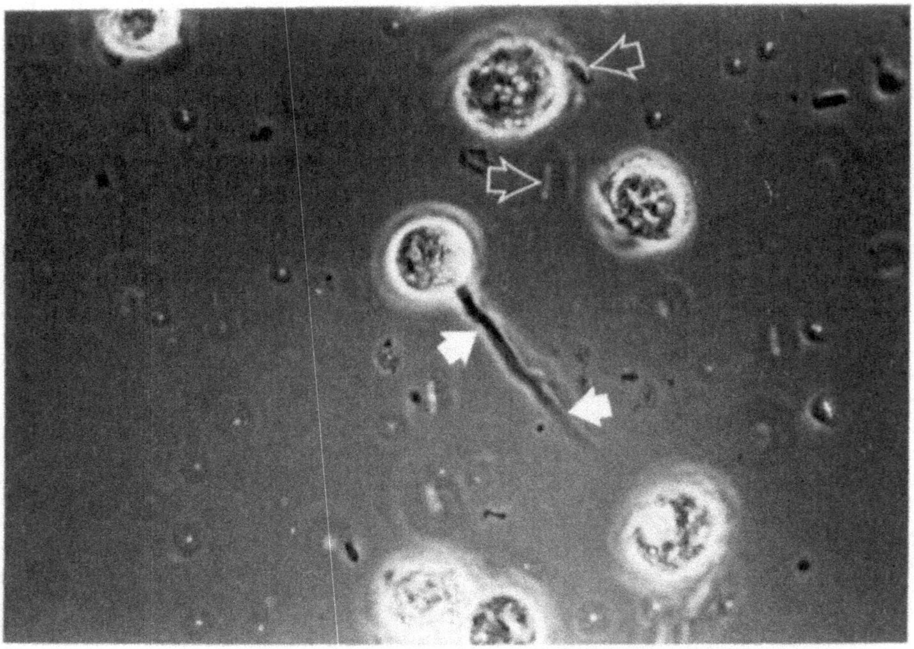

Fig. 4. Phagocytosis of a filamented *K. pneumoniae* (JK) (▲) as a result of growth in the presence of subMIC of CGP 17520 (open arrow). The non-filamented Klebsiella (△) are relatively less engulfed by rabbit PMNs (closed arrow).

We have investigated the phagocytosis of cells grown in the presence of cephalosporins by rabbit polymorphonuclear leukocytes (PMNs) in whole blood. Our preliminary results indicated that filamented bacterial cells were more prone to engulfment by the PMNs in comparison to non filamented cells. Fig. 4 shows a photomicrograph of the engulfment of filamented bacterial cells by rabbit PMN.

The outer membrane protein profiles of *E. coli* during growth under iron sufficient and iron-depleted conditions in the absence and presence of ceftriaxone are shown in Fig. 5. Expression of five OMP with M_r of 83K, 81K, 80K, 78K, and 74K were observed when the bacterial cells were grown under iron-depleted conditions (lane 9 to 18). They were expressed early in the logarithmic phase of growth (at OD 470 nm of 0.3) and are barely detectable in the OM of the cells grown under iron sufficient conditions. Growth of the organisms in the presence of cephalosporins either under iron-sufficient or iron-depleted conditions did not seem to affect the expression of any of the major OMPs.

Discussion

The surface structures and composition of bacteria play an important role in the interaction with host defences during *in vivo* growth [23]. Antibiotics have tra-

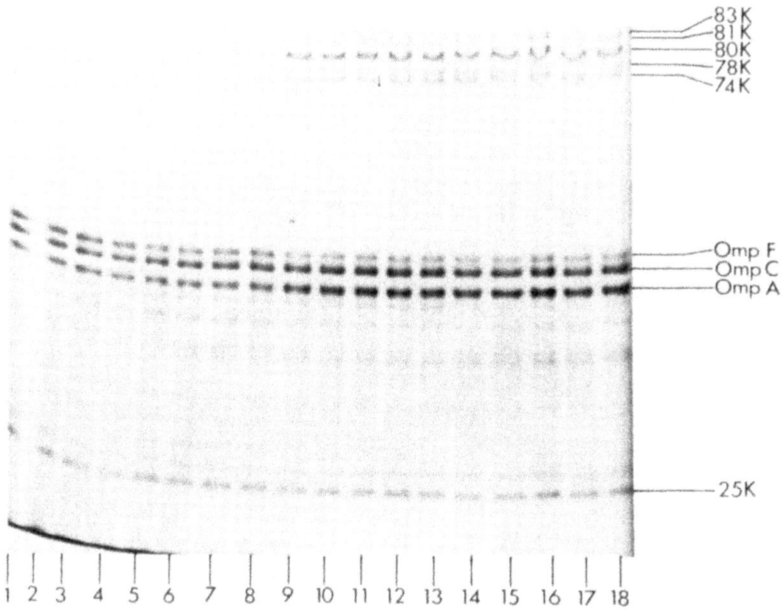

Fig. 5. OMP profiles of *E. coli* col. V^+ during growth under iron-sufficient and iron-depleted conditions. Lanes: 1, Fe + CDM, Optical density of 0.9; 2, Fe + CDM, Optical density of 2.5; 3, Fe + CDM, Optical density of 4.0; 4, Fe + CDM, optical density of 4.8; 5, Fe+CDM+ceftriaxone, optical density of 0.8; 6, Fe+CDM+ ceftriaxone, optical density of 2.3; 7, Fe+CDM+ceftriaxone, optical density of 3.8; 8, Fe+CDM+ ceftriaxone optical density of 4.5; 9, Fe-CDM optical density of 0.7; 10, Fe-CDM, optical density of 2.1; 11, Fe-CDM optical density of 3.8; 12, Fe-CDM optical density of 4.0; 13, Fe-CDM optical density of 4.5; 14, Fe-CDM+ceftriaxone optical density of 0.7; 15, Fe-CDM+ceftriaxone optical density of 2.0; 16, Fe-CDM+ceftriaxone optical density of 3.5; 17, Fe-CDM+ceftriaxone optical density of 3.8; 18, Fe-CDM+ceftriaxone optical density of 4.0. A 150-ml portion of each culture was removed from a 5-liter culture flask at 60-min intervals for OMP preparation

ditionally been developed because of their ability to inhibit or kill the microorganisms *in vitro*. In this study, we have investigated a novel mode of action of cephalosporins which we believe may be an important factor for their success in clinical practice. For infecting bacteria to be successful *in vivo*, they have to overcome the continuous threat enforce by the specific and non-specific host defences. Iron is an essential nutrient for proliferating bacteria. Restriction of iron availability is a non-specific defence against invading microorganisms [10]. However, pathogenic microorganisms have evolved several efficient iron acquisition mechanisms which involve the excretion of iron chelating molecules (siderophores) into the growth environment and expression of receptors (IRMPs) on the OM for the uptake of iron-siderophore complexes [14].

It has been suggested that the production of siderophores *in vivo* may be necessary for the pathogensise of certain microorganisms [10, 16, 22] and certainly that *E. coli* harbouring col V^+ plasmid are more virulent and invasive than plasmid-free strains [16, 35]. There is evidence to demonstrate that aerobactin mediated iron uptake is independent of the chromosomally mediated enterochelin system [35]. Recent studies has shown that enterochelin becomes inactivated after binding to serum proteins [17] and destroyed enzymatically after fulfilling its iron transport function, whereas aerobactin has a superior ability to transfer iron and to recycle following its iron transport without been degraded [4]. In this study we have shown that the production of enterochelin parallels cellular growth, whilst large quantities of aerobactin in culture supernatants of *E. coli* were detected only in late-exponential and stationary phase of the growth cycle. Similar results have been observed by other workers with *E. coli* [26] and *Enterobacter cloacae* [31].

The production of siderophores with the ability to be reused in iron uptake is probably a great advantage for bacterial survival in an iron-restricted environment, especially when cell proliferation has come to a halt. Growth of *E. coli* col V^+ in the presence of ceftriaxone resulted in reduced production of both types of siderophores, yet the antibiotic concentration used had no effect on growth rate.

We observed the induction of 5 new molecular OMP's, when *E. coli* was grown in Fe-CDM: M_r 83K, 81K, 80K, 78K and 74K. These disappeared when the organism was grown in Fe+CDM. The induction of a number of high molecular weight proteins in *E. coli* under iron stressed conditions has been reported [13]. The synthesis of any of the high molecular weight proteins (74-83K) were not affected by growing the organism in the presence of cephalosporins. The 81K and 74K proteins have been identified as the receptors for enterochelin and aerobactin iron uptake systems respectively [10, 35]. The three major bands with M_r values approximately 32K-40K proteins which are known to be Omp A, Omp C and Omp F were always present and appeared to be independent of the growth conditions employed.

The reduction in the synthesis of siderophores caused by the β-lactams may reduce the necessary supply of iron to the bacterial cells in an infection.

The capsular polysaccharides of both Gram-negative and Gram-positive bacteria have been suggested to play in important role in pathogenesis [23]. The encapsulated strains are known to be resistant to phagocytosis and serum killing and are resistant to phagocytosis in the absence of anti-capsular antibodies [34]. Reduction of capsular polysaccharide renders the cells more liable to phagocytosis and complement mediated serum killing [34]. Other workers have found that antibiotics at sub-inhibitory concentrations caused a reduction of capsule of *K. pneumoniae* when the cells were examined microscopically [8]. In this work, we have used immunological techniques to investigate the surface properties of filamented cells induced as a result of growth in the presence of cephalosporins. Rapid agglutination was observed when the filamented cells were reacted with antisera raised against a non-capsulated isogenic strain of *K. pneumoniae* initially isolated by Poxton and Sutherland [21]. No agglutination was observed when the cells were grown in the absence of the antibiotics. This implies that the filamented cells have similar surface components to the non-capsulated strain and different from those grown in the absence of the antibiotics. The explanation for this observation could be that the antibiotics reduce capsular polysaccharide and thus reveal OM components such as OMP and possibly lipopolysaccharide that are normally occluded when the capsule is present. We have undertaken experiments to investigate the surface exposed antigens of cells grown in the presence or absence of cephalosporins. The results indicated that a number of OMP with M_r in the range of 26K to 83K including the

IRMPs were accessible to antibodies when the cells were grown in the presence of cephalosporins. This is in agreement with our observation using a slide agglutination technique that cephalosporins cause reduction of capsular polysaccharide and expression of antigens which are normally occluded in encapsulated strains of *K. pneumoniae* [12]. Other workers had found that subinhibitory levels of antibiotic affected the expression of surface structures such as pili of *Neisseria* species [25].

Surface hydrophobicity has been suggested to play a significant role in the engulfment of bacterial cells by PMNs [30]. In our previous study we have shown that the surface hydrophobicity of *K. pneumoniae* is increased when bacterial cells are grown in the presence of cephalosporins [12]. In this study, we have obtained preliminary evidence that filamented bacteria, because of β-lactam action are more liable to phagocytosis than non filamented cells by rabbit PMNs. Similar observations have been reported by other workers [15].

In addition to any direct lytic activity of β-lactam antibiotics, it appears that in an iron restricted *in vivo* environment some strains of pathogens may be handicapped by an action on surface antigens and iron uptake.

Nutrient limitation, including iron, has been shown to have profound effects on envelope composition [1, 5, 6] including penicillin-binding protein profiles [29] of bacteria with consequent changes in sensitivity to antibiotics [5, 6] and body defense mechanisms [2] as well as virulence factor production [18]. Information about the biological activity of β-lactam antibiotics would be best obtained using cells grown under iron-depleted conditions which closely mimic those grown in vivo [5].

Summary

The production of phenolate and hydroxamate siderophores was reduced when three strains of *Klebsiella pneumoniae* and a strain of *Escherichia coli* col V^+ were grown under iron-depleted conditions in the presence of a number of cephalosporins at sub-inhibitory concentrations which not affecting on the growth rate. Cephalosporins had no significant effect on the outer membrane protein profiles of either *K. pneumoniae* or *E. coli*. However, changes in surface properties were observed in encapsulated strains of *K. pneumoniae* when the bacterial cells were grown in the presence of cephalosporin. Rapid agglutination was observed when cells grown in the presence of antibiotic were reacted with antisera raised against a non-capsulated strain of *K. pneumoniae*. No agglutination was observed when the cells were grown in the absence of antibiotic. The effect of cephalosporins on the surface of encapsulated strains of *K. pneumoniae* was studied at the molecular level. Growth of encapsulated strains of *K. pneumoniae* in the presence of cephalosporins resulted in the exposure on the surface of a number of outer membrane proteins with M_r in the range of 26K to 83K including the iron-regulated membrane proteins.

References

1. Anwar H, Brown MRW, Cozens RM and Lambert PA (1983) Isolation and characterization of the outer and cytoplasmic membranes of *Pseudomonas cepacia*. J Gen Microbiol *129*: 449–507
2. Anwar H, Brown MRW and Lambert PA (1983) Effect of nutrient depletion on sensitivity of *Pseudomonas capacia* to phagocytosis and serum bactericidal activity at different temperatures. J Gen Microbiol *129*: 2021–2027
3. Bjorn MJ, Sokol PA and Iglewski BH (1979) Influence of iron on yields of extracellular products in *Pseudomonas aeruginosa* cultures. J Bacteriol *138*: 143–200
4. Braun V, Brazel-Faisst C and Schneider R (1984) Growth stimulation of *Escherichia coli* in serum by iron (III) aerobactin. Recycling of aerobactin. FEMS Microbios Letts *21*: 99–103
5. Brown MRW and Williams P (1985) Influence of substrate limitation and growth phase on sensitivity to antimicrobial agents. J Antimicrob Chemother *15* (Suppl A): 7–14
6. Brown MRW and Williams P (1985) Influence of environment on envelope properties affecting survival of bacteria in infections. Ann Rev Microbiol *39*: 527–56
7. Crosa JH (1984) The relationship of plasmid-mediated iron transport and bacterial virulence. Ann Rev Microbiol *38*: 69–89
8. Gemell CG, Spear T and Peterson PK (1983) Morphological changes in *Bacteroides fragilis* and *Klebsiella pneumoniae* attributable to growth in the presence of various antibiotics. Eur J Clin Microbiol *2*: 217–218
9. Gibson F and Magrath DI (1969) The isolation and characterization of a hydroxamic acid (aerobactin) formed by *Aerobactar aerogenes*. Biochim Biophys Acta *192*: 175–184
10. Griffiths E (1983) Availability of iron and survival of bacteria in infection. In: Easmon, CS, Jeljaszewicz J, Brown MRW and Lambert PA. Medical Microbiology, Vol. 3, Academic Press, p. 153–178
11. Jann K and Jann B (1983) K antigens of *E. coli*. In: P Kallos et al. Progress in Allergy, Vol. 33, Karger, Basel, p. 53–79
12. Kadurugamuwa JL, Anwar H, Brown MRW and Zak O (1985) Effect of subinhibitory concentrations of cephalosporins on surface properties and siderophore production in iron depleted *Klebsiella pneumoniae*. Antimicrob Agents and Chemother *27: 220–223*
13. Klebba PA, McIntosh MA and Neilands JB (1982) Kinetics of biosynthesis of iron regulated membrane proteins in *Escherichia coli*. J Bacteriol *149*: 880–888
14. Leong J and Neilands JB (1976) Mechanisms of siderophore iron transport in enteric bacteria. J Bacteriol *126*: 823– 830
15. Lorian V and Atkinson B (1984) Bactericidal effect of polymorphonuclear neutrophils on antibiotic-induced filaments of Gram-negative bacilli. J Infect Dis *149*: 719–727
16. Montgomerie JZ, Bindereif A, Neilands JB, Kalmanson GM and Guze CB (1984) Association of hydroxamate siderophore (aerobactin) with *Escherichia coli*: isolated from patients with bacteremia. Infect and Immun *46*: 835–838
17. Neilands JB (1984) Siderophores of bacteria and fungi. Microbiol Sci *1*: 9–14
18. Ombaka EA, Cozens RM and Brown MRW (1983) Influence of nutrient limitation of growth on stability and production of virulence factors of mucoid and non-mucoid strains of *Pseudomonas aeruginosa*. Rev Infect Dis *5*: S880–888
19. Payne, SM (1980) Synthesis and utilisation of siderophores by *Shigella flexnevi*. J Bacteriol *143*: 1420–1424
20. Perry RD and San Clemente CL (1979) Siderophore synthesis in *Klebsiella pneumoniae* and *Shigella sonnei* during iron deficiency. J Bacteriol *140*: 1129–1132
21. Poxton IR and Sutherland IW (1976) Isolation of rough mutants of *Klebsiella aerogenes* and their synthesis of polysaccharides. J Gen Microbiol *96*: 195–202
22. Rogers HJ (1973) Iron binding catechols and virulence in *Escherichia coli*. Infect & Immun *7*: 445–456
23. Smith H (1977) Microbial surfaces in relation to pathogenicity. Bacteriol Rev. *41*: 475–500
24. Spratt BG (1975) Distinct penicillin binding proteins involved in the division, elongation and shape of *Escherichia coli*. Proc Natl Acad Sci USA *72*: 2999–3003

25. Stephens DS, Krebs JW and McGee ZA (1984) Loss of pili and decreased attachment to human cells by *Neisseria meningitidis* and *Neisseria gonorrhoea* exposed to subinhibitory concentrations of antibiotics. Infect and Immun *46*: 507–573
26. Stuart SJ, Greenwood KT and Luke RKJ (1980) Hydroxamate mediated transport of iron controlled by Col. V Plasmids. J Bacteriol *143*: 35–42
27. Swanson J (1981) Surface-exposed protein antigens of gonococcal outer membrane. Infect and Immun *34*: 804–816
28. Towbin H, Stachelin T and Gordon J (1979) Electrophoretic transfer of proteins from polyacrylamide gels to nitrocellulose sheets. Procedure & some applications. Proc Natl Acad Sci USA *76*: 4350–4354
29. Turnowsky F, Brown MRW, Anwar H and Lambert PA (1983) Effect of iron limitation of growth rate on the binding of penicillin G to penicillin binding proteins of mucoid and non-mucoid strains of *Pseudomonas aeruginosa*. FEMS Microbiol. Letts *17*: 243–245
30. van Oss CJ (1978) Phagocytosis as a surface phenomenon. Ann Rev Microbiol *32*: 19–39
31. van Tiel Menkveld GJ, Mentjox-Vervuurt JM, Oudega B and De Graf FK (1982) Siderophore production by *Enterobacter cloacae* and a common receptor protein for the uptake of aerobactin and cloacin DF13. J Bacteriol *150*: 490–497
32. Waxman DJ and Strominger JL (1983) Penicillin binding proteins and the mechanism of action of β-lactam antibiotics. Ann Rev Biochem *52*: 825–869
33. Williams P, Brown MRW and Lambert PA (1984) Effect of iron deprivation on the production of siderophores and outer membrane proteins of *Klebsiella aerogenes*. J Gen Microbiol *130*: 2357–2365
34. Williams P, Lambert PA, Brown MRW and Jones JR (1984) The role of the O & K antigens in determining the resistance of *Klebsiella aerogenes* to serum killing and phagocytosis. J Gen Microbiol *129*: 2181–2191
35. Williams PH (1979) Novel iron uptake system specified by Col V plasmids: an important component in the virulence of invasive strains of *Escherichia coli*. Infect and Immun *26*: 925–932

Discussion

Verhoef: You showed that in the presence of certain drugs iron binding proteins are being produced to a large extent. We showed that between the iron chelator deferoxamme and certain drugs a synergism exists for a number of bacteria (Van Asbeck et al., Eur. J. Clin. Microbiol., 1983). Does that contrast with your data?

Brown: The β-lactams used *exposed* IRMP's on the surface of bacteria grown under iron deprived conditions. These are present in drug-free bacteria but not surface exposed as indicated by our procedure. They are expressed as a response to iron-deprived conditions. This could handicap the bacterium *in vivo* in the presence of antibodies directed against the IRMP's. The *in vitro* synergism referred to is difficult to interpret in *in vivo* terms. I do not think that it contradicts our data.

Vosbeck: Do antibiodies against iron-regulated proteins affect bacterial growth in vitro?

Brown: We have no direct evidence that antibodies directed agains IRMP's cause slow growth *in vitro* or *in vivo*. I speculate that iron restriction *in vivo* may not of itself greatly reduce growth rate, especially when the high affinity iron uptake system is expressed. It seems probable that the slow growth *in vivo* may partly be due to these antibodies directed against the iron uptake system.

The Influence of β-Lactam Antibiotics Including Monobactam on the outer and inner Membrane of *E. coli*

H. J. LEYING, H. KARCH, H. P. KROLL, and W. OPFERKUCH

Medizinische Mikrobiologie und Immunologie, Ruhr-Universität Bochum,
D-4630 Bochum 1, F.R.G.

Introduction

For more than 20 years it is known that β-lactam antibiotics inhibit the transpeptidation of the peptidoglycan [18]. Their primary biochemical targets are some enzymes which are involved in the cross-bridging of the murein peptide sidechains and which are called penicillin binding proteins (PBP's) [12], for review see Martin et al., [8, 13, 19].

The question whether β-lactam antibiotics are able to alter structures of the bacterial cell envelope others than peptidoglycan became relevant because of the observation that some derivatives of this class of antibiotics in subinhibitory concentrations can modulate host-parasite interactions. Bacteria grown in subinhibitory concentrations of β-lactam antibiotics have been reported to exhibit alterations in such surface associated virulence properties as adherence to eucaryotic cells or phagocytosis by polymorphonuclear leucocytes or macrophages [2, 1, 9, 11, 14, 15] and serum resistance [3, 17]. The molecular basis of the drug-induced changes in the host-parasite interaction is unknown until now. The aim of this study was to examine which alterations of the bacterial envelope are caused by β-lactam antibiotics.

Material and Methods

Bacteria

E. coli WF 96 was obtained from St. Mary's hospital, London. This strain belongs to serogroup 07:K1:H6. E. coli strain D 509 was originally obtained from Dundee University, England. It belongs to serogroup 086. Bacteria were grown in Mueller-Hinton Broth (Merck, Darmstadt, Germany) at 37 °C without agitation. For antibiotic treatment appropriate amounts of the drugs were added to the medium. MIC (minimal inhibitory concentration) determination was performed in a microtiter assay according to DIN 58 940, using Mueller-Hinton broth (Merck, Darmstadt, Germany).

The Influence of Antibiotics on the
Host-Parasite Relationship II
Edited by D. Adam, H. Hahn, W. Opferkuch

Antibiotics

The following compounds were used: aztreonam (Squibb), ciprofloxacin (Bayer), imipenem (MSD), mecillinam (Leo-Pharmacia), moxalactam (Lilly), piperacillin (Cyanamid), ticarcillin (Beecham).

Preparation of bacterial membranes

Membranes were prepared according to Kroll et al. [5] by sucrose density ultracentrifugation. Bacteria were grown to an optical density of about 0.5 at 578 nm, centrifuged and washed once with phosphate buffered saline. Subsequently, the bacteria were suspended in 5 ml 0.05 M HEPES (N-2-hydroxyethylpiperazine-N'-2-ethanesulfonic acid) (pH 7.2) containing 0.75 M sucrose and 50 μg of egg white lysozyme (Sigma Chemical Co., St. Louis, Mo., U.S.A.) per ml. After 10 minutes of incubation the plasmolysed bacteria were lysed by transferring them into 120 ml of distilled water containing 1 mg DNase I (Serva Feinbiochemica, Heidelberg, Germany). After this membranes were separated from soluble material by ultracentrifugation (1 h, 130,000 g, 4 °C). Subsequently, outer and inner membrane vesicles were isolated by sucrose density ultracentrifugation (16 h, 120,000 g, 4 °C). During the fractionation of the gradients, the absorbance at 278 nm was continuously monitored with a Uvicord (LKB). The density of the fractions was determined by refractometry.

Pulse-chase labelling of bacteria with H^3-glycerol

Cells (1:100 dilution of an overnight culture) were grown in Mueller-Hinton broth containing ⅛ MIC moxalactam, imipenem, aztreonam, ciprofloxacin, or without antibiotic for 4 h. After this incubation time, an optical density of about 0.5 at 578 nm was reached. For pulse labelling, subsequently 0.4 μCi H^3-glycerol (specific activity 1 mCi/μmol; Amersham, England) per ml were added to each culture, vigorously shaken, and after 10 seconds of incubation glycerol was added (dilution of radioactive glycerol 1:1 000). After different chase times, aliquots were taken from the cultures, rapidly chilled by shaking in a −60 °C ethanol bath and centrifuged (8 000 g, 20 min, 4 °C). From the sedimented cells, inner and outer membranes were separated as described, and the incorporation of H^3-glycerol into the outer membrane in percent of the total radioactive incorporation was measured.

Analytical methods

Protein was estimated according to Markwell et al. (1978). Discontinuous sodium dodecyl sulfate-polyacrylamide slab gel electrophoresis (SDS-PAGE) was performed as described by Lugtenberg et al. [6] with the modification that the running gel contained 11% polyacrylamide.

Results

Membranes obtained after osmotic lysis of E. coli were sedimented by ultracentrifugation, layered on top of a linear sucrose density gradient and again ultracentrifuged for 16 h at 120,000 g. The whole membrane mixtures generated 4 bands within the sucrose gradients (Fig. 1a). The first peak with a buoyant density of about 1.27g/cm^3 (peak I) and the second peak with a buoyant density of about 1.24 g/cm^3 (Peak II) both represent outer membrane vesicles. The bands at 1.17 g/cm^3 (peak III) and 1.14 g/cm^3 (peak IV) represent cytoplasmic membrane (CM) vesicles as ascertained by measurement of the NADH oxidase activity, a marker enzyme of the cytoplasmic membrane.

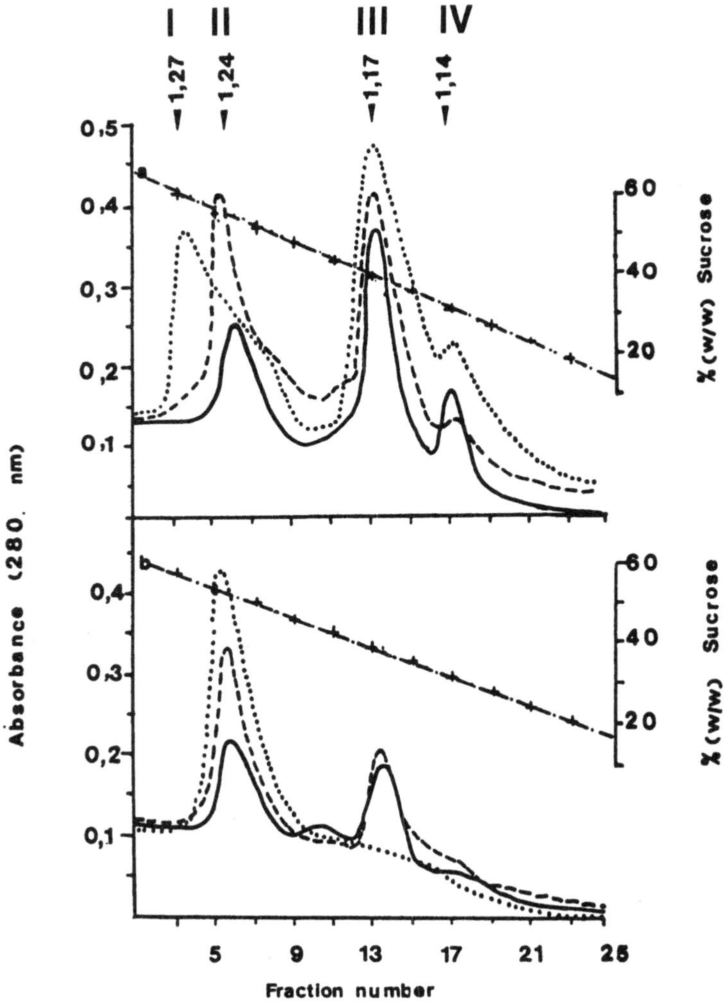

Fig. 1. Distribution of membrane vesicles of E. coli WF 96 grown in the absence (*a*) or presence (*b*) of imipenem for 150 min (-), 210 min (---), 300 min (...) after sucrose density ultracentrifugation

In different growth phases the two first membrane fractions are formed in different amounts (Fig. 1a). The solid line shows the membrane profile of E. coli WF 96 after 150 minutes, the dashed one after 210 minutes, and the dotted one after 300 minutes growth in antibiotic free medium. After 150 minutes, that means in the early log phase, only the lighter outer membrane (OM) fraction could be isolated, while in the early stationary phase the dense OM peak was quantitatively stronger developed. Finally in the late stationary phase almost exclusively the dense peak of the OM was detectable by protein measurement (data not shown).

Figure 1b shows the distribution of membrane vesicles from E. coli WF 96 after incubation in culture medium containing ¼ MIC imipenem. There are two remarkable consequences of the β-lactam antibiotic treatment: first, in all growth phases only the OM peak with the lower buoyant density was formed; second, with

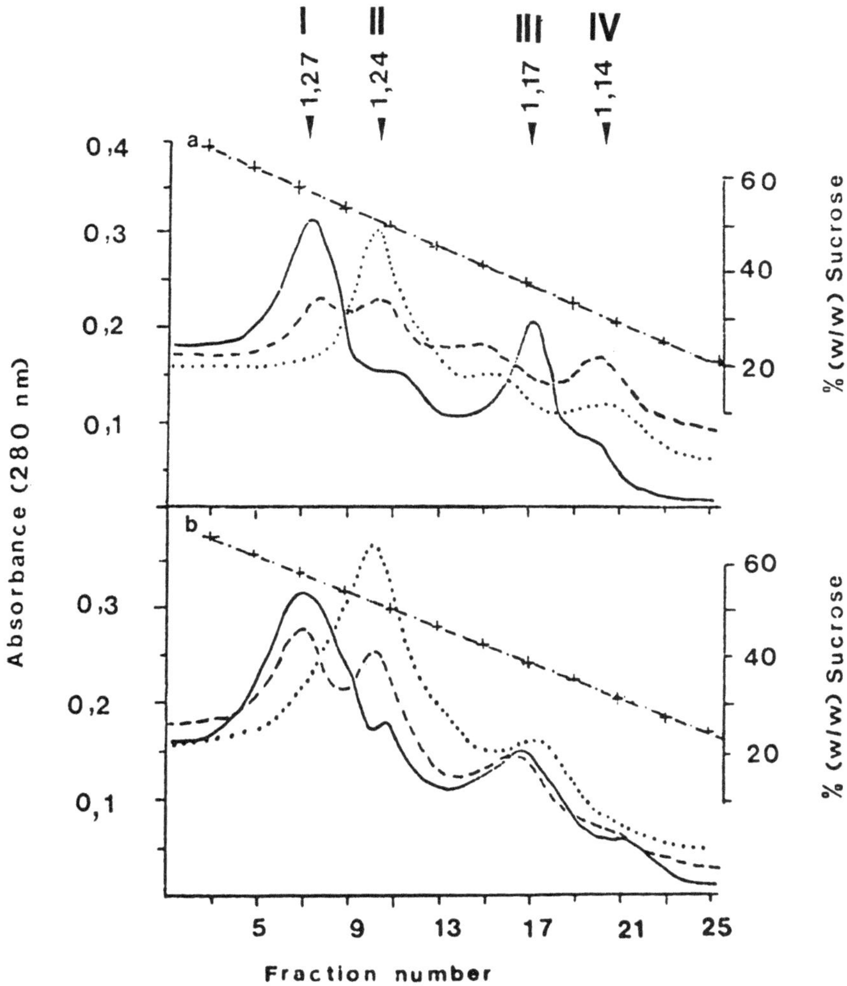

Fig. 2. Distribution of membrane vesicles of E. coli WF 96 grown in the absence (-) or in the presence of 1/12 MIC (---), 1/6 MIC (...) imipenem (*a*) or aztreonam (*b*) after sucrose density ultracentrifugation

increasing incubation times in imipenem containing medium there was a large loss of recoverable CM protein and probably CM material, a process which lead to an almost total reduction of recoverable CM material after 5 h incubation of bacteria in imipenem containing medium. The described effects were principally the same for E. coli strain D 509 and for all β-lactam derivatives used in this study (aztreonam, mecillinam, moxalactam, piperacillin, ticarcillin).

Comparable results were obtained after treatment of E. coli with different subinhibitory concentrations of the drugs. The membrane profiles one obtains under these growth conditions are shown in Fig. 2. The solid line is the membrane profile of the control cells, the dashed one that of bacteria treated with 1/12 MIC imipenem (Fig. 2a) or aztreonam (Fig. 2b) and the dotted one that of cells which were incubated in 1/6 MIC of the two β-lactam derivatives for 5 hours. The β-lactam antibiotics caused a dose dependent quantitative shift from the dense OM peak to the lighter one. In 1/6 MIC of the drugs only the OM fraction corresponding to the buoyant density of 1.24 g/cm^3 was formed within the sucrose gradients. Imipenem also had an influence on the CM profile. It induced a shift from the dense CM peak to the lighter one.

An SDS-PAGE analysis of the proteins present in the four membrane peaks is shown in Fig. 3. The CM peaks contained only traces of the major outer membrane

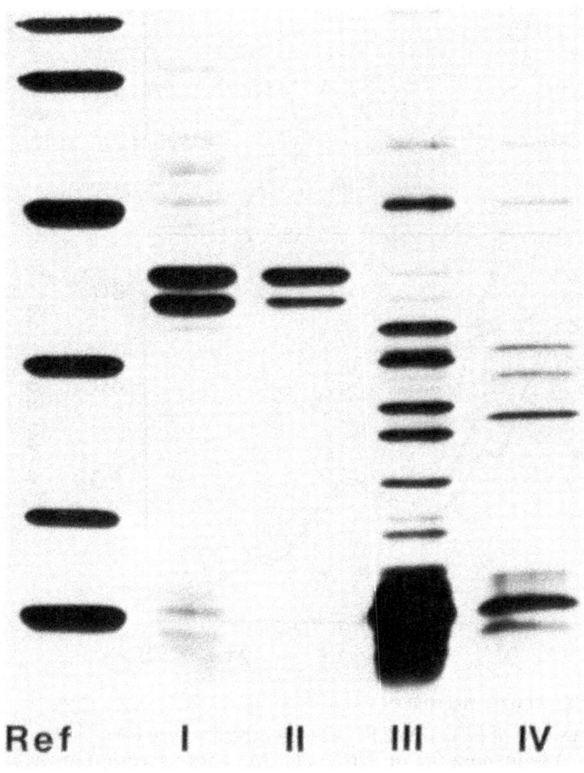

Fig. 3. SDS-PAGE of the four peaks (I, II, III, IV) obtained after sucrose density ultracentrifugation of whole membranes of E. coli WF 96. The following proteins from Pharmacia (electrophoresis calibration kit) were used as molecular weight standards (Ref): phosphorylase b (94,000), bovine serum albumin (67,000), ovalbumin (43,000), carbonic anhydrase (30,000), soy bean trypsin inhibitor (20,100), and alpha-lactalbumin (14,400). The gel was stained with Coomassie brilliant blue R 250

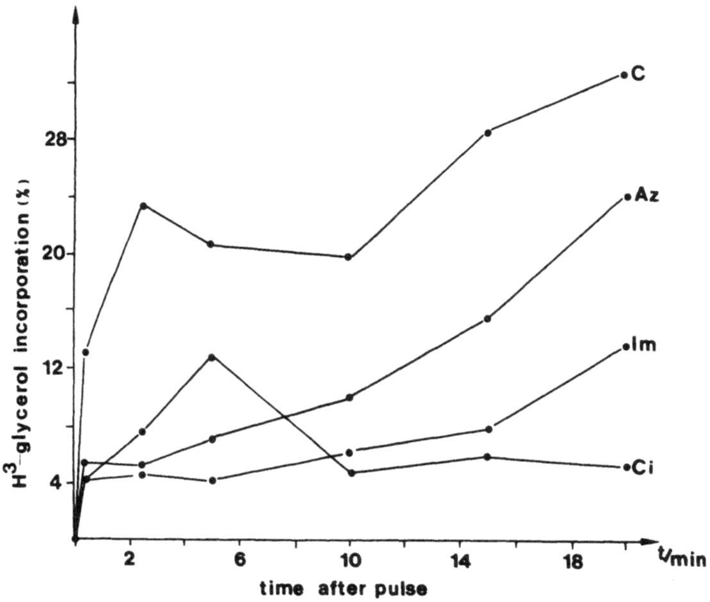

Fig. 4. H^3-glycerol incorporation into the outer membrane in percent of total radioactive incorporation at different chase times. C = control, Az = aztreonam, Ci = ciprofloxacin, Im = imipenem

protein bands. The two OM peaks were very similar with respect to their protein pattern.

At present we are investigating the quantitative contribution of the three major components proteins, lipopolysaccharides and phospholipids to the OM peaks. Although the final picture of the OM-composition of the β-lactam antibiotic treated and untreated bacteria is not clear, there is some evidence that cells exposed to a β-lactam antibiotic concentration of ⅛ MIC have a decreased phospholipid/OM protein ratio as compared with control cells. We could show that some β-lactam derivatives are able to inhibit the translocation of phospholipids from the cytoplasmic to the outer membrane by pulse-chase experiments that were performed as follows: E. coli D 509 were grown in antibiotic-free and ⅛ MIC antibiotic-containing medium for 4 h and subsequently pulsed for 10 seconds with H^3-glycerol as a phospholipid precursor. After different chase times aliquots were taken from the cultures. From these cells, CM and OM were separated as described, and the incorporation of H^3-glycerol into the inner and outer membrane was measured. Fig. 4 shows the H^3-glycerol incorporation into the outer membrane in percent of total radioactive incorporation at different chase times. Ciprofloxacin and imipenem induced a very strong inhibition of the phospholipid translocation over the whole chase period, while aztreonam only delayed the translocation. 20 minutes after the pulse the normal equilibrium for this E. coli strain of about 30% phospholipid incorporation into the OM and 70% into the CM was reached.

Discussion

There is little information in the literature about the influence of β-lactam antibiotics on structures of the bacterial cell envelope others than peptidoglycan. However, some reports indicate that changes may occur in cell envelope macromolecules after treatment of bacteria with subinhibitory concentrations of certain β-lactam antibiotics. For example, James [4] found an effect on the outer membrane protein profile, Taylor et al. [17] on the LPS and capsule content, and Rogers et al. [10] noted alterations in phospholipid biogenesis. We could neither detect any differences in the quality of OM protein, CM protein, and lipopolysaccharide patterns by SDS-PAGE between drug treated and untreated cells nor in the migration properties of CM and OM phospholipids in thin-layer chromatography (manuscript in preparation).

Although the exact quantitative contribution of proteins, lipopolysaccharides, and phospholipids to the OM composition remains to be determined, it seems likely that all β-lactam derivatives used caused alterations in the quantity and arrangement of the membrane constituents. Such alterations may be the reason for the altered membrane distribution within sucrose gradients after treatment of E. coli with subinhibitory concentrations of different β-lactam antibiotics. The question whether the different membrane peaks isolated by sucrose density ultracentrifugation have different functional significance remains to be answered. But the assumption that the different membrane fraction reflect the heterogeneity of the OM and CM as it occurs on living bacteria seems to be likely, because the distinct membrane fractions in pulse-chase experiments with H^3-glycerol as precursor of phospholipids and N-acetyl-D-1-^{14}C-glucosamine as precursor of lipopolysaccharides show different labelling kinetics. Furthermore, the fact that the CM peaks contain only traces of major outer membrane proteins indicates that the procedure of separating the two membranes is very effective.

Summary

The aim of this study was to investigate the influence of subinhibitory concentrations of different β-lactam antibiotics and the quinolone ciprofloxacin on the outer and inner membrane of E. coli. For this purpose membranes were prepared from plasmolyzed bacteria grown in antibiotic containing and antibiotic free medium by osmotic lysis and subsequently separated into outer and inner membranes by sucrose density ultracentrifugation. The membranes of E. coli generated 3 or 4 main fractions within the gradient with buoyant densities of 1.27 g/cm^3, 1.24 g/cm^3 (OM) and 1.17 g/cm^3, 1.14 g/cm^3 (CM). The OM of bacteria which were not treated with β-lactam antibiotics had a buoyant density of 1.24 g/cm^3 at the beginning and of 1.27 g/cm^3 at the end of the log phase. In contrast to this, bacteria grown to late log phase in antibiotic containing medium showed a highly significant dose dependent shift in the buoyant density of their OM from 1.27 g/cm^3 to 1.24 g/cm^3. Furthermore, the amount of recoverable cytoplasmic membrane material decreased after treatment of bacteria with increasing β-lactam antibiotic concentrations and with increasing incubation times. The phospholipid translocation from the inner to the outer membrane was inhibited in the presence of ⅛ MIC of aztreonam, imipenem and ciprofloxacin as assessed by pulse chase labelling of bacteria with H^3-glycerol as phospholipid precursor. Thus it can be established that the antibiotics used affected the structure of the inner and outer membrane although they had no influence on the qualitative composition of their constituents.

References

1. Bar-Shavit Z, Goldman R, Ofek I, Sharon N, Mirelman D (1980) Mannose-binding activity of Escherichia coli: a determinant of the attachment and ingestion of bacteria by macrophages. Infect Immun 29: 417–424
2. Beachey EH, Eisenstein BJ, Ofek I (1980) Sublethal concentrations of antibiotics and bacterial adhesion. In: Elliott K, O'Connor M, Whelan J (eds) Adhesion and microorganism pathogenicity: Ciba foundation symposium 1980. Pitman Medical, London, pp 288–305
3. Friedman H, Warren GH (1976) Antibody-mediated bacteriolysis: enhanced killing of cyclacillin-treated bacteria. Proc Soc Exp Biol Med 153: 301–304
4. James R (1975) Identification of an outer membrane protein of Escherichia coli, with a role in the coordination of deoxyribonucleic acid replication and cell elongation. J Bacteriol 124: 918–929
5. Kroll HP, Bhakdi S, Taylor PW (1983) Membrane changes induced by exposure of Escherichia coli to human serum. Infect Immun 42: 1055–1066
6. Lugtenberg B, Meijers J, Peters R, Van der Hoek P, Van Alpen L (1975) Electrophoretic resolution of the major outer membrane proteins of Escherichia coli K 12 into four bands FEBS Lett 58: 254–259
7. Markwell MAK, Haar SM, Bieber LL, Tolbert WE (1978) A modification of the lowry procedure to simplify protein determinations in membrane and lipoprotein samples. Anal Biochem 87: 206–210
8. Martin HH, Staboulis D, Schilf W (1981) Penicillinbindeproteine der Bakterien als spezifische Wirkorte der β-Laktam-Antibiotika und als Faktoren der Antibiotikaresistenz. Immun Infekt 9: 99–105
9. Ofek I, Beachey EH, Eisenstein BJ, Alkan MC, Sharon N (1979) Suppression of bacterial adherence by subminimal inhibitory concentrations of β-lactam and aminoglycoside antibiotics. Rev Inf Dis 1: 832–837
10. Rogers HJ, Thurman PF (1983) Interrelationships between wall and membrane biosynthesis. In: Hakenbeck R, Höltje JV, Labischinski H (eds) The target of penicillin. Walter de Gruyter, Berlin, New York, pp 637–642
11. Sandberg T, Stenqvist K, Svanborg-Eden C (1979) Effects of subminimal inhibitory concentrations of ampicillin chloramphenicol and nitrofurantoin on the attachment of Escherichia coli to human uroepithelial cells in vitro. Rev Infect Dis 1: 838–844
12. Spratt BG (1975) Distinct penicillin binding proteins involved in the division, elongation, and shape of Escherichia coli K12. Proc Nat Acad Sci USA 72: 2999–3003
13. Spratt BG (1983) Penicillin-binding proteins and the future of β-lactam antibiotics. J Gen Microbiol 129: 1247–1260
14. Svanborg-Eden C, Sandberg T, Stenqvist K, Ahlstedt S (1978) Decrease in adhesion of Escherichia coli to human urinary tract epithelial cells in vitro by subinhibitory concentrations of ampicillin. Infection 6, Suppl 1: 121–124
15. Svanborg-Eden C, Sandberg T, Stenqvist K, Ahlstedt S (1979) Effects of subinhibitory amounts of ampicillin, amoxycillin, and mecillinam on the adhesion of Escherichia coli bacteria to human urinary tract epithelial cells: a preliminary study. Infection 7, Suppl 5: 452–455
16. Taylor PW, Gaunt H, Unger FM (1981) Effect of subinhibitory concentrations of mecillinam on the serum susceptibility of Escherichia coli strains. Antimicrob Agents Chemother 19: 786–788

17. Taylor PW, Kroll HP, Tomlinson S (1982) Effect of subinhibitory concentrations of mecillinam on expression of Escherichia coli surface components associated with serum resistance. Drugs Exp Clin Res 8: 625–631
18. Tipper DJ, Strominger JL (1965) Mechanism of the action of penicillins: a proposal based on their structural similarity to acyl-D-alanyl-D-alanine. Proc Nat Acad Sci USA 54: 1133–1141
19. Waxman DJ, Strominger JL (1983) Penicillin-binding proteins and the mechanism of action of β-lactam antibiotics. Ann Rev Biochem 52: 825–869

Discussion

Martin: The appearance of duplicate bands with different buoyant densities for both outer membrane and cytoplasmic membrane in the density gradient centrifugation may be evidence for an inhomogeneous architecture of the respective membrane mosaics. Have you found chemical evidence for such an inhomogeneity and what could be the reason for the disappearance of one outer membrane band under β-lactam treatment?

Leying: We found evidence for the inhomogeneity of the outer membrane of E. coli by means of pulse-labelling experiments with ^{3}H-glycerol as a phospholipid precursor. In such experiments the ^{3}H-radioactivity at first was detected in the light and several cell divisions later also in the dense outer membrane peak, that means, the radioactivity was in a time dependent fashion incorporated into distinct membrane areas. The reason for the disappearance of one outer membrane band under β-lactam treatment is at present unknown.

Martin: Is there an explanation for the greatly decreased recovery of a cytoplasmic membrane fraction from the β-lactam treated cells?

Leying: No, there is no explanation for this phenomenon. It may be interesting that Dr. Kroll observed a similar decrease of recoverable cytoplasmic membrane material after treatment of serum-sensitive E. coli with serum in the early, prekilling phase of the bactericidal reaction.

Haag: You found two bands each of outer membrane vesicles and cytoplasmic membrane vesicles in sucrose gradients of lysed, non-antibiotic treated E. coli. After treatment with imipenem the number of bands is reduced to one each. On the other hand you showed that the composition of these membrane vesicles does not differ from untreated control cells. My question is, what is the explanation for the differing number of membrane vesicle bands. Do they simply have different but distinct sizes?

Leying: In preliminary experiments we could detect differences in the molecular composition of outer membrane vesicles from β-lactam antibiotic treated and untreated cells; the different composition of membranes is likely to be the reason for the differing buoyant densities of the membranes. The sites of OM vesicles of β-lactam treated and untreated bacteria do not markedly differ as assessed by electron microscopy.

Stephens: The fimbrial (pilus) subunits were not apparently altered by exposure to the β-lactam antibiotics. Were those pilus subunits assembled into pilivisible on the surface of the E. coli?

Leying: Yes, also after β-lactam treatment there were fimbriae visible on the surface of E. coli.

Kadurugamuwa: Do any of the β-lactams used in the study have any effects on the peptidoglycan associated proteins?

Leying: No, the OM-protein patterns of drug treated and untreated bacteria appeared absolutely identical in SDS-PAGE.

Changes in outer Membrane Composition after Treatment of *Proteus mirabilis* with Antibiotics

KATHRYN NIXDORFF, H. H. MARTIN, and NICOLA SIEGMUND-SCHULTZE

Institut für Mikrobiologie, Technische Hochschule Darmstadt, Schnittspahnstrasse 9, D-6100 Darmstadt, FRG

Introduction

In the course of an infection with gram-negative bacteria, surface structures on the bacterial cells play a central role in the interaction of the defense system of the host with the pathogen. The cell surface of gram-negative bacteria is a mosaic of components including lipopolysaccharide (LPS), a few major proteins and phospholipids, which are all complexed with one another in the native outer membrane structure.

When an infection with gram-negative bacteria has been established, antibiotic therapy is usually prescribed. It is known that treatment of gram-negative bacteria with β-lactam antibiotics leads to functional disorganization of the peptidoglycan layer of the cell wall. Bacteria with progressive cell wall defects (from filaments to spheroplasts to protoplasts) can be induced by treating gram-negative bacteria with increasing concentrations of certain antibiotics or by using antibiotics which cause specific defects [6, 7].

On the other hand, very little is known about possible changes in outer membrane composition of gram-negative bacteria treated with β-lactam antibiotics. We therefore considered it essential to investigate such possible changes, as alteration of cell surface components may affect the sensibility of gram-negative bacteria to the bactericidal action of serum [10], an important defense system of the host.

For these studies, we have used a model system consisting of the bacterial form of *Proteus mirabilis* VI, its unstable filament form induced specifically by subinhibitory concentrations of the β-lactam antibiotic cefuroxime and its unstable spheroplast L-form induced by the β-lactam antibiotic cefoxitin. All of these forms have either intact or defective cell walls. A stable protoplast L-form of *P. mirabilis* VI, which is completely devoid of a cell wall and grows in the absence of antibiotics [2], was also included in the model system.

A comparison was made of the outer membrane composition of the bacterial form, its filament form and its spheroplast L-form. The results of these studies are

---

Abbreviations used

$C_{14\text{-}OH}$, β-hydroxy myristic acid; C_{16}, palmitic acid; $C_{16:1}$ palmitoleic acid; $C_{17\ cyclo}$, C_{17} cyclopropane fatty acid; C_{18}, stearic acid; $C_{18:1}$, vaccinic acid; $C_{19\ cyclo}$, C_{19} cyclopropane fatty acid; O.D., optical density; LPS, lipopolysaccharide; rpm, revolutions per minute; SDS, sodium dodecyl sulfate

The Influence of Antibiotics on the
Host-Parasite Relationship II
Edited by D. Adam, H. Hahn, W. Opferkuch

presented in the present report, along with a summary of our previous findings [2, 3] concerning the composition of the cytoplasmic membrane of the stable protoplast L-form.

Materials and Methods

Bacteria

P. mirabilis VI and its stable protoplast L-form were from the collection of H.H. Martin (this institute). The stable protoplast L-form was originally obtained from U. Taubeneck (Jena, GDR).

Antibiotics

The β-lactam antibiotic cefuroxime (Hoechst AG, Frankfurt/Main, FRG) was used for the induction of the filament form of *P. mirabilis* VI. This antibiotic was chosen because of its ability to specifically induce filament formation [7]. The β-lactam antibiotic cefoxitin (MSD-Sharp & Dohme GmbH, Munich, FRG) was used to induce spheroplast formation in *P. mirabilis* VI.

Cultivation of the bacterial form, the unstable filament form and the unstable spheroplast L-form of P. mirabilis VI

For cultivation of the bacterial form in large amounts an overnight culture of *P. mirabilis* VI at 37°C in L-medium (6) was added to 8 flasks containing 400 ml L-medium to an optical density (O.D.) at 578 nm of 0.1–0.2 (Eppendorf Photometer, Netheler & Hinz GmbH, Hamburg, FRG). This culture was incubated at 37 °C with shaking at 120 rpm until an $O.D._{578\ nm}$ of 1.0 was reached (logarithmic phase of growth). The culture was then rapidly cooled in an ice bath and the cells were harvested by centrifugation at 4°C.

For cultivation of the filament form in large amounts, 1.0 ml of an overnight culture of the bacterial form was added to each of several flasks containing 50 ml L-medium plus 0.1 µg cefuroxime/ml and incubated overnight at 37°C with shaking at 120 rpm. From this culture, 8 flasks containing 400 ml L-medium plus 0.1 µg cefuroxime/ml were inoculated to an $O.D._{578nm}$ of 0.1–0.2. This culture was incubated at 37°C with shaking at 120 rpm until an O.D. of 1.0 was reached (logarithmic phase of growth). The cells were harvested as above.

P. mirabilis VI was induced to spheroplast formation 2–3 days before cultivation in large amounts by addition of 60 µg cefoxitin/ml L-medium. On the first day of induction, the culture was incubated overnight at 37 °C without shaking. On the second day, 1.0 ml of the culture from the first day was inoculated into each of several flasks containing 50 ml L-medium plus 60 µg cefoxitin/ml. This culture was incubated at 37°C with increasing degrees of shaking during the day (70 rpm, 90 rpm) and finally with shaking at 120 rpm overnight. On the third day, the overnight culture from the second day was added to 8 flasks containing 400 ml L-medium plus 60 μg cefoxitin/ml to an $O.D._{578\ nm}$ of 0.1–0.2 and incubated at 37°C with shaking (120 rpm) to an $O.D._{578\ nm}$ of 0.9 (logarithmic phase of growth). Cells were harvested as above.

Isolation of cell walls

Cell walls of *P. mirabilis* VI free of cytoplasmic membranes were obtained by shaking aqueous suspensions of bacteria with glass beads (0.17 mm diameter) in a cooled cell mill (E. Bühler, Tübingen, FRG) for 20 min in the presence of 0.4% sodium dodecyl sulfate (SDS) as previously described [6, 8]. Approximately 10 g bacteria (wet weight) were mixed with 90 ml of 0.4% SDS, pH 7.0, and 150 g of glass beads. To obtain intact cell walls of the filament form, treatment for 5 min in the cell mill was sufficient. The spheroplast L-form was treated for only 5 sec in the cell mill.

SDS-polyacrylamide gel electrophoresis

SDS-polyacrylamide gel electrophoresis of proteins in cell walls was carried out in a slab gel apparatus GE4 (Deutsche Pharmacia, Freiburg, FRG) according to Lugtenberg *et al.* [4] as previously described [8].

Analytical methods

Protein content of cell walls was determined by a modification of the Lowry technique [5]. Fatty acids of phospholipids and LPS were analyzed by gas-liquid chromatography as described [2].

Results

Comparative analyses of outer membrane components and fatty acid composition of phospholipids in cell walls of the bacterial and the unstable filament forms of P. mirabilis VI

The results presented in Table 1 show that there were no significant differences in the total amounts of protein, phospholipids or LPS in the cell walls of the bacterial and the filament forms of *P. mirabilis* VI when these two forms were cultivated under identical conditions and harvested in the same growth phase.

However, we did detect a slight increase in the ratio of saturated:unsaturated+cyclopropane fatty acids in the outer membrane phospholipids of the filament form (Table 2).

The filament form also showed a different outer membrane protein pattern from the bacterial form when proteins extracted from cell walls with 2% SDS at 100°C for 5 min were analyzed by SDS-polyacrylamide gel electrophoresis. Results are pre-

Table 1. Quantitative analyses of amounts of protein, phospholipids and lipopolysaccharide in cell walls of the bacterial and the filament forms of *Proteus mirabilis* VI

Sample	Percent Protein	Phospholipids $\frac{\text{nmol } C_{16}^{a}}{\text{mg sample}}$	Lipopolysaccharide $\frac{\text{nmol } C_{14\text{-}OH}^{b}}{\text{mg sample}}$
Bacterial form	55±5	96.03±23.22	126.03±28.66
Filament form	53±4	66.68± 8.92	121.41± 3.53

[a] C_{16} = palmitic acid
[b] $C_{14\text{-}OH}$ = β-hydroxy myristic acid
Values in the table represent the means ± standard errors from two determinations each on two separate samples.

Table 2. Fatty acid analyses of phospholipids in cell walls of the bacterial form, the filament form and the spheroplast L-form of *Proteus mirabilis* VI

Fatty acid[a]	Bacterial form $\frac{\text{% Fatty acid}}{\text{Total fatty acids}^{b}}$	Filament form $\frac{\text{% Fatty acid}}{\text{Total fatty acids}^{b}}$	Spheroplast L-form $\frac{\text{% Fatty acid}}{\text{Total fatty acids}^{b}}$
C_{16}	35.79±2.75	43.67±2.96	39.90±0.13
$C_{16:1}$	45.86±1.61	31.30±1.29	20.55±0.93
$C_{17\ cyclo}$	4.13±2.09	4.51±0.49	1.90±0.13
C_{18}	0.89±0.26	1.16±0.24	1.28±0.23
$C_{18:1}$	10.51±1.63	18.27±0.41	19.60±0.68
$C_{19\ cyclo}$	2.82±0.38	1.05±0.57	3.44±0.21
$\frac{\text{Saturated}}{\text{Unsaturated + cyclo}}$	0.58	0.81	1.19

[a] C_{16} = palmitic acid; $C_{16:1}$ = palmitoleic acid; $C_{17\ cyclo}$ = C_{17} cyclopropane fatty acid; C_{18} = stearic acid; $C_{18:1}$ = vaccinic acid; $C_{19\ cyclo}$ = C_{19} cyclopropane fatty acid.
[b] Mean values ± standard errors taken from two separate determinations.

sented in Fig. 1. There was an apparent decrease in the amount of the outer membrane protein with an apparent molecular weight of 39,000 daltons (39KD protein) of the filament form (lane C), which was consistent from batch to batch. This decrease was also observed in cell walls of the filament form of *P. mirabilis* strain 211 cultivated under the same conditions (data not shown). In previous studies [9], we did not detect a loss of the 39KD protein in cell walls of the filament form of *P. mirabilis* 211. In these studies, however, the filament form was cultivated at 30 °C instead of 37 °C, the temperature used to cultivate the bacterial form. Thus, loss of this protein may be in part temperature dependent, but it was, however, observed only in the filament form and not in the bacterial form when both were cultivated at 37 °C.

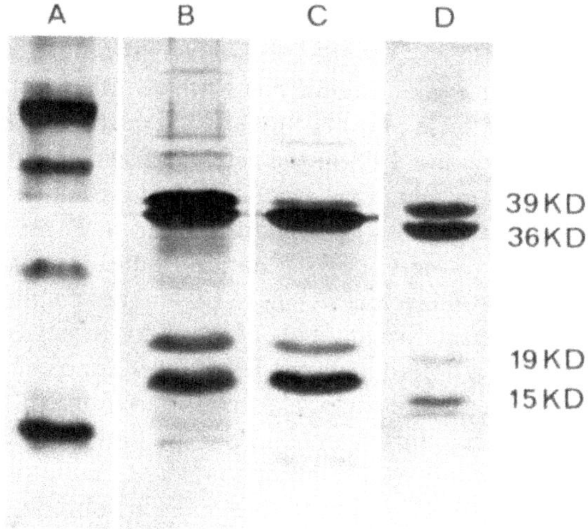

Fig. 1. SDS-polyacrylamide gel electrophoresis of proteins from cell walls of the bacterial form, the filament form and the spheroplast L-form of *Proteus mirabilis* VI. In each case, samples of cell walls containing the same amounts of protein were extracted with 2% SDS for 5 min at 100°C. **A,** standard proteins: bovine serum albumin (67KD), egg albumin (45KD), chymotrypsinogen (25KD) and cytochrome C (12.5KD). **B,** 45 µg protein from cell walls of the bacterial form. **C,** 45 µg protein from cell walls of the filament form. **D,** 45 µg protein from cell wall fragments of the spheroplast L-form

Comparative analyses of outer membrane components of the bacterial form and the unstable spheroplast L-form of P. mirabilis VI

The results presented in Table 3 show that there were no significant differences in the total amounts of proteins and LPS in the cell walls of the bacterial form and in cell wall fragments isolated from the spheroplast L-form of *P. mirabilis* VI induced by cefoxitin. Indeed, the analyses of these components of the bacterial form are not significantly different from those obtained in other determinations made at a different time (see Table 1). However, we did detect a significant increase in phospholipids in cell wall fragments of the spheroplast L-form, as indicated by an increase in palmitic acid content of phospholipids. If phospholipids are indeed increased, then a concomitant decrease in one of the other components of the outer

Table 3. Quantitative analyses of amounts of protein, phospholipids and lipopolysaccharide in cell walls of the bacterial form and the spheroplast L-form of *Proteus mirabilis* VI

Sample	Percent Protein	Phospholipids $\frac{\text{nmol } C_{16}{}^{a}}{\text{mg sample}}$	Lipopolysaccharide $\frac{\text{nmol } C_{14\text{-OH}}{}^{b}}{\text{mg sample}}$
Bacterial form	60±5	118.30±11.39	153.91±11.58
Spheroplast L-form	66±7	187.83± 8.75	169.09±28.40

[a] C_{16} = palmitic acid
[b] $C_{14\text{-OH}}$ = β-hydroxy myristic acid
Values in the table represent the means ± standard errors from two determinations each on two separate samples.

membrane would be expected. Our preliminary determinations of β-hydroxy myristic acid ($C_{14\text{-}OH}$) in the spheroplast form were subject to a relatively high degree of standard error. Taking the extreme cases into account (165 nmol $C_{14\text{-}OH}$ in the bacterial form and 141 nmol $C_{14\text{-}OH}$ in the spheroplast L-form), there may actually be slightly less LPS in the cell wall fragments of the spheroplasts. In addition, the $C_{14\text{-}OH}$ determinations may not reflect the actual weight of LPS in the sample. In this regard, two types of LPS are present in *P. mirabilis,* one with very short and one with very long O-polysaccharide chains designated LPS I and LPS II, respectively [1]. Thus, the actual weight of LPS in the sample could vary considerably, depending upon the relative amounts of the two types of LPS, which would not be reflected in the $C_{14\text{-}OH}$ determinations. It is known, for example, that about one third of the LPS molecules in the bacterial form of *P. mirabilis* VI are LPS II [1], while the protoplast L-form contains mainly LPS I in the cytoplasmic membrane [2]. Whether the observed increase in phospholipids is actual or only relative to a decrease in the weight of LPS in spheroplast outer membranes will have to be determined.

Analyses of fatty acids in phospholipids of cell wall fragments of the spheroplast L-form (Table 2) also showed an increase in the ratio of saturated:unsaturated+cyclopropane fatty acids over the ratios of these components in phospholipids of the bacterial and the filament forms. Thus, a tendency toward an increase in rigidity of phospholipids in the outer membrane with progressive cell wall defects is indicated.

On the other hand, no differences in outer membrane protein patterns of the bacterial form and the spheroplast L-form could be observed (Fig. 1, lanes B and D). From these patterns it can also be seen that the cell wall fragments of the spheroplast L-form were not contaminated with cytoplasmic membranes.

Composition of the cytoplasmic membrane of the stable protoplast L-form of P. mirabilis VI

The stable protoplast L-form of *P. mirabilis* VI, which grows and retains its form in the absence of antibiotics, is completely devoid of all cell wall components, except for small amounts of LPS that are present in the cytoplasmic membrane [2].

In a previous study [3], we performed quantitative analyses of the components in protoplast membranes isolated from this form. The results yielded 44% protein, 179 nmol palmitic acid/mg dry weight (phospholipids) and 24 nmol β-hydroxy myristic acid/mg dry weight (LPS). Compared with cell walls of the bacterial form, the protoplast membrane contains slightly less protein, less LPS and more phospholipids.

Discussion

Results presented in this report show various changes in the composition of outer membrane components of *P. mirabilis* VI after induction to filament formation with cefuroxime or induction to spheroplast formation with cefoxitin.

The question naturally arises as to the significance of these changes for the survival of the various cell forms in the presence of serum. In an accompanying paper (Siegmund-Schultze, Martin and Nixdorff, this volume), we investigated the sensitivity of the bacterial, the filament, the spheroplast and the protoplast forms of *P. mirabilis* VI to the bactericidal action of normal human serum. We found that the bacterial and the filament forms were relatively resistant. Sensitivity to serum was first observed after induction of this organism to the spheroplast L-form. The most sensitive of all was the stable protoplast L-form.

Obviously, the apparent decrease in the 39KD outer membrane protein and the slight increase in the ratio of saturated:unsaturated+cyclopropane fatty acids in outer membrane phospholipids after induction to the filament form had little effect on conversion to serum sensitivity. Whether the apparent increase in the amount of phospholipids in the outer membrane of the spheroplast L-form contributed to the increase in serum sensitivity is difficult to assess. In this case, we analyzed only the cell wall fragments which survived contact with SDS during isolation. There remains the possibility that the spheroplasts contain areas of defective outer membrane composition, or even areas completely devoid of outer membrane material, which we did not detect using the present methods of isolation. It has been determined, for example, that the spheroplast L-form of *P. mirabilis* 19 induced by 50 μg/ml cefoxitin contained only 39% of the peptidoglycan of the bacterial form (E. Ruttkowski and H. H. Martin, unpublished data). If such areas exist, they are very likely to be more accessible for the action of complement in serum. Indeed, the protoplast L-form, which is completely devoid of cell wall components other than small amounts of LPS in the cytoplasmic membrane [2, 3], was the most sensitive of all forms to the bactericidal action of serum.

In order to resolve such questions, studies employing more gentle methods of isolating spheroplast surface structures are now underway.

Acknowledgements. This work was supported by the Deutsche Forschungsgemeinschaft. We thank Heide-Marie Rostek and Gesche Bauer for excellent technical assistance.

Summary

The effects on outer membrane composition of the serum resistant gram-negative bacterium *Proteus mirabilis* VI after treatment of this organism with antibiotics were investigated. Comparison of the outer membrane composition of the bacterial form and its filament form induced by subinhibitory concentrations of cefuroxime showed no significant differences in the total amounts of protein, phospholipids and lipopolysaccharide (LPS). There was, however, an apparent decrease in the amount of the 39KD outer membrane protein in the filament form. In addition, analyses of fatty acids showed a slight increase in the ratio of saturated:unsaturated+cyclopropane fatty acids in phospholipids from the outer membrane of the filament form. Cell wall fragments free of cytoplasmic membranes and cell contents were isolated from *P. mirabilis* VI after treatment of this bacterium with cefoxitin, which led to spheroplast formation. Analyses of these fragments showed no significant differences in the amounts of proteins and LPS, or in the protein patterns obtained upon electrophoresis, as compared with analyses of cell walls of the bacterial form. We did, however detect a significant increase in the ratio of saturated:unsaturated + cyclopropane fatty acids of outer membrane phospholipids, as well as an apparent increase in the amounts of phospholipids in the spheroplast L-form.

References

1. Gmeiner J (1975) The isolation of two different lipopolysaccharide fractions from various *Proteus mirabilis* strains. Eur J Biochem 58: 621–626
2. Gmeiner J, Martin HH (1976) Phospholipid and lipopolysaccharide in *Proteus mirabilis* and its stable protoplast L-form. Difference in content and fatty acid composition. Eur J Biochem 67: 487–494
3. Karch H, Nixdorff K (1980) Comparison of quantitative and qualitative antibody-producing cell responses to lipopolysaccharide in cell walls of the bacterial form and in membranes of the protoplast L-form of *Proteus mirabilis*. Infect Immun 30: 349–352
4. Lugtenberg B, Meijers J, Peters R, van der Hoek P, van Alphen L (1975) Electrophoretic resolution of the "major outer membrane protein" of *Escherichia coli* K12 into four bands. FEBS Lett 58: 254–258
5. Markwell MAK, Haar SM, Bieber LL, Tolbert E (1978) A modification of the Lowry procedure to simplify protein determination in membrane and lipoprotein samples. Anal Biochem 87: 206–210
6. Martin HH (1964) Composition of the mucopolymer in cell walls of the unstable and stable L-form of *Proteus mirabilis*. J Gen Microbiol 36: 441–450
7. Martin HH, Staboulis D, Schilf W (1981) Penicillinbindeproteine als spezifische Wirkorte der β-Laktam-Antibiotika und als Faktoren der Antibiotikaresistenz. Immun Infekt 3: 99–105
8. Nixdorff K, Fitzer H, Gmeiner J, Martin HH (1977) Reconstitution of model membranes from phospholipid and outer membrane proteins of *Proteus mirabilis*. Role of proteins in the formation of hydrophilic pores and protection of membranes against detergents. Eur J Biochem 81: 63–69
9. Siegmund-Schultze N, Nixdorff K (1985) Modulation of the interaction of bacteria and their membrane components with host defense systems by antibiotics. In: Pulverer G, Jeljaszewicz J (eds) Chemotherapy and immunity. Zentralbl Bakteriol Mikrobiol Hygiene I. Abteilung, supplement 13. Gustav Fischer Verlag, Stuttgart New York, pp 231–242
10. Taylor PW (1983) Bactericidal and bacteriolytic activity of serum against gram-negative bacteria. Microbiol Rev 47: 46–83

Discussion

Stephens: What is the function of the 39,000 m.w. protein in the outer membrane of Proteus?

Nixdorff: In earlier studies we investigated the function of the 39 KD protein in reconstituted model membrane vesicles *in vitro*. The protein did act as a porin in this system; it is, however, not associated with peptidoglycan. The 36 KD protein is also a porin, and it is associated with peptidoglycan.

Weight and Morphology of Bacteria Exposed to Antibiotics

V. LORIAN[1], W. TOSCH[2], and D. JOYCE[1]

[1] Section of Microbiology and Epidemiology, Department of Pathology, The Bronx Lebanon Hospital Center, Bronx, New York
[2] Present Address: Ciba Geigy, K–125.3.09, CH–4002 Basel, Switzerland

Introduction

Exposure of bacteria to sub-MIC of various beta-lactam and other antibiotics results in morphologic alterations including a substantial increase in size. Staphylococci change into large cells with a diameter of 2–3 μm [1] and gram-negative bacilli grow into long filaments [2, 3, 4].

This investigation explored changes in weight in strains of *Escherichia coli*, *Pseudomonas aeruginosa*, and *Staphylococcus aureus* following exposure to several beta-lactam antibiotics and to ciprofloxacin.

Materials and Methods

Five strains of bacteria were used in the study: *E. coli* 4298 (from Dr. Lowell Young); *P. aeruginosa* 3073, *S. aureus* Fitz and 10B (from the culture collection of Ciba Geigy, Basel, Switzerland), and *S. aureus* 209P (from the FDA). These organisms were exposed to the following antibiotics: ampicillin (Bristol); amoxicillin (Beecham); amdinocillin (mecillinam) (Hoffmann-La-Roche); mezlocillin (Bayer); cefsulodin (Ciba Geigy); CGP-17520, a new cephalosporin (Ciba Geigy); cloxacillin (Beecham); and ciprofloxacin (Miles).

The bacteria were first grown overnight in Mueller-Hinton Broth (MHB). Then filter membranes (0.3 u pore, PHWP04SO, Millipore Corporation, Bedford Mass.) placed on Mueller-Hinton Agar (MHA) were inoculated with 0.1 ml of a 1:10 dilution of the MHB culture and incubated for 90 minutes at 37 °C. These membranes were transferred to MHA alone and MHA containing the various antibiotics at concentrations known to produce morphologic alterations [5] where they were incubated at 37 °C for four hours.

For each weight determination, each strain of bacteria was cultured on 20 membranes. The organisms present on the 20 membranes were eluted into 10 ml of distilled water in a pre-weighed 50 ml centrifuge tube [5]. An aliquot of 0.1 ml was taken from the suspension, serially diluted, and planted on agar for a colonyforming unit (CFU) count. One drop of the suspension was examined by interference phase contrast microscopy and used to prepare a smear.

After centrifugation at 1,500 × g for 30 minutes, the supernatant was removed. Then the tubes were dried in an incubator at 37 °C for five hours and placed in a

The Influence of Antibiotics on the
Host-Parasite Relationship II
Edited by D. Adam, H. Hahn, W. Opferkuch

desiccator until constant weights were obtained. With gram-negative bacilli, two separate weight determinations were done and the weight/CFU ratio calculated. One determination was done for each strain of *S. aureus*. To show the bacterial nuclear material in filaments, the smear was flamed and immersed in 10% v/v hydrochloric acid in water at 95–100 °C for 2–3 minutes. After being washed with tap water, it was stained for 15 minutes with Giemsa at pH 7.6 [6, 7].

Results

As expected from previous experience, the gram-negative bacilli exposed to ampicillin, amoxicillin, mezlocillin, cefsulodin, CGP-17520 and ciprofloxacin grew into filaments, and *E. coli* exposed to amdinocillin (mecillinam) produced round cells [6]. *S. aureus* exposed to cloxacillin developed into round and oval cells 2–3 µm in diameter [1], hereafter referred to as large staphylococci.

Figures 1 through 6 show the weight/CFU ratios and the predominant morphology of gram-negative bacilli, both those exposed to antibiotics and those not exposed. The close resemblance of the two curves, representing the two experiments in each group, demonstrates the reproducibility of the method.

Provided one morphologic form predominates, the weight/CFU ratio of a bacterial population sample is comparable to the weight of a single organism in that population. In this study, the approximative mean weight of a normal *E. coli* bacillus

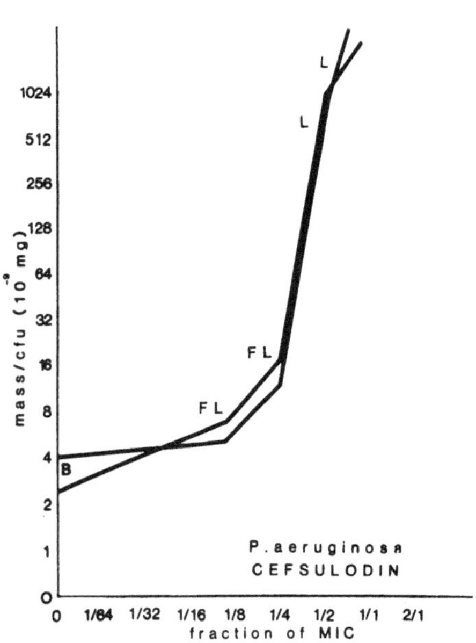

Fig. 1

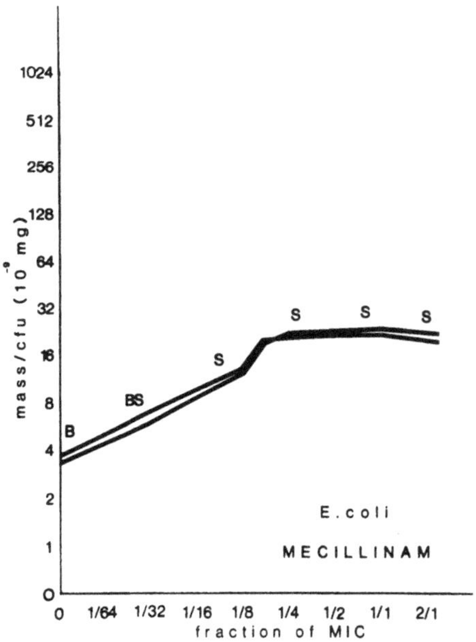

Fig. 2

Fig. 1–6. Weight/CFU ratio of organisms according to antibiotic concentrations. The MICs (µg/ml) are: cefsulodin, 2; mecillinam (amdinocillin), 3.2; mezlocillin, 4; amoxicillin, 25; CGP, 2; ampicillin, 12; ciprofloxacin, 0.02. (B = bacilli, F = filaments, L = lysis, S = spheroid-like)

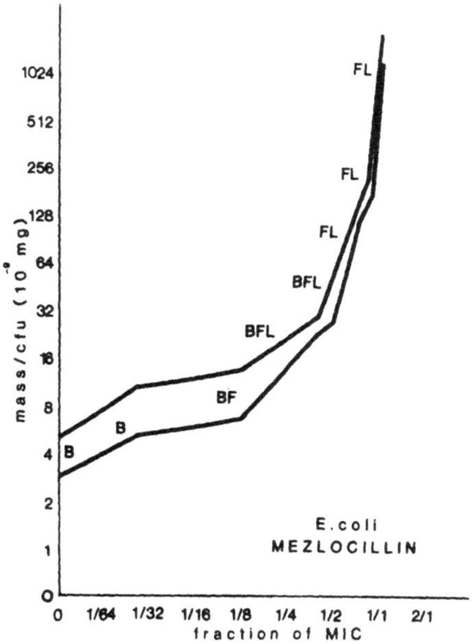

Fig. 3

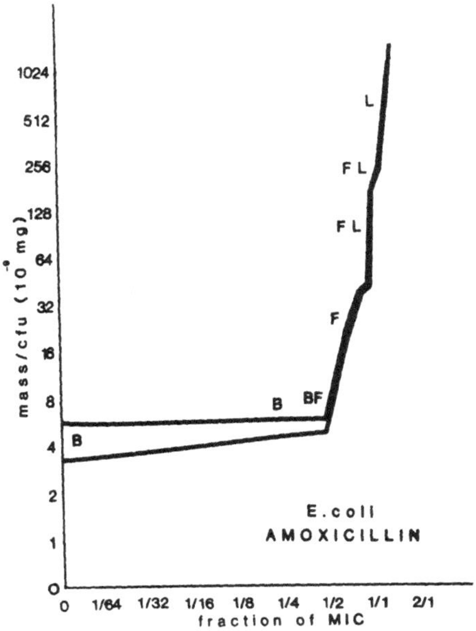

Fig. 4

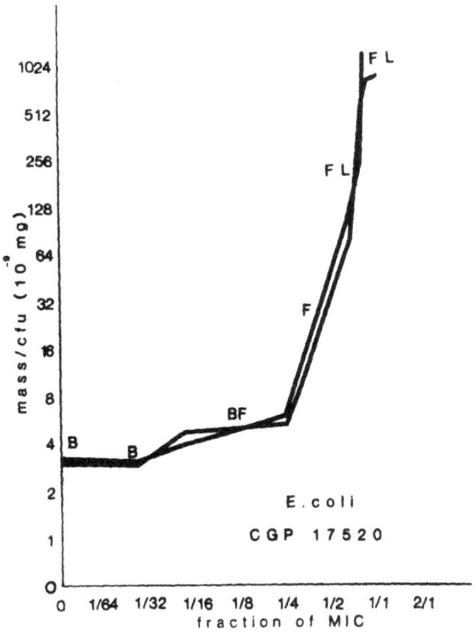

Fig. 5

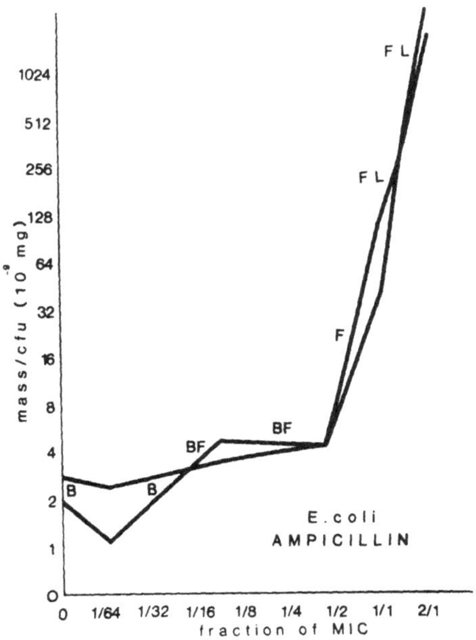

Fig. 6

was calculated to be 3.67×10^{-9} mg or 3.67 picograms and that of a normal *P. aeruginosa* bacillus was calculated to be 4.5×10^{-9} mg or 4.5 picograms.

The change from bacilli to filaments does not occur at the same time or at the same ratio for all the organisms in a given population of bacteria exposed to an antibiotic. Therefore, in most instances a sample contains one predominant form such as filaments but some normal size bacilli or elongated bacilli might be present too. The authors definition of an elongated bacillus is a rod that is longer than the double its length during the mid division cycle; when longer than 10 μm, it is termed a filament. When the bacterial population is exposed to antibiotic concentrations equal to or higher than the MIC lysis takes place but it is not installed in the entire bacterial population at the same time. This results in most cases in a mixture of filaments and lysed filaments. When *E. coli* is exposed to mezlocillin (Fig. 3) at concentrations between ⅛ and ½ the MIC, all three forms, bacilli, filaments and lysed filament can be observed in the same sample at the same time.

At low antibiotic concentrations— 1/64 to ⅛ of the MIC—the weight/CFU ratio for gram-negative bacilli was 2–3 times that of a bacterial population consisting of normal bacilli. At concentrations up to ½ of the MIC, the ratio increased 10–20 times. An antibiotic concentration at or above the MIC resulted in a large increase in the weight/CFU ratio. However, this increase coincided with significant lysis of filaments observed microscopically and, therefore, the ratio cannot be used for the weight estimation of a single organism.

Three patterns were observed with beta-lactam antibiotics to which the gram-negative bacilli were exposed:

(a) With amdinocillin (mecillinam) a low antibiotic concentration up to ¼ of MIC caused a weight/CFU ratio increase which then remained practically unchanged up to two times the MIC, with no lysis observed microscopically;

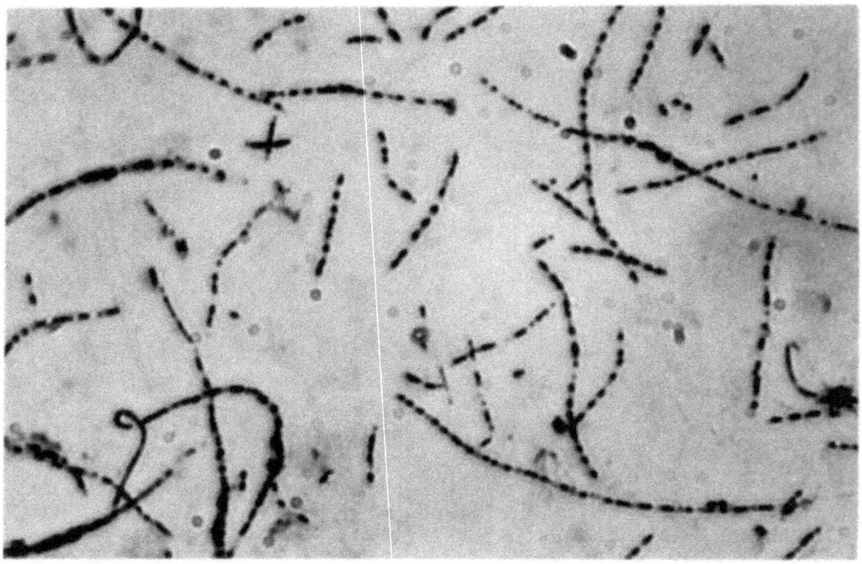

Fig. 7. The nuclear material (dots) is homogeneously dispersed throughout the filaments

Table 1. Weight/CFU Ratios of Three Strains of *S. aureus*

Strain	Control			Cloxacillin			Clox/Control Ratio
	Total CFU Number	Total Weight	Weight/ CFU Ratio	Total CFU Number	Total Weight	Weight/ CFU Ratio	
209P	8.8×10^{9}	19 mg	2.2×10^{-9}* mg	7.6×10^{8}	16 mg	21×10^{-9} mg	9.54
Fitz	2.1×10^{10}	37 mg	1.8×10^{-9} mg	1.8×10^{9}	26 mg	14×10^{-9} mg	7.78
10B	4.2×10^{10}	71 mg	1.7×10^{-9} mg	7.3×10^{8}	13 mg	18×10^{-9} mg	10.58

* 10^{-9} mg = 1 picogram

(b) With low concentrations (up to ½ of MIC) of ampicillin, amoxicillin, mezlocillin, and CGP-17520, the weight/CFU ratio increased proportionalo to the antibiotic concentration;

(c) With cefsulodin, concentrations up to ¼–½ of the MIC resulted in a small weight/CFU ratio increase followed by an abrupt, large weight/CFU ratio increase with significant lysis. The stained slide showed nuclear material evenly dispersed throughout the entire filament (Fig. 7).

Table 1 shows the weight/CFU ratio of the three strains of *S. aureus* with and without exposure to cloxacillin. The mean weight of an *S. aureus* was 1.9×10^{-9} mg or 1.9 picograms; the mean weight of a large staphylococcus grown with cloxacillin was 17×10^{-9} mg or 17 picograms, a weight about nine times the normal.

Discussion

Bacterial mass can be measured by electron microscopy [8]. However, when the mass of different bacterial sizes must be determined, measurement of direct weight is the more accurate method [9].

A filament, which appears to be a simple elongated bacillus, actually consists of a series of bacilli, the result of replication without separation. A filament of *E. coli* produced by ampicillin usually weight 30–60 picograms and contains multiple genomes (Fig. 7). In drug-free medium the filament separates into its constituent normal bacilli [5]. In contrast, a round cell produced by amdinocillin weight 18–20 picograms, but is the result of abnormal growth of a single bacillus and contains a single genome [10, 11].

The large staphylococci are similar to the filaments of gram-negative bacilli in that they contain several genomes. On drug-free medium, large staphylococci separate into multiple staphylococci of normal size and growth rate [1].

An increase in the weight/CFU ratio is evidence of morphologic alteration and, therefore, of antibacterial activity. Since the ratio is quantitative, it expressed the magnitude of change more accurately than microscopic observation alone. A

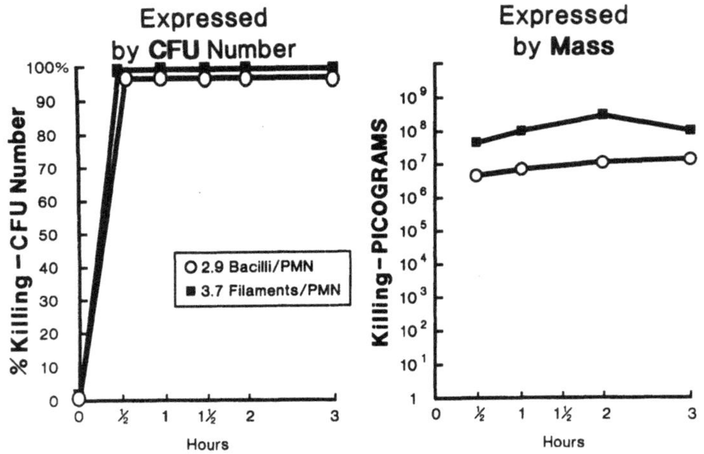

Fig. 8. Killing by PMNs of *E.coli* bacilli and filaments produced by exposure to ½ MIC of ciprofloxacin. Left: Killing as determined by CFU numbers; Right: Killing as determined by mass (picograms)

decrease in the weight/CFU ratio after the antibiotic altered organisms are transferred to drug-free agar, could be indicative of recovery (separation into bacteria of normal size and mass).

In experiments with large bacteria, the recognition of bacterial weights (mass) could lead to conclusions different from those based on bacterial numbers. In a recent study, the bactericidal effect of PMNs on gram-negative bacilli and their respective filaments was evaluated by two criteria: the number of killed organisms/PMN and the weight of organisms killed/PMN [12]. Only the comparison of the weight of killed bacilli and the weight of killed filaments revealed that filaments are significantly more susceptible than normal bacilli to the bactericidal activity of PMNs. To illustrate this point please examine Fig. 8. When the CFU criteria is applied, the killing by PMNs of both the *E. coli* bacilli and the filaments produced with ciprofloxacin appear to be equal. When, however, the weight (mass) of the organisms killed is determined, the PMNs killed a significantly larger mass of filaments than of bacilli. Regardless of its size or weight, when planted on agar, each bacterium is assumed to produce a single colony. Therefore, the correct criterion for comparison of large bacteria with their normal counterparts appears to be their respective weight rather than their number.

The chance that one PMN will kill 20 bacilli or 10 normal staphylococci is better when the bacilli or the cocci are together in the form of a filament or a large cell than when they are dispersed as usual. Since PMNs are able to kill filaments and large staphylococci remarkably well (in fact, much more efficiently than they kill a comparable mass of normal size organisms, their presence is a favorable consequence of antibiotic treatment in immunocompetent patients.

Summary

One strain of *Escherichia coli*, one strain of *Pseudomonas aeruginosa*, and three strains of *Staphylococcus aureus* were grown in the presence of various beta-lactam antibiotics. Morphologic alterations such as filamentation of gram-negative bacilli and substantial enlargement of staphylococci resulted.

For both normal and antibiotic-exposed organisms, the number of colony-forming units (CFU) and the dry weight were determined, then the approximative dry weight of an individual organism was calculated. The weight of one normal *E. coli* bacillus ranged from 2.8×10^{-9} to 4.2×10^{-9} mg, whereas the weight of an *E. coli* filament resulting from exposure to a beta-lactam antibiotic ranged from 20×10^{-9} to 120×10^{-9} mg. The weight of a normal staphylococcus ranged from 1.7×10^{-9} to 2.1×10^{-9} mg, whereas the weight of one large staphylococcus resulting from exposure to cloxacillin ranged from 14×10^{-9} to 21×10^{-9} mg.

It was concluded that recognizing bacterial weight during experiments with these large bacteria could lead to interpretations different from those based on bacterial numbers alone.

References

1. Lorian V (1975) Some effects of subinhibitory concentrations of penicillin on the structure and division of staphylococci. Antimicrob Agents Chemother 7: 864–870
2. Gardner AD (1940) Morphological effects of penicillin on bacteria. Nature 146: 837–838
3. Zimmerman SB, Stapley O (1976) Relative morphological effects induced by cefoxitin and other beta-lactam antibiotics in vitro. Antimicrob Agents Chemother 9: 318
4. Lorian V, Sabath LD (1972) Penicillins and cephalosporins; differences in morphologic effects on *Proteus mirabilis*. J Infec Dis 125: 560–564
5. Lorian V (1980) Effects of subminimum inhibitory concentrations of antibiotics on bacteria. In: Lorian V (ed.) Antibiotics in laboratory medicine. Williams and Wilkins, Maryland, p 342–408
6. Robinow CF (1944) Cytological observations on Bact. coli, *Proteus vulgaris* and various aerobic spore-forming bacteria with special reference to the nuclear structures. J Hygiene 43: 413–423
7. Everson-Pearse AG (1972) Histochemistry theoretical and applied. Williams and Wilkins, Maryland, p 1186–1192
8. Bahr CF, Engler WF, Mazzone HM (1976) Determination of the mass of viruses by quantitative electron microscopy. Quarterly Revs Biophys 9: 459–489
9. Scullard G, Meynell E (1966) Bacterial mass measured with the MRC greywedge photometer. J Pathol Bacteriol 91: 608–612
10. Lorian V, Atkinson B (1977) Comparison of the effects of mecillinam and 6-aminopenicillanic acid on *Proteus mirabilis*, *Escherichia coli*, and *Staphylococcus aureus*. Antimicrob Agents Chemother 11: 541–552
11. Amaral L, Viano I, Atkinson B, Lorian V (1981) Development, replication, and reversion of mecillinam-induced ovoid form of *Salmonella typhimurium*. Curr Microbiol 5: 207–211
12. Lorian V, Atkinson B (1984) Bactericidal effect of polymorphonuclear leukocytes on antibiotic induced filaments of gram-negative bacilli. J Infect Dis 149: 719–727

Discussion

Opferkuch: I have some doubts about counting of CFU of antibiotic treated bacteria because of the dependence on the medium used.

Lorian: Growing bacteria on membrane placed on Trypticase Soy Agar containing antibiotics gave us quite reproducible results.

Daschner: How do you get rid of the antibiotic present in the test medium, used for production of "masses". Is it possible that you weigh not only the bacterial mass, but also the antibiotic.

Lorian: The bacteria are grown on filter membranes placed on agar containing antibiotics. Before drying, the filters are "washed" on drug free agar and floated on water which, as documented, removes all the antibiotic present.

Ginsburg: The problem of mass/number of organisms may have important bearing on the survival of peptidoglycan-polysaccharide (PPG-PS) complexes in macrophages in chronic inflammatory sites.
a) activation of macrophages to release lysosomal enzymes
b) activation of the O_2 metabolism
c) immunogenicity
d) blocking of macrophages by excess PPG-PS complexes, PPG-PS complexes
e) transaction of PPG-PS to other tissue sites
f) resistance to degradation by lysosyme other lysosomal enzymes

Lorian: I certainly agree with all points.

Pulverer: When you prepare to calculate the mass of bacteria instead of CFU this is to my opinion very well possible in an in vitro system. But how can you calculate the mass of bacteria when you are performing in vivo experiments?

Lorian: The microscopic examination of a specimen from an "in vivo" subject will indicate the approximate size of filaments present. An approximative mass of the filament can be calculated by multiplying the length of the filaments (µm) with 3 which will approximate the weight in picograms of the respective filament. Multiply them with the CFU number per gram of tissue and you obtain a very approximate mass of bacteria present.

Voigt: Would you please tell something about the weighing of bacteria.

Lorian: The bacteria are grown on filter membranes placed on agar containing antibiotics. Before eluting the bacteria, the filters are "washed" on drugfree agar and floated on water which, as documented, removes all the antibiotic present. The organisms are eluted in water, dried, desiccated to constant weight and then weighed 3 times and mean taken.

II. The Influence of Antibiotics on Immunogenicity

Immunogenicity of Bacteria Treated in vitro with Subinhibitory Concentrations of Antibiotics

H. FRIEDMAN

Department of Medical Microbiology and Immunology, University of South Florida College of Medicine, Tampa, FL USA

Introduction

Studies in this and other laboratories over the last few years have pointed to the possible synergistic relationship between the host immune defense system and antibiotics in protection and/or treatment against a variety of infections agents, especially those caused by opportunistic bacteria [1–4]. Studies in this laboratory have shown that a semi-synthetic penicillin, such as cyclacillin or ampicillin, but not Penicillin G itself, may alter the susceptibility of susceptible Gram negative bacteria such as Escherichia coli to host immune defenses, both humoral and cellular. E. coli pretreated in vitro with a subinhibitory concentration of cyclacillin or ampicillin became much more susceptible to the bactericidal activity of specific antibody and complement. Such antibiotic pretreated bacteria, even after extensive washing, also showed increased susceptibility to phagocytosis by normal peritoneal or splenic macrophages [5, 6]. In addition, the antibiotic pretreated bacteria were even more readily phagocytized by peritoneal macrophages from adjuvant treated mice. Thus antibiotic pretreatment appeared to result in marked increases in the susceptibility of the microorganism to phagocytosis. The bacteria also became much more susceptible to antibody mediated lysis in the presence of complement.

It appears that the antibiotic, even at subinhibitory concentration, may have altered the surface of the microorganism and resulted in the formation and/or release of increased amounts of surface antigens such as lipopolysaccharide which reacted with cellular or humoral components of the immune system. As shown in the present study, as well as earlier ones in this laboratory, the antibiotic pretreated E. coli induced larger numbers of antibody forming cells in the spleen of normal mice, as well as specifically antigen primed mice. Macrophages from these mice, after ingesting the antibiotic pretreated and washed bacteria, contained increased immunogenic moieties of the bacteria as shown by subsequent injection of cell-free extracts of the cells into recipient animals.

General Experimental Procedures

For these studies a penicillin resistant strain of E. coli was used exactly as described previously [2,4–6]. The organisms were cultured in vitro on brain heart infusion agar and after 18 hrs, were harvested, washed with sterile saline, and resuspended to a

The Influence of Antibiotics on the
Host-Parasite Relationship II
Edited by D. Adam, H. Hahn, W. Opferkuch

concentration of 10^8 viable bacteria per ml as determined by standard agar plate count. Cyclacillin and ampicillin, as well as penicillin G, were obtained from Wyeth Laboratories, Radnor, PA and dissolved in sterile physiological saline before use. For antibody studies, pooled sera from rabbits immunized by several courses of injections with 10^8 heat killed E. coli LPS were utilized. The rabbits were given 4–5 injections of the bacterial LPS at weekly intervals over a period of 4–6 weeks and 10 days after the last injection, serum was obtained. The titer of pooled sera was generally 1:120 or greater by microhemagglutination procedure in which serial 0.025 ml dilutions of serum were incubated in 96 well microtiter plates (Nunc), to which were added equal volume of a 10^8 suspension of heat killed E. coli. After 1 hour incubation at 37 °C and 18 hrs in 4 °C, the titer was determined as the reciprocal of the highest dilution resulting in complete agglutination of the bacteria. For bacteriolytic assays, standard amounts of serum in 0.1 ml of saline, plus 0.1 ml of a 1:10 dilution of guinea pig complement, were added to the microwell plates. To each well was added 10^6 or 10^7 viable E. coli. After 2 hrs incubation at 37 °C the number of viable E. coli present in individual wells was determined by quantitative plate counts.

For phagocytic assays, peritoneal cells were obtained from normal BALB/c mice, weighing 16 gms each, which were injected intraperitoneally with 5 ml saline 15 min previously. Cells were pooled from individual mice, pelleted by centrifugation at $1000 \times g$ at 4 °C and resuspended in medium 199 containing fetal calf serum, washed twice with medium, and allowed to adhere to plastic petri plates for 1 hour at 37 °C. The nonadherent cells were decanted and the adherent cells harvested by scraping with a rubber policeman. The resulting cells were resuspended to a concentration of 10^8 nucleated cells per ml. Microscopic examination indicated that 85%–90% or more of the cell were morphologically similar to macrophages and over 95% of the cells readily phagocytized latex particles within 15–25 minutes [2,7]. In some experiments, adherent splenic cells were similarly prepared from the mice and resuspended to a concentration of 10^8 cells per ml; over 80% of these cells phagocytized latex particles.

For the phagocytic assay, suspensions of 10^6 nucleated cells in 0.5 ml medium were added to individual wells of 20 well Linbro plates. To each well was then added 0.1 ml of a fresh suspension of E. coli, either untreated or previously treated with antibiotic as indicated. The cultures were incubated in an atmosphere of 95% air and 5% CO_2 for 2 hours. Samples of each culture was placed on microscopic slides which were then stained with Giemsa and examined microscopically. At least 100 macrophages were examined per sample. The number of E. coli present per macrophage was scored and the percent of macrophages containing two more ingested bacteria was determined. For killing experiments, the macrophage–E. coli suspensions were pelleted by centrifugation at $10{,}000 \times g$ for 10 min. The number of bacteria in the supernatants was determined by plate counting. The sedimented macrophages were resuspended in saline, disrupted by freezing and thawing three times and the number of viable E. coli determined by plate counting. In some experiments, macrophages were obtained from mice injected intraperitoneally with 1.0 mg Bacillus calumette guarin (BCG) as an immunologic adjuvant. Alternatively the mice were injected with 50 μg MDP 2 days earlier.

Experimental Results

Pretreatment of E. coli with either cyclacillin or ampicillin, but not penicillin, rendered them much more susceptible to bacteriolysis by the rabbit antisera and normal guinea pig complement (Table 1 and 2). For example, pretreatment of the E. coli with 1–5 μg of either cyclacillin or ampicillin increased the apparent titer of the serum when these treated bacteria were used as the target for the antibody (Table 1). On the other hand, penicillin G, at a concentration of 20 U or even more (data not shown) had little if any effect on the titer of the bacteriolytic antibody. The optimal increase in the killing ability of the antibody against the ampicillin or cyclacillin pretreated E. coli occurred at about 2 hours. However, even a 1 hr

Table 1. Effect of pretreatment of E. coli with graded amounts of antibiotics on bacteriolytic titer with antiserum and complement

Bacterial treatment[a]			Bacteriolytic titer[b]
None (control)			1:160
Antibiotic pretreated	Cyclacillin	5.0 μg	1:1280
		2.5	1:960
		1.0	1:360
	Ampicillin	5.0	1:960
		2.5	1:900
		1.0	1:2180
	Penicillin	G 20.0 U	1:240

[a] Washed 18 hr cultures of E. coli (10^8 bacteria/ml) incubated at 4 °C for 2 hr with graded amounts of indicated antibiotic before washing and incubation for 18 hrs with serial two fold dilutions of anti E. coli rabbit serum and 1:10 dilution of sterile guinea pig complement.

[b] Average bacteriolytic titer of serum and complement using untreated or treated E. coli.

Table 2. Effect of pretreatment of E. coli with antibiotics for varying times on susceptibility to bacteriolysis by antiserum.

Bacterial treatment[a]		Serum dilution[b] 1:80	1:60	1:320	1:640	1:1250	1:2500
None (control)		0	0	2+	4+	4+	4+
Ampicillin	1 hr	0	0	1+	2+	4+	4+
	2 hr	0	0	0	±	2+	4+
	4 hr	0	0	0	±	1+	4+
Cyclacillin	1 hr	0	0	4+	3+	4+	4+
	2 hr	0	0	0	±	2+	4+
	4 hr	0	0	0	±	1+	4+
Penicillin	4 hr	0	±	2+	3+	4+	4+

[a] Suspension of 10^8 E. coli/ml medium exposed to 2.0 μg ampicillin or cyclacillin, or 20 U Penicillin G, for indicated lengths of time at 4° before washing and testing for susceptibility to bacteriolysis by anti-E. coli LPS serum and complement.

[b] Growth of E. coli after 18 hr incubation at 37° in presence of serial dilution of rabbit anti E. coli serum plus 1:10 dilution of sterile guinea pig complement; 0 = No growth and 4+ = maximum growth.

pretreatment resulted in an increase in the susceptibility of the bacteria to the antibody (Table 2). No killing occurred in the absence of specific antibody (i.e., normal serum) or when complement was not used.

Pretreatment of E. coli with similar quantities of ampicillin or cyclacillin also increased their uptake and destruction in liver and spleen. As seen in Table 3, the bacteria, after treatment with subinhibitory concentrations of the antibiotics, followed by extensive washing, were much more rapidly removed from the blood of injected mice within 30 minutes, and fewer viable organisms were present in the liver and spleen at 24 hrs as compared to the number of untreated E. coli or the E. coli pretreated with penicillin G. Similar results occurred in terms of in vitro phagocytosis, i.e., pretreatment of the E. coli with ampicillin or cyclacillin at subinhibitory concentrations increased their ability to be phagocytized by mouse peritoneal exudate cells (Table 4). More macrophages had two or more E. coli after

Table 3. Uptake of antibiotic pretreated bacteria by tissue of normal mice

Antibiotic treatment of E. coli[a]		Blood (30 min)	E. coli in tissue[b] Liver (24 hr)	Spleen (24 hr)
None (control)		2.1×10^2	8.3×10^3	3.9×10^5
Ampicillin	5.0 μg	0.9×10^2	3.1×10^3	1.8×10^3
	1.0	1.0×10^2	2.1×10^3	7.3×10^2
Cyclacillin	5.0	1.0×10^2	1.1×10^3	6.9×10^4
	1.0	1.2×10^2	8.6×10^2	6.5×10^3
Penicillin G	G 20.0 U	3.8×10^2	5.1×10^4	2.1×10^5

[a] Suspension of 10^8 E. coli pretreated for 2 hr at 4 °C with indicated doses of antibiotic prior to washing and injection into BALB/c mice.
[b] Average number of virulent E. coli CFU per ml blood or gram tissue determined by bacterial plate count at indicated time after injection into normal test mice.

Table 4. Effect of antibiotic pretreatment on phagocytosis of E. coli by normal mouse peritoneal cells

Antibiotic pretreatment of E. coli[a]		E. coli presence Macrophage culture Whole cell[b] extract[c]	Cell free ($\times10^2$)	Cell free medium ($\times10^3$)[c]
None (control)		7.5±1.3	14.8±21.0	36.5±2.3
Ampicillin	5.0 μg	26.5±2.5	2.1±0.4	12.5±1.2
	1.0	21.3±2.7	1.3±0.5	18.6±2.5
Cyclacillin	5.0	28.2±3.4	3.2±0.7	14.2±2.1
	1.0	22.1±2.5	2.0±0.5	15.5±3.1
Penicillin G	20 U	9.3121.2	10.3±0.9	32.5±3.0

[a] Suspensions of 10^6 washed E. coli pre-treated for 2 hr with indicated E. coli before exposure to 10^6 ml normal PE cells
[b] Percent normal cells with 2 or more E. coli after 60 minutes incubation of cells with untreated or antibiotic pretreated bacteria
[c] Presence of viable E. coli (number ± S.D.) for 0.5 ml cell-free macrophage extract or culture supernatant

incubation with ampicillin or cyclacillin pretreated E. coli as compared to macrophages incubated with non-treated bacteria or those which had been pretreated with penicillin G. Furthermore, assessment of the number of viable bacteria per ml cell free extract from these macrophages showed many fewer living microorganisms present when antibiotic pretreatment occurred prior to washing of the bacteria and exposure to the phagocytes (Table 4). In addition, assessment of the number of extra-cellular bacteria in the medium after phagocytosis showed that many more bacteria were taken up by the macrophages incubated with cyclacillin or ampicillin pretreated organisms as compared to those which had been incubated with penicillin G treated bacteria.

Enhanced phagocytosis occurred when ampicillin or cyclacillin pretreated bacteria were exposed to macrophages from mice which had been injected with either BCG or MDP to activate the macrophages (Table 5). Although more phagocytosis occurred with macrophages from adjuvant treated animals incubated in vitro with non-antibiotic pretreated E. coli or those which were pretreated with penicillin G, there was a 2–3 fold greater increase in the percentage of macrophages which ingested the ampicillin or cyclacillin pretreated E. coli.

It appeared that antibiotic pretreatment made the microorganisms much more susceptible to antibody mediated bacteriolysis or phagocytosis. Thus experiments were performed to determine whether antibiotic pretreatment of the bacteria before their exposure to phagocytes resulted in more immunogenic bacterial material in the cells as compared to phagocytes containing bacteria which were not pretreated with an effective antibiotic. As is apparent in Table 6, ingestion of ampicillin or cyclacillin pretreated E. coli, but not control E. coli or those which had been pretreated with penicillin, resulted in more immunogenic components which, when injected into normal mice, induced a higher level of agglutinating antibody to the E. coli. Furthermore, these extracts, when injected into normal mice, resulted in increased numbers of anti-E. coli PFCs. Even greater responses were apparent when these macrophage extracts were injected into mice which had been primed four weeks earlier with E. coli vaccine (data not shown). Thus the ampicillin and/or cyclacillin pretreated organisms exposed to macrophages not only were taken up in greater

Table 5. Effect of antibiotic pretreatment of E. coli on subsequent phagocytosis by peritoneal macrophages from normal or adjuvant treated mice

Antibiotic pretreatment of E. coli[a]		Mouse macrophage tested[b] Normal	BCG	MDP[c]
None (control)		6.5±2.1	13.2±4.1	11.3±2.3
Ampicillin	5.0 μg	21.4±3.4	36.1±4.5	30.1±2.8
	1.0	19.3±2.1	30.3±2.6	27.3±3.1
Cyclacillin	5.0	21.5±4.2	39.3±4.0	36.5±3.8
	1.0	18.4±3.1	36.2±2.8	30.3±4.0
Penicillin G	20 U	6.9±1.9	10.6±2.8	11.9±2.4

[a] Indicated antibiotic used to pretreated 10^8 E. coli for 2 hrs before assay with PE cells

[b] Average percentage of PE cells, ± SD, phagocytizing 2 or more E. coli after incubation for 2 hr at 37 °C with bacteria

[c] Donor mice injected 7 days earlier with 1.0 mg BCG or 2 days earlier with 50 μg MDP

Table 6. Effect of antibiotic pretreatment of E. coli on the ability of phagocytized bacteria to induce anti-E. coli PFC in recipient mice

Antibiotic pretreatment of E. coli[a]		Antibody response[b] Agglutinin titer	Anti E. coli LPS PFC
None (control)		1:480	383±72
Ampicillin	5.0 μg	1:1280	650±48
	1.0	1:1280	495±32
Cyclacillin	5.0	1:1280	586±47
	1.0	1:960	542±30
Penicillin G	20.0 U	1:480	310±60

[a] Indicated antibiotic used to pretreated 10^8 E. coli for 2 hrs prior to phagocytosis with 10^6 normal mouse peritoneal cells.

[b] Average antibody response for 3–4 mice per group injected 8–10 days earlier with 1.0 ml of cell-free extract of PE cells which phagocytized E. coli; control mice given extract of 10^6 normal peritoneal cells had anti-E. coli titer of 1:20 and fewer than 100 PFC per spleen

numbers, but also more immunogenic components were present in the macrophages as shown by the ability of extracts of these cells to induce increased antibody responses to E. coli in recipient animals, either normal or antigen primed. Control experiments showed that these macrophage extracts did not induce increased antibody responses to other bacteria such as Salmonella, so that the immunogenic component appeared to be antigen specific.

In order to study in more detail whether antibiotic pretreated organisms as compared to control or penicillin G treated organisms were indeed more immunogenic, they were injected directly into mice and specific anti-E. coli PFCs determined at various times thereafter (Table 7). The untreated E. coli, as controls, induced a significant numbers of PFCs when injected directly into normal BALB/c mice. Injection of similar numbers of E. coli which had been pretreated with cyclacillin or ampicillin resulted in even greater numbers of PFCs to E. coli LPS, indicating that the bacteria, although not inhibited in their growth capacity by

Table 7. Effect of antibiotics on immunogenicity of E. coli as shown by ability of bacteria to induce bacteriolytic PFC

Antibiotic pretreatment of E. coli[a]		Anti E. coli PFC on day[b] +3	+6	+10	+15
None (control)		121±15	865± 32	652± 42	210±16
Ampicillin	5.0 μg	265±40	1830±240	1445± 62	325±40
	1.0	210±32	1750±138	1230±130	428±60
Cyclacillin	5.0	242±19	1563±120	1230± 41	652±30
	1.0	152±20	1410±193	1140±160	570±48

[a] Indicated antibiotic used to pretreat 10^8 E. coli for 2 hrs before washing and injection into recipient mice

[b] Average number of anti E. coli PFC, ± SD, on indicated day for spleen cells from 3–4 mice per group

cyclacillin or ampicillin, nevertheless appeared to develop more potent immunogenic capacity as shown by their ability to induce larger numbers of specific antibody forming cells.

Discussion and Conclusions

There is now much evidence that antibiotic pretreated bacteria become much more susceptible to the host defense system, either cellular or humoral, as compared to untreated bacteria [2,6,9–11]. This probably occurs because cell surface components of the bacteria may be altered after the exposure to an antibiotic, even in subinhibitory concentrations. It has been noted that filimentous forms of gram negative bacteria often develop after exposure of the organisms to even subinhibitory concentrations of antibiotics. Such doses of the antibiotics do not appear to interfere with the viability of the bacteria in vitro. Similar numbers of colony forming units are usually observed with such antibiotic pretreated bacteria as compared to untreated ones. Moreover, a number of previous studies from many laboratories, including this one, have shown that antibiotic pretreatment of some bacteria with subinhibitory doses of an antimicrobial agent resulted in increased susceptibility to phagocytosis by macrophages, both in vivo and in vitro [2,4–6,11]. As shown in the present study, as well as in previous ones, peritoneal cells derived from normal mice or even mice stimulated with an adjuvant showed an increased ability to phagocytize and kill a bacterium such as E. coli after pretreatment with subinhibitory doses of semi-synthetic antibiotics such as cyclacillin and ampicillin, but not with an antibiotic such as penicillin G to which they were resistant. This effect appeared due to a direct interaction of the antibiotic on the bacterium, since the organisms were washed extensively before injection into animals or addition to macrophages from normal or immune animals or even to antibody.

Macrophages appeared to be able to degrade more readily E. coli into potent immunogens when the bacteria were pretreated with the subinhibitory doses of cyclacillin or ampicillin as compared to the bacteria which had not been pretreated. For example, incubation of peritoneal cells from normal mice or even mice which had been pretreated with an adjuvant such as BCG or MDP resulted in more rapid phagocytosis of cyclacillin or ampicillin pretreated E. coli and showed increased immunogenicity in the cell-free extracts as evident by increased numbers of specific anti-E. coli antibody forming cells upon subsequent injection into test mice. There is some question, however, as whether or not the increased immunogenicity of the macrophage extracts was due to increased amounts of antigen because of larger numbers of bacteria present in each macrophage or because of the degradation or release of increased amounts of E. coli surface antigens. Since each macrophage exposed to the cyclacillin or ampicillin pretreated bacteria contained more bacteria, it is possible that this could account for an apparent increased immunogenicity. It is also possible, however, that the bacteria were not only much more readily phagocytized but more readily degraded into larger amounts of immunogenic moieties when pretreated with the antibiotics.

Further studies quantitating the amount of immunogen present in an average macrophage should be performed, including experiments with radiolabelled E. coli.

Nevertheless, it appears that pretreatment of the E. coli with ampicillin or cyclacillin, but not penicillin, not only makes them more susceptible to phagocytosis, i.e., greater uptake of the bacteria and larger numbers of bacteria per phagocyte, but also increases the amount of immunogen per macrophages. It should be noted that the in vitro serologic tests also indicated that antibiotic pretreatment of the bacteria makes them more susceptible to lysis by antiserum and complement. Thus it is probable that these microorganisms are also more readily lysed or destroyed by intracellular components of the phagocytes. It is also possible that the organisms could be so readily degraded in vivo or in vitro that they are digested to less immunogenic or non-immunogenic components. Thus, it does not seem likely that greater induction of antibody forming cells to E. coli by injection of such macrophage extracts can permit at this time a definitive conclusion as to whether more or less bacterial antigen or other components become available as an immunogen when phagocytes ingest and/or degrade antibiotic pretreated bacteria.

It does seem likely that antibiotic pretreatment makes the organism more susceptible to host immune factors, including humoral ones, i.e. antibody, and/or phagocytes. The ability of the antibiotic pretreated organisms to stimulate enhanced antibody formation may also be important, at least as a secondary feature of antibiotic treatment. Although it is apparent that in many infections by opportunistic bacteria, such as gram negative ones, the eventual outcome of the disease process occurs quite early after exposure of an individual to the microorganism, development of antibody forming cells, as well as cell mediated immunity dependant upon macrophages and T cells, plays an important role in resistance to further infection. The exposure of larger amounts of immunogenic components of the bacterium to the host's specific defense system may be important in secondary resistance to infection. The ability of the antibiotic pretreatment to make a microorganism more susceptible to antibody as well as phagocytes may be important in the early events following infection and antibiotic treatment, even if a subinhibitory dose of the antibiotic is utilized. Certainly more studies are warranted to investigate these possibilities in greater depths.

Summary

Pretreatment of Escherichia coli with subinhibitory concentrations of the semisynthetic penicillin cyclacillin or ampicillin, but not with penicillin G, resulted in increased susceptibility of the bacteria to phagocytosis as well as to increased bacteriolysis mediated by antibody and complement. Such increases in susceptibility to phagocytosis and bacteriolysis correlated with previous observations of increased susceptibility to the host immune system. Enhanced phagocytosis also occured when the bacteria were pretreated with subinhibitory concentrations of cyclacillin or ampicillin and, following extensive washing, incubated in vitro with macrophages from mice stimulated with the nonspecific immunomodulators BCG or MDP. Furthermore, pretreatment of E. coli with cyclacillin or ampicillin, followed by phagocytosis by normal macrophages, resulted in increased amounts of immunogenic components in the macrophages as indicated by subsequent injection of cell-free extract of these cells into normal or E. coli primed recipient mice. Larger numbers of antibody forming cells to the E. coli LPS and increased antibody titers occurred in mice given cell-free extracts from macrophages which had phagocytized the cyclacillin and ampicillin pretreated E. coli as compared to those which had phagocytized untreated E. coli or those which had been pre-incubated with penicillin G. These results support the view that antibiotic pretreated bacteria, even with subinhibitory doses, become much more susceptible to the host immune defense system.

References

1. Adam D, Schaffert W, Marget W (1974) Enhanced in vitro phagocytosis of Listeria monocytogens by human monocytes in the presence of ampicillin, tetracycline, and chloramphenicol. Infect Immun 9: 811
2. Friedman H, Warren GH (1974) Enhanced susceptibility of penicillin resistant staphylococci to phagocytosis after in vitro incubation with low doses of nafcillin. Proc Soc Exp Biol Med 146: 207–211
3. Gillissen G (1980) Antibiotics and immune response, concomitant effect of chemotherapy. Immunität und Infektion 8: 79–88
4. Friedman H, Warren GH (1982) Increased phagocytosis of Escherichia coli pretreated with subinhibitory concentrations of cyclacillin or ampicillin. Proc Soc Exp Biol Med 169: 301–305
5. Friedman H, Warren GH (1976) Antibody-mediated bacteriolysis: Enhanced killing of Cyclacillin treated bacteria. Proc Soc Exp Biol Med 153: 301–304
6. Friedman H, Warren GH (1977) Cyclacillin-induced potentiation of Escherichia coli immunogenicity in vivo and in vitro. Chemotherapy 23: 324–336
7. Friedman H, Warren GH (1984) Antibacterial activity of cyclacillin in vivo and in vitro: Effects on Gram negative organisms. Prog Chemother 2: 119–126
8. Eickenberg U-U, Hahn H, Opferkuch W (eds) (1982) The influence of antibiotics on the host-parasite relationship. Springer
9. Lorian V, Atkinson B (1975) Abnormal forms of bacteria produced by antibiotics. Am. J. Clin Pathol 64: 678
10. Ofek I, Beachey EH, Eisenstein BJ, Aklan MC, Sharon N (1979) Suppression of bacterial adherence by subminimal inhibitory concentrations of Beta-lactam and aminoglycoside antibiotics. Rev Inf Dis 1: 832
11. Gemmell CG, Peterson PK, Schmeling D, Mathews J, Quie PG (1983) Drug-induced modification of Bacteroides fragilis and its susceptibility to phagocytosis by human PMN. Eur J Clin Micro-biol 2: 327

Discussion

Lorian: Since organsims exposed to the antibiotic increase in size and in *mass*, is it possible that the increased immunity is the result of a larger mass of bacteria?

Friedman: Immunity may be a result of a larger mass of bacteria. This would be due to a larger amount of LPS in each bacteria. That is one of the postulates we have, i.e., that the antibiotic causes increased amounts of LPS to appear on the surface of the bacterium. Obviously if there were a greater amount of the bacteria, one would expect to have more LPS antigen or determinants per unit bacterium. The question can be raised, however, whether the same weight of bacteria is made up of larger bacteria or to many more smaller bacteria. Would one larger bacterium have the same or different amounts of LPS? This should be investigated in greater detail.

Influence of Subminimal Inhibitory Concentrations of Antibiotics on Antigen Expression and Metabolic Activity of Salmonella Species

G. GILLISSEN and B. MELZER

Department of Medical Microbiology, Medical Faculty, Pauwelsstraße, 5100 Aachen (FRG)

Introduction

Subminimal inhibitory concentrations (subMICs) of antibiotics may induce changes in bacterial morphology, ultrastructure, biochemistry and in multiplication rate [15, 19, 21, 22, 23, 26, 31, 36]. SubMICs of antibiotics may also increase susceptibility to phagocytosis and intracellular killing [8, 9, 30], or affect the ability of bacteria to adhere to epithelial cells [1, 34]. Considering that the morphological and structural changes produced by subMICs of antibiotics are different from those induced by ≧ MICs, *Lorian* [19] postulated that the biochemical changes responsible for them must also be significantly different. It should, therefore, be expected that bacteria exposed to subMICs of antibiotics show also alterations in antigen expression with changes in their pathogenicity and the related immune response.

Only few studies are concerned with this problem. M-protein synthesis of streptococci was shown to be inhibited by chloramphenicol and tetracycline but only with concentrations that inhibit bacterial growth [4]. An alteration of the expression of virulence factors of bacteria could, however, also be demonstrated when grown in subMICs of antibiotics [11, 17, 27]. Immunization of rabbits with vaccines prepared with group A – streptococci grown in ½ of the MIC (but not in ⅛ of the MIC) of erythromycin resulted in a slower and weaker development of protective antibodies than in controls [6]. This was suggested to be due to a partial loss of antigenic material, particularly of M-protein, by the streptococci exposed to subMICs of the antibiotic [6]. Experiments performed to examine the influence of subMICs of fosfomycin on immunogenicity of S. wien showed higher antibody titers in rabbits both concerning response to somatic and flagella antigens than in controls [37]. According to the methods described, the differentiation of the two groups of specificities is, however, difficult to explain.

Modifications of antigenic structures have first been described in Salmonella by tetracyclines [35]. The loss of the ability of streptococci to adhere to epithelial cells [1] was explained biochemically by the associated loss of lipoteichoic acid, the ligand that binds the germ to cell membrances. In E. coli, the nonseptate filaments developed in subMICs of penicillin are devoid of pili and therefore of adherence ability (cit.[19]). According to these observations a change in antigenic composition by subMICs of antibiotic should be suggested. Particularly in case of antibacterial agents that affect peptidoglycan and produce morphological and biochemical changes in the cell wall, alterations of antigenic structure of gramnegative bacilli

The Influence of Antibiotics on the
Host-Parasite Relationship II
Edited by D. Adam, H. Hahn, W. Opferkuch

should be expected [19]. Experiments with S. typhimurium performed in this respect showed, however, that organisms grown in subMICs of antibiotics as ampicillin and others had lower agglutination titers when the antigen was prepared in alcohol and higher titers when prepared by heat treatment [20, 24]. Comparably, no decisive difference in antigen structure was seen, when organisms were visualized by specific fluorescent antibody staining.

There are several observations on the influence of antibiotics on enzyme activities of microorganisms [12; rev. 10, 14]. In most of these experiments, however, ≧ MICs of antibacterial agents have been assayed but relatively few publications are concerned with the effect of subMICs of antibiotics. Examples are the inhibition of D-alanin-carboxypeptidase of Bac. subtilis [2] and of E. coli [33] by penicillin or of acid phosphatase activity of Staphylococci by penicillin and neomycin [13].

The aim of the experiments presented here was to investigate in parallel the effects of subMICs of the three antibiotics chloromycetin (C), ampicillin (A) and of a tetracycline (T) on morphology and structure of a given Salmonella strain and its antigenicity as well as on its metabolic activities.

Materials and Methods

Salmonella strains

S. paratyphi A ATCC 11511 (1, 2, 12: a:-), S. paratyphi A mutant (1, 2, 12: a-) no. 256K, S. I 16: a:- (no. IP 903/71). The last 2 strains were kindly made available by Dr. S. Aleksić (Salmonella-Center, Hamburg, FRG).

Antibiotics

Chloramphenicol (Boehringer, Mannheim, FRG) (C), ampicillin-sodium (Grünenthal, Stolberg, Rhld., FRG) (A) and tetracycline-hydrochloride (Achromycin, Cyanamide Inc., Wolfratshausen, FRG) (T). The minimum inhibitory concentrations (MICs) of these antibiotics for S. paratyphi A ATCC 11511 were 0.39 (C), 0.63 (A) and 1.56 μg per ml (T).

Culture media

Standard broth ('Standard I', Merck, Darmstadt, FRG) without glucose and complemented with 1 per cent glucose. For subcultures, D.S.T.-Agar (Oxoid, Basingstoke, Hants., GB) and to facilitate H-antigen-expression Sven-Gard-Agar (Pasteur Institute, Paris, France) were used.

Diagnostic sera

Anti-O2 (ORNA 02/03) and anti-Ha-(ORWA 02/03)-sera (Behringwerke, Marburg, FRG).

Flagella staining

Flagella-staining was carried out by a mordant (solution of tannic acid and Victoria Blue R – Aldrich, Milwaukee, USA) and subsequent staining with Victoria Blue alone as described by Tabeyi (see [16]). A loopful of germs of a 18–20 hours broth culture sedimented by centrifugation was transferred into approximately 6 ml of sterile tap water. After an incubation time of 1½ hours at 37 °C without agitation 20 μl were brought onto a slide and stained after air-drying.

Immunization

For all vaccine preparations germs of 20 hours broth cultures were used. The sediment obtained by centrifugation was washed with a 0.5 per cent solution of phenol in saline, resuspended in the same medium up to 2.5×10^9 cells per ml and incubated for 30 min at 56°C. Three male outbred rabbits of 2.5–3.0 kg b.w. for each vaccine were injected on day 1, 4, 10, 14, 18 and 22 intravenously with 20 μl-–0.1–0.3–1.0-1.5 ml of vaccine respectively in a total volume of 2.0 ml complemented with saline and on day 22 with 2 ml of undiluted vaccine. Serum was obtained on day 30 by bleeding.

Metabolic activity

All 30 parameters of metabolic activities of Salmonella mentioned by *Cowan* [5] and by *Braude* et al. [3] were assayed. Media contained the antibiotics C, A and T in ¼ and 1/16 of the MICs.

Under quantitative aspects, the influence of subMICs of antibiotics was examined on acid phosphatase activity as described previously [13] using p-nitrophenyl-disodium-phosphate (NPP), mw 262.66, (Carl Roth, Karlsruhe, FRG) as substrate at a concentration of 1523 mM. S. paratyphi A cultured for 18 hours in standard and in glucose broth containing ¼ and 1/16 of the MICs were washed once with saline and then adapted to 5×10^8 cells per ml of distilled water. To 1.0 ml of this suspension were given 1.5 ml of NPP adapted to pH 5.6 with 0.1 n HCl, 2.5 ml Michaelis buffer of pH 5.6 and 1.0 ml distilled water of pH 6.0. After an incubation time of 30 min in a waterbath of 37 °C, 4.0 ml of 0.1 n NaOH and 0.5 ml glycocoll buffer of pH 12.0 were added and centrifuged. The quantity of p-nitrophenol liberated was evaluated by the O.D. at 405 nm using a calibration curve and taken for acid phosphatase activity. O.D. was measured against the same mixture without bacteria. All assays were performed in triplicate. Protein content of the bacteria used in each assay was evaluated as described by Markwell [25] with a calibration curve established with bovine serum albumin (Behringwerke, Marburg, FRG).

Results

In a first series of experiments S. paratyphi A was grown for 20 hours in broth containing ¼ and ¹⁄₁₆ of the MIC of C, A and T. T induced in ¼ of the MIC but not in the lower concentration a loss of agglutinability for anti-Ha-antibodies (Table 1). This phenomenon could only be observed if glucose containing medium was used but not in standard broth. The vaccine prepared with the Ha-deficient bacteria was also Ha-negative. The other two antibiotics had no influence on 02- and Ha-agglutinability.

The loss of Ha-expression is apparently due to a loss of flagella formation as shown in Fig. 1. Bacteria exposed to ¼ of the Mic of T had no flagella or only rudimentary forms. 'Recovery' to normal forms were observed after one subculture in medium without antibiotic.

The same result was obtained if the bacteria were cultured successively 4 times in glucose containing broth with ¼ and ¹⁄₁₆ of the MIC of C, A and T (Table 1). If, however, the strain was repeatedly cultured in increasing concentrations of antibiotics (Table 2) and taken out of antibiotic containing media for agglutination the Ha-agglutinability was not only lost in case of T but additionally in case of C-exposure.

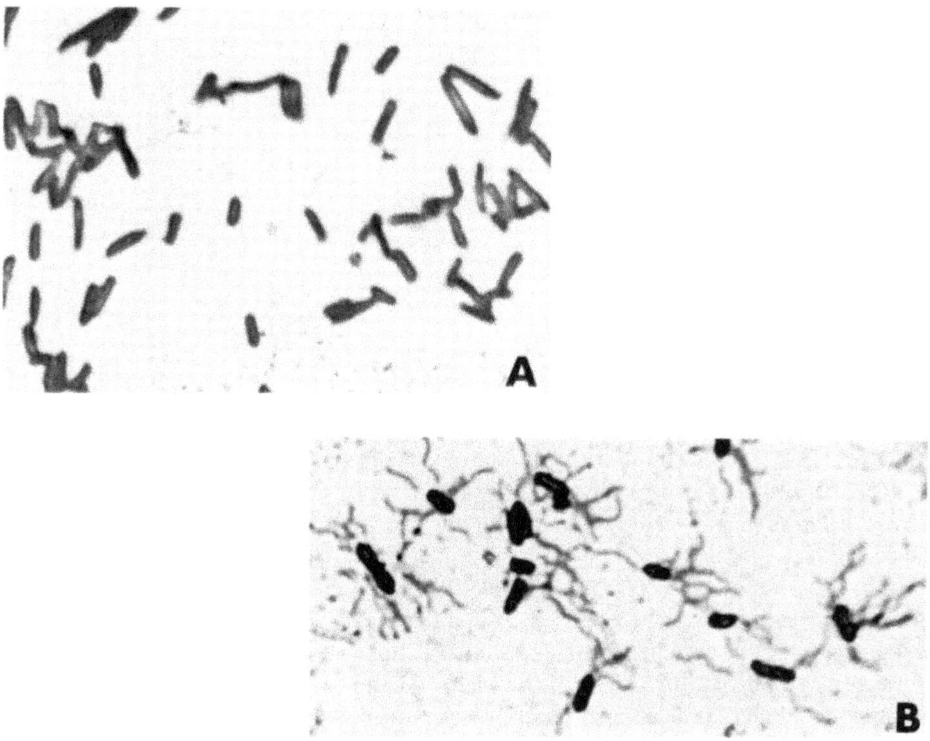

Fig. 1. Flagella staining of S. paratyphi A grown in glucose broth with ¼ of the MIC of tetracycline (**a**); Bacterial growth in the absence of antibiotics (**b**)

Table 1. Agglutination of S. paratyphi A ATCC 11511 after growth in glucose broth with subMICs of antibiotics[1)]

Antibiotics	MIC	20 hours O2	20 hours Ha	20 hours Saline	4×24 hours O2	4×24 hours Ha	4×24 hours Saline
Control	–	+	+	+	+	+	O
C	1/4	+	+	O	+	+	O
	1/16	+	+	O	+	+	O
A	1/4	+	+	O	+	+	O
	1/16	+	+	O	+	+	O
T	1/4	+	O[2)3)]	O	+	O	O
	1/16	+	+	O	+	+	O

[1] When cultured in subMICs of antibiotics diluted in standard broth the antigen expression is like in the controls.
[2] Subcultured once for 18 hours in normal medium without antibiotics: $02^{+}Ha^{+}$.
[3] Vaccine: $O2^{+}Ha^{-}$.
OOO = no agglutination

Table 2. Agglutination behaviour of chemoresistant variants of S. paratyphi A grown in subMICs of antibiotics (glucose broth)

Antibiotics	No. of subcultures	MIC µg/ml	subMIC[1)] µg/ml	O2	Ha	Saline
C	18	37.5	7.8	+	O[3)]	O
A	8	40	12.5	+	±[2)]	O
T	30	50	31.2	+	O[3)]	O

[a] 20 × original MIC-values.
[b] Subcultured in normal medium: $O2^{+}Ha^{+}$.
Subcultured once in glucose broth or twice on D.S.T.-Agar: $O2^{+}Ha^{-}$;
[c] Subcultured once in glucose broth and additionally on Sven-Gard-plate: $O2^{+}Ha^{+}$

Flagella staining of these strains showed (Fig. 2 T and C) that both were devoid of flagella. In contrast, filaments of ampicillin resistent bacteria had developed flagella, however, less regularly and not so numerous as did normal cells (Fig. 2A).

'Recovery' of Ha-antigen expression was not obtained after 2 successive subcultures on D.S.T-agar but only if grown in glucose broth and subsequently on Sven-Gard-plates.

Immunization of rabbits with a vaccine of bacteria cultured in glucose broth with ¼ of the MIC of T resulted as expected in an appreciable titer against O-antigen of S. paratyphi A but without anti-Ha-antibody (Table 3). In contrast, using a vaccine of bacteria equally cultured in ¼ of the MIC of T, however in standard broth, 2 of 3 animals developed anti-Ha-agglutinating antibodies additionally to an anti-O-titer. One of the animals was anti-Ha-negative. The reason for this might be explained by a less pronounced expression of Ha-antigen but not a complete loss under the culture conditions of bacteria used for the vaccine preparation.

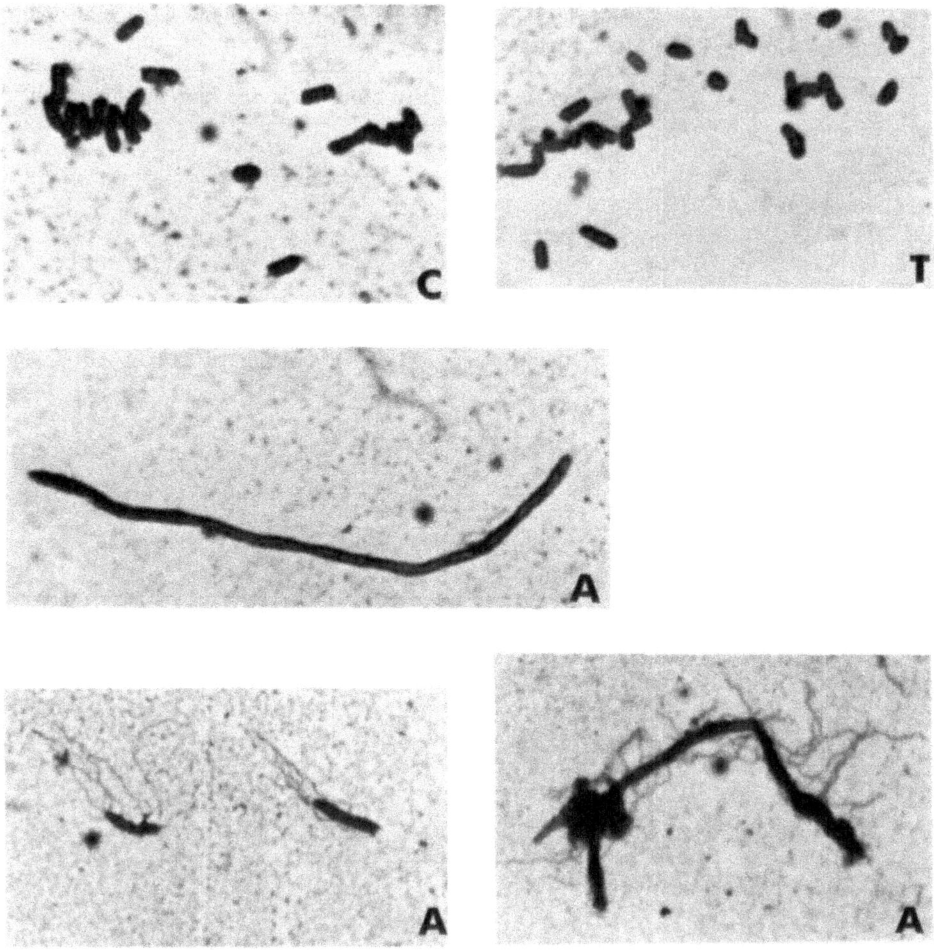

Fig. 2. Flagella staining of chemoresistant variants of S. paratyphi A grown in glucose broth containing antibiotics. C = Chloramphenicol, T = Tetracycline, A = Ampicillin

Most of the usual parameters for testing metabolic activity of Salmonellae were not influenced by subMICs of antibiotics. Only the citrate reaction (citrate as carbon source) was negative in the presence of $\frac{1}{4}$ as well as of $\frac{1}{16}$ of the MIC of C and A and in case of T only in $\frac{1}{4}$ of the MIC containing medium (Table 4). Moreover, acid formation from dulcitol was delayed in case of C and T as well as H_2S formation by A.

Acid phosphatase activity related to bacterial protein is inhibited by $\frac{1}{4}$ of the MIC of A whether the germs were cultivated in standard or in glucose containing broth (Fig. 3). In contrast, C and T – equally in a $\frac{1}{4}$ of the MIC — enhanced the acid phosphatase activity. In any case, lower concentrations were less effective. The protein content of bacteria related to 5×10^8 cells is particularly high if bacteria were grown in $\frac{1}{4}$ of the MIC of A (Fig. 4) reflecting an increase in cell volume. If acid phosphatase activity is related to the cell number instead to cell protein the difference of enzyme activity in per cent of controls is necessarily inverse.

Table 3. Antibody titer (reciprocal value) of rabbit sera obtained by vaccination with S. paratyphi A grown in 1/4 MIC of tetracycline

	Vaccine from standard broth			Vaccine from glucose broth		
Antigen	1	2	3	1	2	3
S. paratyphi A ATCC 11511 1, 2, 12: a	320	320	80	640	320	320
Same strain grown in standard broth with antibiotic	640	320	320	1280	320	160
Same strain grown in glucose broth with antibiotic	160	160	320	640	320	160
S. paratyphi A (mutant) 1, 2, 12: —	40	20	80	320	80	80
S.I 16: a No. IP 903/71	O (<20)	40	20	O (<20)	O (<20)	O (<20)

Table 4. Influence of subMICs of antibiotics on metabolic activity of S. paratyphi A (only results diverging from normal are listed)

Antibiotics	MIC	Citrate	Dulcitol	H_2S
Control	–	+	+	+
C	1/4	O	×	+
	1/16	O	×	+
A	1/4	O	+	×
	1/16	O	+	×
T	1/4	O	×	+
	1/16	×	×	+

× = Delayed reaction (48 hours)

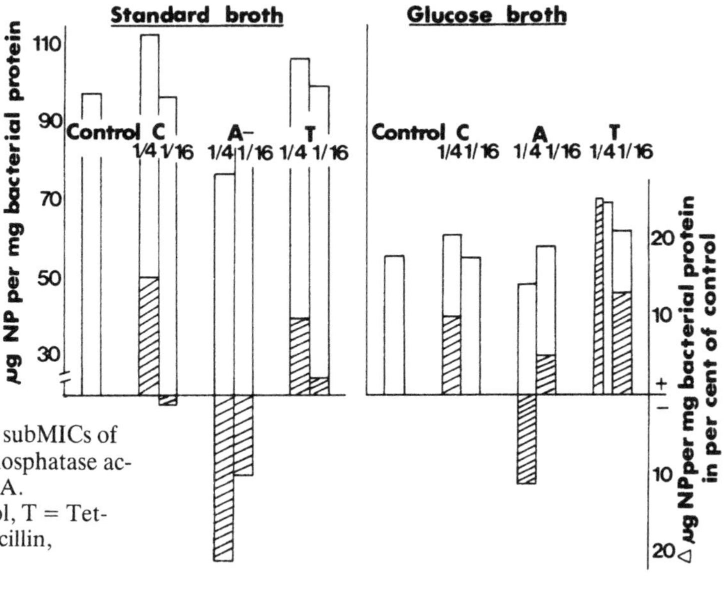

Fig. 3. Influence of subMICs of antibiotics on acid phosphatase activity of S. paratyphi A. C = Chloramphenicol, T = Tetracycline, A = Ampicillin, NP = p-nitrophenol

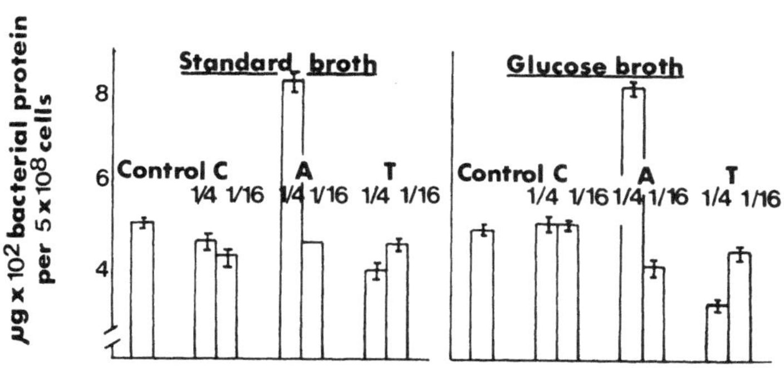

Fig. 4. Protein content of $5x10^8$ cells of S. paratyphi A grown in subMICs of antibiotics. C = Chloramphenicol, T = Tetracycline, A = Ampicillin

Discussion

S. paratyphi A lost its ability to agglutinate with Ha-specific antisera when grown for 18–20 hours in ¼ of the MIC of T containing broth in the presence of glucose. Smaller concentrations of 1/16 of the MIC were ineffective as well as subMICs of C and A. The loss of Ha-agglutinability is confirmed by a loss of flagella formation and reflected by a failure of anti-Ha-antibody production in rabbits after immunization with a vaccine prepared with these germs. If rabbits were vaccinated with bacteria grown in ¼ of the MIC of T in a glucose free standard broth only 2 out of 3 animals produced anti-Ha-antibody, but the other one did not or at least not to a detectable extent. This is not due to an unresponsiveness of this animal, because the anti-O-titers were high. It might be due to a certain but not complete inhibition of flagella formation induced by T. The loss of Ha-agglutinability favoured in glucose containing broth may be explained by the observation that the uptake of T is dependent on the presence of glucose [7].

In other experiments, S. typhimurium grown in subMICs of A or T reacted similarly as the control bacteria obtained from antibiotic free media when examined with fluorescent antibody staining or agglutination [19]. This must not necessarily be contradictory to our results because O- and H-agglutinability seem not be examined separately. On the other hand, it should also be considered that different bacterial species may behave differently. Assays with S. typhimurium (not presented in this paper) showed that subMICs of antimicrobial agents might also influence–at least partially–O-antigen formation.

A recovery of Ha-defective antigen expression to normal was observed after a single subculture on antibiotic free medium. This corresponds to the reversibility of abnormal forms of Staph. aureus exposed to subMICs of penicillin [18] or of P. mirabilis exposed to mecillinam [15, 23, 26, 29] or fosfomycin [22, 32].

The same change in Ha-expression was observed when the strain was cultured 4 times successively in ¼ of the MIC of T. If, however, the strain was repeatedly cultured in increasing concentrations of the antibiotics, the Ha-antigen expression examined with slide agglutination was not only lost in case of T but also in that of C. Therefore, it seems that the two antibiotics inhibiting protein synthesis have the same effect which is more pronounced in case of T than in case of C. 'Recovery' to normal Ha-agglutinability was only observed after two subsequent cultures on normal media and on Sven-Gard-plates. The reason why bacterial cells exposed for only one culture turn to subMICs of T recover 'easier' than those exposed to increasing concentrations of antibiotics is difficult to explain. It can be taken for sure that subMICs of antibiotics do not induce a change in the genetic repertoire, so that the 'delayed recovery' to normal forms were due to different generation times of different variants. On the other hand, it could be shown that not all bacterial cells of a population exposed to subMICs of antibiotics recover simultaneously when grown on normal media [18]. This could possibly be explained by different local antibiotic concentrations on/in the bacterial cells together with the assumption of *Lorian* [19] that the mode of action might differ with the concentration.

There are only few observations on the influence of subMICs of antibiotics on bacterial metabolic enzyme activities (see above). Tetracycline inhibits alkaline phosphatase [12] or citrate brake down [28], however, assayed using rat kidney extracts or beef liver respectively. Besides the inhibitory effect of penicillin in subMICs on carboxy-peptidase activity of Staphylococci [2, 33], subMICs of penicillin reduced acid phosphatase activity of a penicillin resistant strain of Staphylococcus aureus significantly but not that of a penicillin sensitive one. Neomycin in subMICs was inhibitory for both strains [13]. In contrast, the experiments described above showed that in S. paratyphi A subMICs of C and T, i.e. the antibiotics acting by inhibiting protein synthesis, increased acid phosphatase activity, whereas A acting on cell wall synthesis was inhibitory. These assays are of course not comparable because different strains and antibiotics were used. It seems rather likely that inhibition of protein synthesis and influence on phosphatase activity must not correlate and that the mode of action of $\geqq$ MICs of antibiotics may be different from subMICs [19].

From other metabolic parameters only citrate utilization was clearly inhibited by ¼ and $\frac{1}{16}$ of the MICs of C and A whereas T showed this effect only at ¼ of the MIC. The latter effect is in some respect comparable with that of an inhibition of citrate brake down in beef liver by aureomycin [28]. Otherwise, subMICs of C and T delayed acid formation from dulcitol and A the H_2S-production. These assays were performed only on a qualitative level. More detailed information on the mode of action of subMICs of antimicrobial agents on bacterial metabolism might possibly be obtained by investigating the kinetic of these reactions quantitatively.

Summary

The influence of subminimal inhibitory concentrations (subMICs) of ampicillin, chloramphenicol and tetracycline on antigen expression of S. paratyphi A and the metabolic activities of this strain was examined. The experiments showed that ¼ of the MIC of tetracycline in glucose containing broth inhibited the agglutinability with an anti-Ha-serum reflected by a loss of flagella. This phenomenon was reversible when bacteria were grown on media free of antibiotics. If the strain was repeatedly subcultured in incrasing concentrations of these antibiotics, the Ha-agglutinability was also lost in case of chloramphenicol, but not in that of ampicillin. Acid phosphatase activity per mg cell protein increased when the strain was grown in subMICs of chloramphenicol and tetracycline but was inhibited by ampicillin. Citrate utilization was inhibited by chloramphenicol and ampicillin in ¼ and 1/16 of the MIC; tetracycline was inhibitory only in ¼ of the MIC, whereas 1/16 of the MIC induced a delayed reaction.

References

1. Alkan ML, Beachey EH (1978) Excretion of lipoteichoic acid by group A streptococci. Influence of penicillin on excretion and loss of ability to adhere to human oral mucosal cells. J Clin Invest. 61: 671–677
2. Blumberg PM, Strominger JL (1971) Inactivation of d-alanine carboxypeptidase by penicillins and cephalosporins is not lethal in Bacillus subtilis. Proc Natl Acad Sci USA 68: 2814–2817
3. Braude AI, Davis CE, Fierer J (eds) (1981) Medical Microbiology and infectious diseases. W.B. Saunders Company, Philadelphia, London, Toronto, Mexico City, Sydney, Tokyo
4. Brock TD (1963) Effect of antibiotics and inhibitors on M protein synthesis. J Bacteriol 85: 527–531
5. Cowan ST (1974) Manual for the identification of medical bacteria. Cambridge University Press (2nd edition)
6. Ferne M, Schlossberg M, Michel J (1984) Elaboration d'anticorps chez le lapin après vaccination par Streptococcus pyogenes cultivé en présence de concentrations sub-inhibitrices d'érythromycine. Pathol Biol 32: 369–371
7. Franklin TJ, Godfrey A (1965) Resistance of Escherichia coli to tetracyclines. Biochem J 94: 54–60
8. Friedman H, Warren HG (1974) Enhanced susceptibility of penicillin resistant staphylococci to phagocytosis after in vitro incubation with low doses of Nafcillin (38177). Proc Soc Exp Biol Med 146: 707–711
9. Friedman H, Warren HG (1976) Enhanced killing of cyclacillin-treated bacteria. Proc Soc Exp Biol Med 153: 301–304
10. Gale EF, Cundliffe E, Reynolds PE, Richmond MH, Waring MJ (eds) (1981) The molecular basis of antibiotic action. John Wiley & Sons, London, New York, Sidney, Toronto (2nd edition)
11. Gemmel CG (1982) Expression of virulence factors of bacteria when grown in the presence of subinhibitory concentrations of various antibiotics. Drug Exp Clin Res 8: 245–250
12. Ghatak S, Krishna Murti CR (1953) Enzyme inhibition studies in relation to drug action. IV. Action of certain antibiotics on alcaline phosphatase. J Sci Ind Research (Ind.) 12B: 160
13. Gillissen G, Ruda M (1958) Die Beeinflussung der Phosphatasen von M pyogenes var. aureus durch Penicillin und Neomycin. Zeitschr f Hygiene 144: 401–406
14. Gottlieb D, Shaw PD (eds) (1967) Antibiotics. I. Mechanism of action. Springer Verlag Berlin, Heidelberg, New York
15. Greenwood D, O'Grady F (1973) FL 1060: a new β-lactam antibiotic with novel properties. J Clin Pathol 26: 1–6
16. Hallmann L (1955) Bakteriologie und Serologie. Georg Thieme Verlag, Stuttgart
17. Kaplan L, Anthony BF, Chapman SS, Ayonb EM, Wannamaker LW (1970) The influence of the site of infection on the immune response to group A streptococci. J Clin Invest 49: 1405–1414
18. Lorian V (1975) Some effects of subinhibitory concentrations of penicillin on the structure and division of staphylococci. Antimicrob Agents Chemother 7: 864–870
19. Lorian V (1980) Effects of subminimum inhibitory concentrations of antibiotics on bacteria. In: Lorian V (ed) Antibiotics in Laboratory Medicine. Williams & Wilkins, Baltimore/London, pp 342–408

20. Lorian V, Atkinson B (eds) (1974) Agglutination changes of enterobacteria exposed to ampicillin. Abstracts. Annual Meeting of the American Society for Microbiology, Chicago, p 129
21. Lorian V, Atkinson B (1975) Abnormal forms of bacteria produced by antibiotics. Am J Clin Pathol 64: 678–688
22. Lorian V, Atkinson B (1976) Effects of subinhibitory concentrations of fosfomycin on bacteria. G Ital Chemioter 23: 65–74
23. Lorian V, Atkinson B (1977) Comparison of the effects of mecillinam and 6-aminopenicillanic acid on Proteus mirabilis, Escherichia coli and Staphylococcus aureus. Antimicrob Agents Chemother 11: 541–552
24. Lorian V, Atkinson B, Ewing W (1976) Agglutination with O sera of salmonella exposed to antibiotics. Am J Clin Pathol 66: 1004–1011
25. Markwell MAK, Hass SM, Bieber LL, Tolbert NE (1978) A modification of the lowry procedure to simplify protein determinations in membrane and lipoprotein samples. Anal Biochem 87: 206
26. Melchior NH, Blom J, Tybring L, Birth-Andersen A (1973) Light and electron microscopy of the early response of Escherichia to a 6β-amidinopenicillanic acid (FL 1060). Acta Pathol Microbiol Scand 81: 393–407
27. Michel J, Ferne M, Rotenberg M (1983) Effects de concentrations sub-inhibitrices d'antibiotiques sur la virulence de streptocoques hémolytiques du groupe A. Pathol Biol 31: 528–532
28. Mulli K, Uhlenbrook K, Ludwig L (1953) Zum Wirkungsmechanismus des Aureomycins. Arzneim Forsch 3: 559–561
29. Neu H (1976) Mecillinam, a novel penicillanic acid derivative with unusual activity against gram-negative bacteria. Antimicrob Agents Chemother 9: 793–799
30. Nishida M, Mine Y, Nonoyama S, Yokota Y (1976) Effect of antibiotics on the phagocytosis and killing of Pseudomonas aeruginosa by rabbit polymorphonuclear leukocytes. Chemotherapy 22: 203–210
31. Rolinson GN (1977) Subinhibitory concentrations of antibiotics. J. Antimicrob. Chemother 3: 111–113
32. Stapley EO, Hendlin D, Mata JM, Jackson M, Wallick H, Hernandez S, Mochales S, Currie SA, Miller RM (1969) Phosphomycin I discovery and in vitro biological characterization. Antimicrob Agents Chemother 1968: 284–290
33. Strominger JL (1969) Penicillin sensitive enzymatic reactions in bacterial cell wall synthesis. Harvey Lect 64: 179–213
34. Svanbord Eden C, Sandberg T, Ahlstedt K (1978) Decrease in adhesion of E. coli to human urinary tract epithelial cells in vitro by subinhibitory concentrations of ampicillin. Infection 6 (suppl. I): 121–124
35. Tanner J (1960) Contribution à l'étude de l'action des antibiotiques sur l'immunité. Action de la tetracycline sur les antigènes de S. typhi. Ann Inst Pasteur 98: 772
36. Tybring L, Melchior NH (1975) Mecillinam (FL 1060) a 6β-amidinopenicillanic acid derivative: bacterial action and synergy in vitro. Antimicrob Agents Chemother 8: 271–276
37. Viano I, Martinetto P, Valtz A, Santiano M, Barbaro S (1979) Variability of immune response induced by bacteria treated with subminimal inhibitory concentrations of fosfomycin. Rev Infect Dis 1: 858–861

Discussion

Karch: I found your observation on the effect of tetracycline on flagella expression by *Salmonella typhimurium* very interesting. We have found that flagella contribute to the virulence of *S. typhimurium*. Does tetracycline also prevent flagella formation by other species of enterobacteriaceae, for example *E. coli* or *Salmonella typhi?*

Gillissen: Several other Salmonella strains, but not E. coli, have been assayed. According to these for the moment only preliminary results it seems that the effect of sub MIC's of tetracycline on flagella formation is not completely the same in all strains as shown in S. paratyphi A.

Beachey: Although your antibiotic treatment was sufficient to suppress the expression of visible flagellae, it is possible that flagellar subunits are still being made. Did you examine the treated bacteria for the presence of flagellar subunits?

Gillissen: Tetracycline in ¼ of the MIC inhibited H-agglutinability indicative of a loss of visible flagella. We did not examine the treated bacteria for the presence of flagellar subunits. We think, however, that the treated bacteria are also devoid of such subunits, because rabbit sera obtained by immunization with this strain did not exhibit any agglutinating ability for (normal) H-antigen.

Ghione: The disappearance of flagella in Salmonella cultured in the presence of tetracyclin can be attributed to a preferential inhibition of flagellin synthesis or to the isolation of a preexistent atrichous mutant? Is there a quick reversion of these bacteria to the flagellate form during subculturing in plain medium?

Gillissen: Bacteria grown for 18 hours in glucose containing broth with ¼ of the MIC of tetracycline "recovered" normal flagellar agglutinability after a subculture of another 18 hours in broth free of antibiotics. On the other hand, chemoresistant bacteria taken out of medium with antibiotic showed a "delayed recovery", i.e. 2 to 3 subcultures were needed. In this case, the possibility of mutants with different generation times have, indeed, to be considered. Previously, it was however described by Dr. Lorian that even in short time cultures reversion in morphology does not appear simultaneously in all cells of a culture in a log growth phase, possibly induced by different concentrations of the antibiotic bound in/on the germs.

Influence of Ciprofloxacin on Specific Interactions on Listeria-specific T Cells with Antigen in vitro

H. HAHN, BEATE WOS, and U. SPERLING

Department of Medical Microbiology; Freie Universität Berlin, Berlin, FRG

Introduction

The decision for the application of antimicrobial drugs should not only be based on the activity of the drug on the microorganism itself, but must take into consideration the reciprocal effects between agent and host organism [2, 7]. In particular, possible influences on the immune system should be considered–either on components of the immune system directly or indirectly *via* alterations of the bacteria which might result in a different handling by the immune system [9]. 1982, a first meeting with international attendance was held in Munich which was entirely devoted to this subject [2].

There are now many published studies which report on various antimicrobial drugs having various influences on the immune system. For instance, tetracyclines are reported to negatively influence phagocytosis [4], chemotaxis [5, 6], antibody formation [1], and specific interactions between T cells and antigen [11]. Others, such as β-lactam antibiotics have so far been shown to exert no, or minor, effects on the immune system [2].

Of particular interest seemed to us to investigate possible influences antimicrobials might exert on specific immune reactions relevant to the defense of bacterial diseases. Since interactions between antigen-presenting macrophages and specific T cells represent a key function in T cell-dependent immunity to facultatively intracellular bacteria, we have started to investigate aspects of these interactions in vitro [11].

Ciprofloxacin is a newly developed antibacterial agent with broad antibacterial activity. It is to be expected that this and similar agents will, in the near future, come into widespread therapeutic use.

The following study was done with the aim of investigating whether Ciprofloxacin influences specific T cell/antigen interactions in vitro. As a model, we chose the in vitro system which measures reactivity of Listeria-specific murine T cells against Listeria antigen presented by murine macrophages.

The Influence of Antibiotics on the
Host-Parasite Relationship II
Edited by D. Adam, H. Hahn, W. Opferkuch

Materials and Methods

Antibiotics

Ciprofloxacin (laboratory standard) was a gift from Bayer AG, Leverkusen.

Penicillin G (laboratory standard) was a gift from Grünenthal, Stolberg, and Doxycycline (laboratory standard) from Pfizer, Karlsruhe.

Solutions were freshly prepared for each experiment by dissolving the agents in RPMI 1640 medium supplemented with 5% FCS and 2 mM L-glutamine.

Bacteria

Listeria monocytogenes, strain EGD, was kept virulent by continuous mouse passage. Bacteria were grown in trypticase-soy broth for 18 h, centrifuged, washed, and heat-killed. Heat-killed Listeriae (HKL) were used as antigen.

T cell/antigen interactions

Specific interactions between T cells and antigen were measured according to Farr et al. [3].

In short, syngeneic (B6×DBA2)F1(B6D2) mice from our own breeding facilities were injected i. v. with 4×10^3 living Listeriae. Seven d later, peritoneal exudates were induced by the i.p. injection of 2 ml of a 10% protease peptone solution. Exudate cells were washed out and the T cells enriched by passaging them over nylon wool. These cells are called PETLEs (peritoneal exudate T lymphocytes enriched).

5×10^3 peritoneal macrophages from normal syngeneic (B6D2) mice were preincubated with 2×10^7 HKL for 5 h. Subsequently, the macrophages were cultured in 96 well flat bottom tissue culture dishes (Nunc) together with $2{,}5\times10^5$ PETLEs from Listeria-immune mice. Final volume was 0,2 ml per well. Tissue medium used was RPMI 1640 supplemented with 5% FCS and 2mM L-glutamine. The substance to be tested was added at concentrations indicated. Incubation was done in 5% CO_2 at 37 °C. After 2 d of incubation, 1 μ Ci ^{3}HTdR (specific activity: 2 Ci/mM, Radiochemical Center, Amersham, U.K.) was added to the wells and incubation continued for another 18 h. Cells were then collected on glass fiber filters (Skatron Multiple Cell Culture Harvester, Flow Laboratories) and ^{3}HTdR incorporation measured using a liquid scintillation counter. ^{3}HTdR is a measure of T cell proliferation and, therefore, in this system, of a secondary immune response of Listeria-specific T cells in vitro.

Results

Dose-response relationship between Ciprofloxacin and T cell proliferation

When Ciprofloxacin, Penicillin G, or Doxycyclin, respectively, were added in various concentrations to the incubation mixture, there was no inhibition of T cell proliferation by Ciprofloxacin up to 12,5 mcg/ml.

At a concentration of 25 mcg/ml, there was a slight (20%) reduction of proliferation, and at 50 mcg/ml, proliferation was reduced markedly.

On the other hand, Penicillin G at 50 mcg/ml did not inhibit proliferation at all, whereas Doxacyclin at 12 mcg/ml markedly, and at 25 mcg/ml completely, suppressed T cell proliferation (Fig. 1)

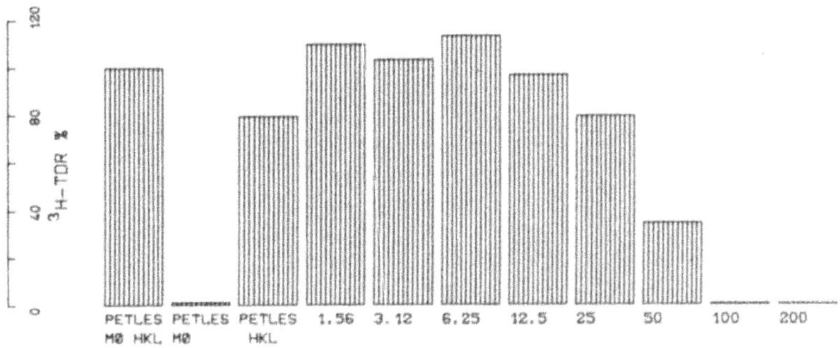

a

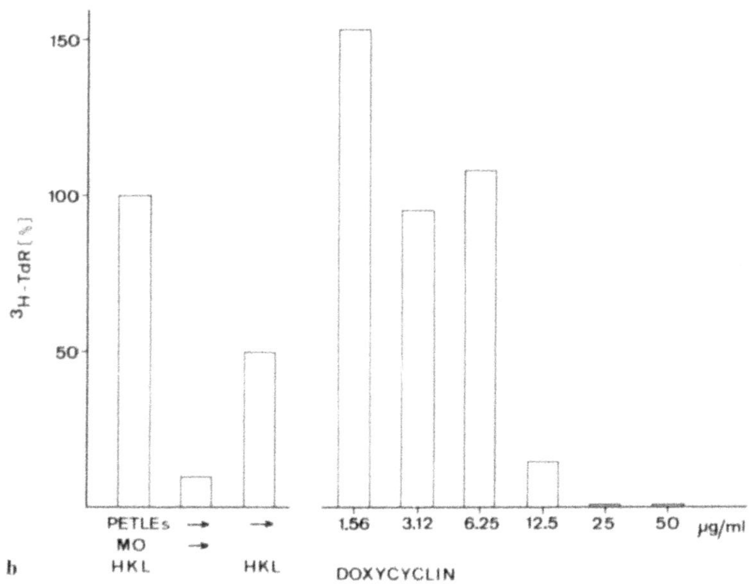

b

Fig. 1a–c. Dose-response relationship of the effect of **a**) Ciprofloxacin, **b**) Doxycyclin, and **c**) Penicillin G on Listeria-specific T cell proliferation in vitro

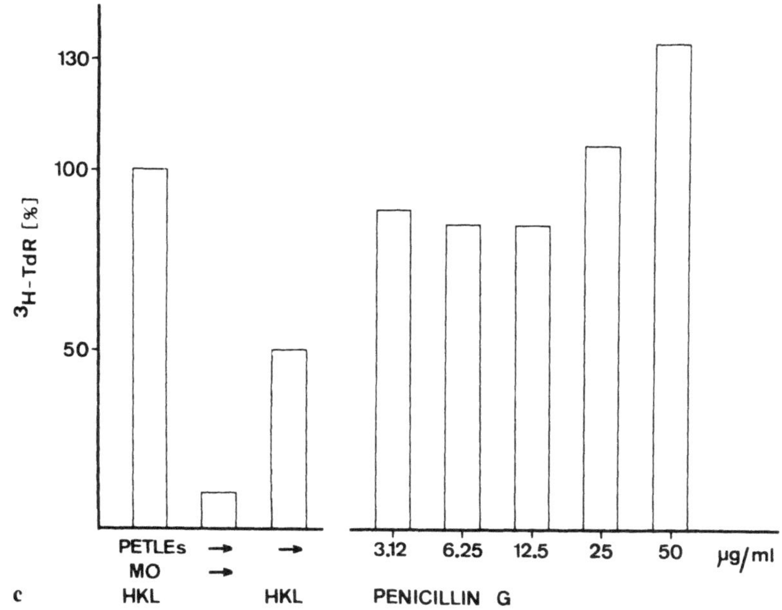

c

Reversibility of the influence of Ciprofloxacin

In order to test whether the suppressive effect of Ciprofloxacin at 50 mcg/ml is reversible, Listeria-specific PETLEs were incubated in the presence of macrophages, HKL, and antibiotic (50 mcg/ml) for 14 h at 37 °C in 5% CO_2. Subsequently, cells were pooled, washed, and resuspended at $2{,}5 \times 10^5$ cells/0,2 ml, and incubation continued in the absence of the test substance. ^{3}HTdR incorporation was measured after a total of 66 h of incubation.

As can be seen from Fig. 2, the suppressing effect of Ciprofloxacin was not reversible as the proliferation rate reached only 40% of control values.

Influence of Ciprofloxacin, Penicillin G, and Doxycyclin when present in the pulsing phase only

In antigen-specific T cell proliferation, the phases of antigen uptake and presentation can be separated from the phase of T cell proliferation. In the following experiment, the question was investigated whether Ciprofloxacin, when present during the antigen pulsing phase only, exerts any influence on subsequent T cell proliferation. Antigen was added to normal macrophages. Ciprofloxacin or, as control substances, Doxycyclin and Penicillin G, respectively, were added and kept in the mixture for 4 h. After 4 h, antigen as well as test substances were washed out and replaced by fresh T cells and fresh medium. T cell concentration was $2{,}5 \times 10^5$/well, incubation time 66 h. ^{3}H-TdR was added 18 h before the end of incubation and proliferation measured at the end of the incubation period.

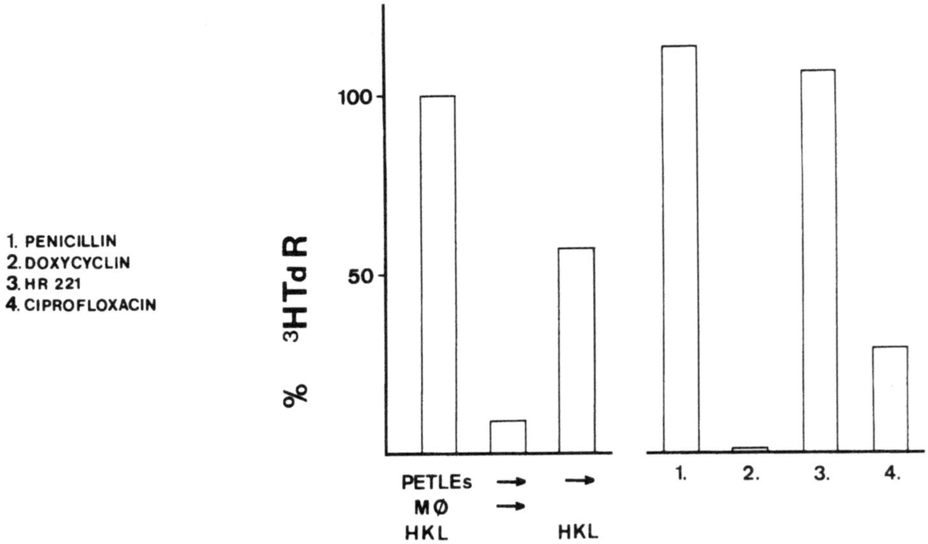

Fig. 2. Reversibility of the influence of Ciprofloxacin, Penicillin G, HR221, and Doxycyclin on Listeria-specific T cell proliferation in vitro

The results of this experiment are depicted in Fig. 3. They show that the presence of Ciprofloxacin as well as Doxycyclin during the antigen uptake/presentation phase does not influence the capacity of macrophages to subsequently stimulate T cell proliferation. Penicillin G and Doxycyclin did not influence the subsequent T cell response, either.

Discussion

A variety of assay systems measuring the immune reactivity have recently been employed to measure the influence of antibiotics on parameters of the immune response. However, rather little appears to be known on the influence of antimicrobial agents in systems which are measuring interactions of T cells with specificity to bacterial antigens and antigen, although a possible influence by antimicrobials on these systems might have the greatest relevance to the actual situation in patients.

We therefore used an assay originally described by Farr et al [3] to test Ciprofloxacin, a newly developed antimicrobial in the class of quinolones, with respect to its influence on the in vitro interaction between antigen-specific T cells and antigen. In this system, Listeria monocytogenes is being used as an antigen. It is presented by macrophages to T cells, and the latter react by proliferation. It is commonly thought that this system reflects the in-vivo antigen specific interaction between T cells of helper type and antigen presenting macrophages.

It was found that Ciprofloxacin at concentrations of 25 mcg/ml and higher, irreversibly reduced T cell proliferation. Penicillin G, on the other hand, did not

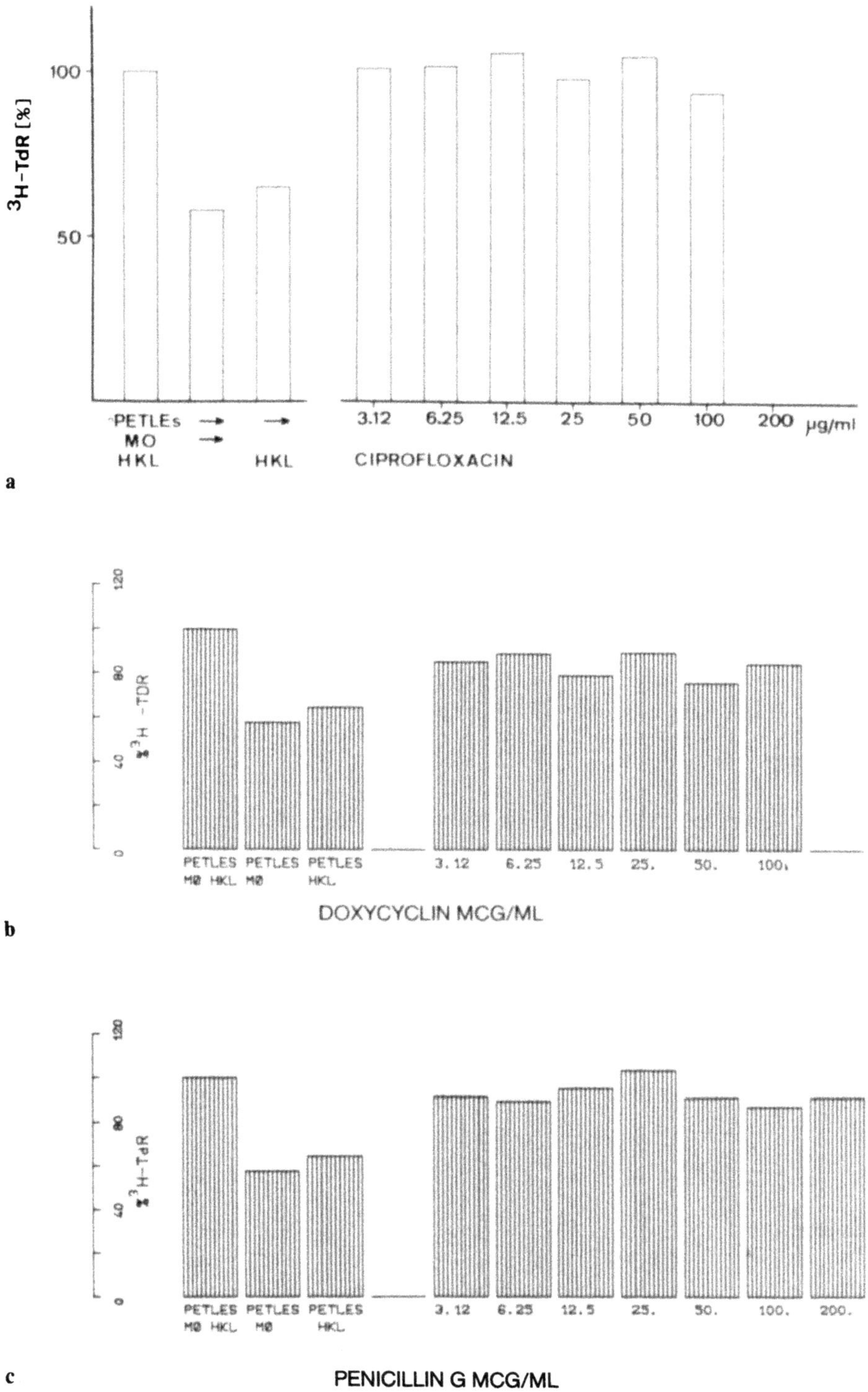

Fig. 3a–c. Proliferation of Listeria-specific T cells after antibiotic had been present during the antigen uptake phase only. **a**) Ciprofloxacin, **b**) Doxacyclin, **c**) Penicillin G

suppress at all, whereas Doxycyclin, as has been shown previously by us [11] at concentrations of 12,5 mcg/ml and higher, suppressed the response.

The observed reduction of antigen specific T cell proliferation by Ciprofloxacin in vitro occurred only when Ciprofloxacin was present in the phase of proliferation of T cells. When, on the other hand, Ciprofloxacin was present during antigen uptake and presentation by macrophages, no effect could be observed.

It should be stressed that the concentrations at which Ciprofloxacin reduced specific T cell proliferation were way beyond concentrations reached in the body after therapeutic dosages. Thus, any effect on antibacterial defenses of patients by Ciprofloxacin appears unlikely. Furthermore, in vivo studies on the influence of Ciprofloxacin on delayed hypersensitivity reactions in mice (Hahn et al., unpublished) have yielded no evidence for in vivo immune suppressive effects of Ciprofloxacin.

The manner in which Ciprofloxacin at concentrations exceeding 25 mcg/ml reduced antigen-specific T cell proliferation appears to have been by a non-specific cytotoxic effect resulting in irreversible cell damage or cell death.

Taken together, the results suggest that Ciprofloxacin when given in therapeutic dosages exerts no specific negative effects on the interaction of Listeria-specific T cells with antigen. The reduced reactivity observed when supratherapeutic concentrations are present in the assay mixture is best explained by assuming that non-specific effects are responsible, such as cell death.

Summary

In this paper, the influence of Ciprofloxacin, a newly developed antibiotic from the class of quinolones was tested for its effects on specific T cell/antigen interactions using a system of Listeria specific murine helper T cells, macrophages, and heat killed Listeria monocytogenes (HKL) as antigen. ^{3}HTdR incorporation was used as the final readout.

When Ciprofloxacin was present in the incubation medium, a slight reduction (20%) of the T cell proliferative response was seen at 25 mcg/ml, while at 50 mcg/ml, T cell proliferation was reduced markedly. This effect of Ciprofloxacin was only partially reversible. The effect of Ciprofloxacin was restricted to the proleferation phase, whereas the preceding antigen uptake by macrophages was unaffected.

References

1. Banck G, Forsgren A (1979) Antibiotics and suppression of lymphocyte function in vitro. Antimicrob Agents and Chemother 16: 554
2. Eickenberg U-U, Hahn H, Opferkuch W (eds) (1982) The influence of antibiotics on the hostparasite relationship. Springer Verlag Berlin, Heidelberg, New York, Tokyo
3. Farr AG, Kiely J-M, Unanue ER (1979) Macrophage-T cell interactions involving Listeria monocytogenes — Role of the H-2 gene complex. J Immunol 122: 2395
4. Forsgren A, Schmeling D, Quic PG (1974) Effect of Tetracycline on the Phagocytic

5. Forsgren A, Schmeling D, Banck G (1978) Effect of Antibiotics on Chemotaxis of Human Polymorphonuclear Leukocytes in Vitro. Infection 6 (Suppl. 1), 102–106
6. Forsgren A, Banck G, Beckmann, Bellahsene A (1980) Antibiotic-Host Defense Interaction in Vitro and in Vivo. Scand J Infect Dis 24, 195-203
7. Hauser WE, Remington JS (1982) Effect of antibiotics on the humoral and cell-mediated immune responses. In: Action of antibiotics in patients. LD Sabath (ed). Hans Huber, Stuttgart, Bern, Wien
8. Mackaness GB (1969) The influence of immunologically committed lymphoid cells on macrophage activity in vivo. J Exp Med 129: 973
9. Munster AM et al (1977) The effect of antibiotics on cell-mediated immunity. Surgery 81:692
10. North RJ (1973) Importance of thymus-derived lymphocytes in cell-mediated immunity to infection. Cell Immunol 7: 166
11. Hahn H, Mielke M (1985) Influence of Antibiotics on Interactions Between Listeria-Specific T Cells and Antigen in vitro. Zbl Bakt Suppl 13: 143–152

Discussion

Friedman: This is a very excellent model for studying the effects of antibiotics in vitro on the interaction of T lymphocytes and macrophages responding to a bacterial antigen. Can you use a similar model for studying the effects of an antibiotic, at its subMIC concentration, on the interaction or processing of Listeria by macrophages?

Hahn: This is an important suggestion which we shall follow.

Ghione: What concentration of tetracyclin (or derivatives) did you employ in the test?

Hahn: 1,0 to 25 mcg/ml.

Ghione: Did Tetracyclin (or derivatives) display a dose-related effect in your test?

Hahn: The effects are dose-dependent.

Pulverer: What are the conclusions from the results achieved in your excellent in vitro test system for the in vivo situation?

Hahn: Our in vivo data (delayed type hypersensitivity in mice) so far have confirmed our in vitro findings (Press and Hahn, manuscript in preparation).

Adam: Did you look also for higher concentrations of β-lactam antibiotics, e. g. more than 200 mcg? What was the effect influencing T-cell-antigen-interaction?

Hahn: Maximum concentrations used were: 200 μg/ml for β-lactams; 50 μg/ml for gyrase inhibitors.

III. The Influence of Antibiotics on the Interaction of Bacteria with Host Defence Mechanisms

Effects of Antibiotics on the Sensitivity of *Proteus mirabilis* to the Bactericidal Action of Normal Human Serum

NICOLA SIEGMUND-SCHULTZE, H.H. MARTIN, and KATHRYN NIXDORFF

Institut für Mikrobiologie, Technische Hochschule Darmstadt, Schnittspahnstrasse 9, D-6100 Darmstadt, FRG

Introduction

Soon after the introduction of antibiotic therapy it became apparent that antibiotics could work synergistically with the complement system to exert bactericidal effects that were greater than those produced by either component alone [2, 13]. It was also shown that gram-positive bacteria, which are normally resistant to the bactericidal action of antibody and complement, could be made sensitive after induction to protoplast formation by treatment of the organism with penicillin [9]. In the case of gram-negative bacteria treated with subinhibitory concentrations of antibiotics, which leads to filament formation, no synergistic effects of antibiotics and serum could be detected [5].

It is known that β-lactam antibiotics induce functional disorganizations in the peptidoglycan layer of bacterial cell walls, which are expressed in the conversion of bacterial forms to filaments, spheroplast L-forms and protoplast L-forms [7]. Progressive defects leading from filament to spheroplast to protoplast formation can be produced by either increasing the concentration of certain antibiotics or using antibiotics which exert specific effects [5, 6, 7, 8].

To investigate the effects of antibiotics on the sensitivity of a serum resistant strain of *Proteus mirabilis* to the bactericidal action of serum, we have used a model system consisting of the bacterial form of *P. mirabilis* VI, its unstable filament form induced by subinhibitory concentrations of the β-lactam antibiotic cefuroxime and its unstable spheroplast L-form induced by the β-lactam antibiotic cefoxitin, all of which have either intact or defective cell walls. A stable protoplast L-form of *P. mirabilis* VI, which is completely devoid of a cell wall and grows in the absence of antibiotics [3], was also included in the model system.

P. mirabilis was chosen because of the advantageous ability of this organism to survive and multiply in the presence of antibiotics after induction to cell wall defective forms [7]. In this case the requirement for growing cultures in bactericidal assay systems [12] was satisfied. With this model system we could detect at which stage of cell wall defectiveness the conversion to serum sensitivity occurs.

Abbreviations used

GPS, guinea pig serum; HA, hemagglutinin; NHS, normal human serum; O.D. optical density; LPS, lipopolysaccharide; rpm, revolutions per minute; SDS, sodium dodecyl sulfate

The Influence of Antibiotics on the
Host-Parasite Relationship II
Edited by D. Adam, H. Hahn, W. Opferkuch

Preliminary studies with this model suggested that relatively extensive cell wall defects are required for conversion of the serum-resistant strain of *P. mirabilis* to serum sensitivity. The bacterial and the filament forms of this strain were completely resistant to the bactericidal action of antibody and complement, while approximately 50% of the spheroplast population and 90% of the stable protoplast population were sensitive [11].

In these studies [11] we used rabbit antibodies directed against cell surface structures of *P. mirabilis* and commercial guinea pig serum as a source of complement. In the present report, we have extended these studies by testing the sensitivities of the various forms of *P. mirabilis* VI to the action of normal human serum, a bactericidal system which is clinically more relevant.

Materials and Methods

Bacteria

P. mirabilis VI and its stable protoplast L-form were from the collection of H.H. Martin (this institute). The stable protoplast L-form was originally obtained from U. Taubeneck, Jena, GDR.

Antibiotics

The β-lactam antibiotic cefuroxime (Hoechst AG, Frankfurt/Main, FRG) was used for the induction of the filament form of *P. mirabilis* VI. This antibiotic was chosen because it specifically induces filament formation [8]. The β-lactam antibiotic cefoxitin (MSD-Sharp & Dohme GmbH, Munich, FRG) was used to induce spheroplast formation of *P. mirabilis* VI. This antibiotic induces complete spheroplast formation in *P. mirabilis* in concentrations which do not inhibit growth of the organism.

Cultivation of the bacterial form of P. mirabilis VI

An overnight culture of *P. mirabilis* VI at 37 °C in L-medium [6] was added to 50 ml L-medium to an optical density (O.D.) at 578 nm of 0.3 (Eppendorf Photometer, Netheler & Hinz GmbH, Hamburg, FRG). This culture was incubated for 90 min at 37 °C with shaking at 120 rpm (logarithmic phase culture). The culture was then adjusted to an O.D.$_{578\ nm}$ of 0.8 with L-medium before addition to the bactericidal assay system.

Cultivation of the unstable filament form of P. mirabilis VI

P. mirabilis VI was induced to filament formation one day before bactericidal assay by addition of 0.1 μg cefuroxime/ml L-medium. Cultivation of the logarithmic phase culture to be used in the bactericidal assay was carried out as described for the bacterial form, except that 0.1 μg cefuroxime/ml L-medium was always included. The logarithmic phase culture was adjusted to an 0. $D._{578}$ nm of 0.8 with L-medium + 0.1 μg cefuroxime/ml before addition to the bactericidal assay system.

Cultivation of the unstable spheroplast L-form of P. mirabilis VI

P. mirabilis VI was induced to spheroplast formation two days before bactericidal assay by addition of 60 μg cefoxitin/ml L-medium. On the first day of induction, the culture was incubated overnight at 37 °C without shaking. On the second day, a fresh culture was incubated at 37 °C with increasing degrees of shaking during the day (70, 90 rpm) and finally overnight at 120 rpm. On the third day, the overnight culture from the second day was added to 50 ml L-medium + 60 μg cefoxitin/ml to an $O.D._{578\ nm}$ of 0.3 and incubated at 37 °C for 180 min with shaking at 120 rpm (logarithmic phase culture). This culture was adjusted to an $O.D._{578\ nm}$ of 0.8 with L-medium + 60 μg cefoxitin/ml before addition to the assay system.

Cultivation of the stable protoplast L-form of P. mirabilis VI

Cultivation of the stable protoplast L-form of *P. mirabilis* VI began 3–4 days before bactericidal assay in L-medium. Each day a fresh culture was initiated from the culture of the previous day and incubated at 37 °C with shaking at 120 rpm. The logarithmic phase culture was prepared by adding an overnight culture to 50 ml L-medium to an $O.\ D._{578\ nm}$ of 0.3 and incubating this culture for 180 min at 37 °C with shaking at 120 rpm. This culture was adjusted to an $O.D._{578\ nm}$ of 0.8 with L-medium before addition to the bactericidal assay system.

Normal human serum

Blood was collected aseptically from healthy volunteers and allowed to clot at room temperature for 1 h and 3–4 h thereafter at 4 °C. Clots were removed and the serum was collected by centrifugation. Sera were stored in aliquots of 1.0 ml at −80 °C until use.

Bactericidal assay

A modification of the photometric assay of Muschel and Treffers [10] was used. All reagents and glassware were sterile and kept in an ice bath before addition to the bactericidal assay. Normal human serum in various amounts ranging from 0.03 to

2.25 ml was added to test tubes (16×100 mm) and brought to 2.25 ml with L-medium. Thereafter 0.75 ml of a logarithmic phase culture (O.D.$_{578\ nm}$ of 0.8) was added and contents were mixed. Concentrations of normal human serum in the assay system thus ranged from 1–75%. From these tubes, 0.8 ml was removed for measuring O.D. at time 0. The tubes were then incubated at 37 °C in a shaker water bath for 60 min (kill period). Again, 0.8 ml was removed from each tube for O.D. measurement, 1.0 ml L-medium was added and the tubes were incubated under the same conditions for an additional 90 min, after which time another O.D. measurement was made (outgrowth of surviving organisms). Controls containing the respective amounts of heat-inactivated serum were included. O.D.$_{578\text{-}\ nm}$ was measured in each case against a blank consisting of the respective concentration of serum without bacteria. The percent survival was calculated by dividing the O.D. of tubes containing active human serum by the O.D. of tubes containing the respective concentration of heat-inactivated human serum and multiplying by 100. Percent kill was calculated by subtracting the percent survival from 100.

Hemagglutination assay

Hemagglutinin titers of normal human serum and of lyophilized, reconstituted guinea pig serum (Behringwerke, Marburg, FRG) were measured against alkali-treated lipopolysaccharide from *P. mirabilis* VI coupled to sheep red blood cells as described [1].

Results and Discussion

Bactericidal reaction of normal human serum with the bacterial and the unstable filament forms of P. mirabilis VI

The results of bactericidal assays determining the sensitivities of the bacterial and the unstable filament forms of *P. mirabilis* VI to the action of normal human serum are reported in Table 1.

Both the bacterial and the filament forms were relatively resistant to the bactericidal action of human serum. In the case of the bacterial form, only 10% killing was achieved with the highest concentration of serum tested (75%). A general decrease in the amount of killing with decreasing concentrations of serum followed, although the decrease was not always proportional to the serum concentration. The standard errors were also high, which is most likely a reflection of the generally weak reactions. In the case of the filament form, 23% of the population was killed by the highest concentration of human serum. Thus, the filament form appeared to be slightly more sensitive to the bactericidal action of normal human serum. However, considering the high standard errors obtained, this slight increase in sensitivity was not significant.

In our previous studies using rabbit antibodies to cell surface components of *P. mirabilis* and guinea pig serum as a source of complement, both the bacterial and the filament forms were totally resistant to this bactericidal system [11].

Table 1. Bactericidal reaction of normal human serum (NHS) with the bacterial and the filament forms of *Proteus mirabilis* VI

Cell form	% NHS	% Kill[a]
Bacterial	75	10.6±6.3
	60	7.8±4.5
	50	2.8±2.8
	40	9.2±6.5
	25	3.2±3.2
	10	2.5±2.2
	3	0
Filament[b]	75	23.5±8.6
	60	17.1±4.3
	50	13.3±7.6
	40	12.8±2.1
	25	2.0±1.1
	10	2.6±2.6
	3	0

[a] Values represent the means ± standard errors taken from 4 separate experiments. % kill was calculated from measurements made 90 min after the kill period (outgrowth of surviving bacteria).

[b] Filament formation was induced by cultivation of the bacterial form in the presence of 0.1 μg/ml cefuroxime.

Bactericidal reaction of normal human serum with the unstable spheroplast L-form of P. mirabilis VI

The sensitivity of the unstable spheroplast L-form of *P. mirabilis* VI to the bactericidal action of various concentrations of normal human serum is presented in Fig. 1

Up to 90% killing of the spheroplast L-form was observed with the highest concentrations of normal human serum, when optical densities of the test reactions were measured 90 min after the actual kill period (curve B). If measurements were made directly after the kill period (curve A), only 65% killing was observed with the highest concentrations of serum. It may be that the differences in the optical densities between the test reactions (active serum) and the control tubes (inactivated serum) are more pronounced when cultures are allowed to grow out beyond the actual kill period.

To compare the activity of normal human serum with the activity of normal guinea pig serum, we also tested the sensitivity of the spheroplast L-form in the present assay system using commercial guinea pig serum. In this case only 16% killing of the spheroplasts, population was achieved with 75% guinea pig serum (data not shown). In our previous studies using rabbit antibodies to surface components of *P. mirabilis* and guinea pig serum as a source of complement, the best combination gave only 40–50% killing of the spheroplast form [11].

Thus, even when antibodies were added to guinea pig serum, normal human serum was still more active against the spheroplast L-form.

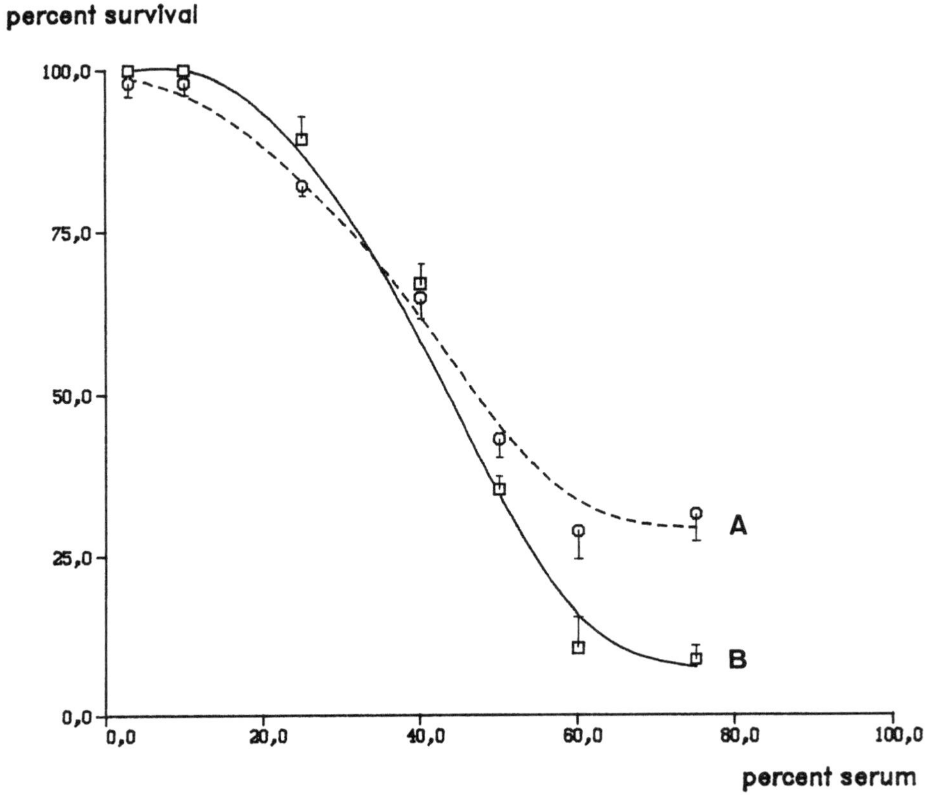

Fig. 1. Bactericidal reaction of normal human serum with the spheroplast L-form of *Proteus mirabilis* VI. **A**, percent survival measured directly after the kill period (60 min). **B**, percent survival measured 90 min after the kill period. Points on the curves represent the means of values taken from 4 separate experiments. Bars represent standard errors of the means; for each point, either the plus or the minus values are given for optical simplicity

Bactericidal reaction of normal human serum with the stable protoplast L-form of P. mirabilis VI

The sensitivity of the stable protoplast L-form of *P. mirabilis* VI to the bactericidal action of various concentrations of normal human serum is presented in Fig. 2.

The protoplast L-form was clearly the most sensitive of all forms of *P. mirabilis* VI. In this case, essentially 100% killing was achieved with 30% normal human serum, regardless of whether reactions were measured after the 60 min kill period (curve A) or 90 min after the kill period (curve B). In contrast, only 21% killing was observed in the same assay system using 30% normal guinea pig serum (data not shown).

Thus, normal human serum was much more active than guinea pig serum in killing all forms of *P. mirabilis* VI tested. That the higher activity of human serum was not due to a higher content of antibodies to lipopolysaccharide (LPS) can be seen from the results of hemagglutination assays of normal human serum and of normal guinea pig serum measured against LPS coupled to sheep red blood cells (*Table 2*). There

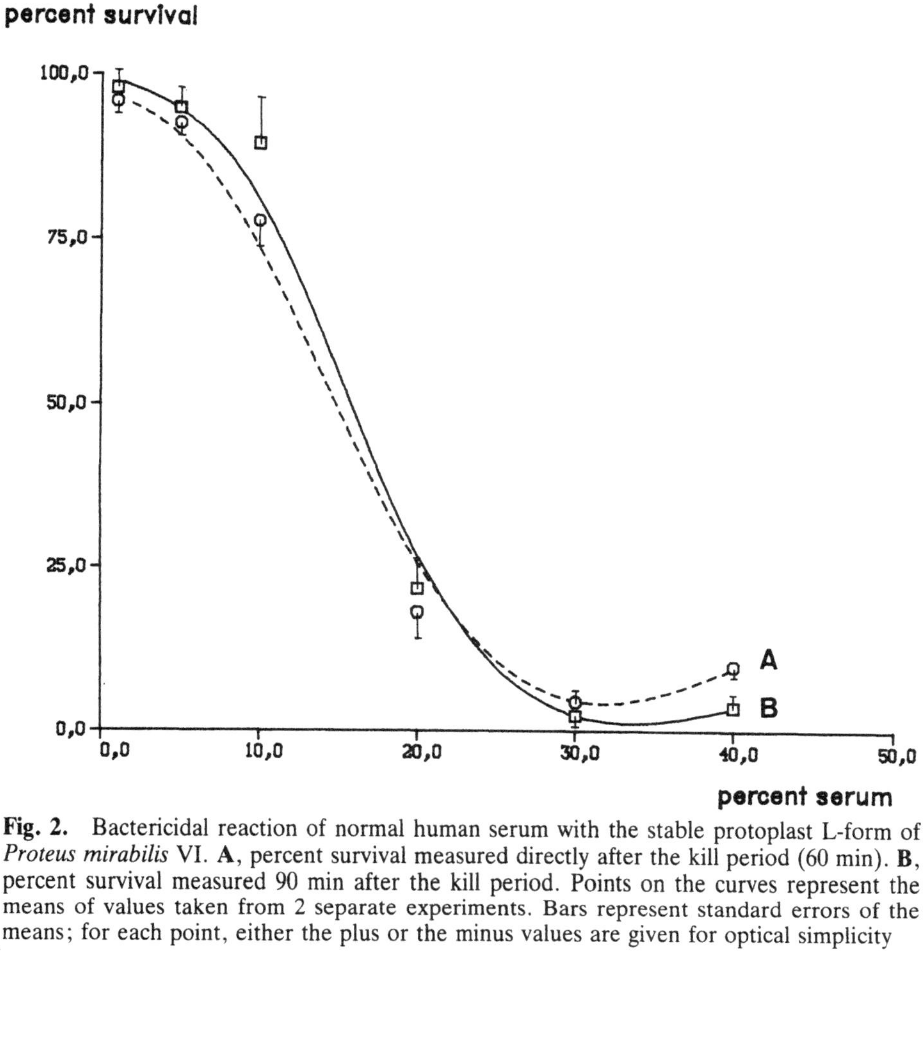

Fig. 2. Bactericidal reaction of normal human serum with the stable protoplast L-form of *Proteus mirabilis* VI. **A**, percent survival measured directly after the kill period (60 min). **B**, percent survival measured 90 min after the kill period. Points on the curves represent the means of values taken from 2 separate experiments. Bars represent standard errors of the means; for each point, either the plus or the minus values are given for optical simplicity

Table 2. Hemagglutinin (HA) titers in normal human serum (NHS) and in lyophilized, reconstituted guinea pig serum (GPS) measured against *Proteus mirabilis* VI lipopolysaccharide coupled to sheep red blood cells

Serum	Source	HA titer[a]
NHS	Donor 1, sample 1	16
NHS	Donor 1, sample 2	16
NHS	Donor 1, sample 3	16
NHS	Donor 2, sample 1	5.6
NHS	Donor 2, sample 2	4
NHS	Donor 2, sample 3	4
NHS	Donor 3	64
GPS	Behringwerke, Marburg, FRG	32

[a] Values represent the geometric means taken from two separate determinations

was no correlation between titers of antibodies to LPS and bactericidal activity of the sera. Whether bactericidal antibodies to other surface components (e.g. outer membrane proteins) are present in these sera will have to be tested.

In general, our results indicate that relatively extensive cell wall defects are required for conversion of the serum resistant strain of *P. mirabilis* to serum sensitivity. Conversion to serum sensitivity was first observed with the spheroplast L-form, and the stable protoplast L-form was even more sensitive.

In an accompanying paper (Nixdorff, Martin and Siegmund-Schultze, this volume), we detected certain changes in outer membrane composition of *P. mirabilis* VI after treatment of this organism with cefuroxime (induction of filament formation) or cefoxitin (induction of spheroplast L-forms). In cell walls of the filament form, we detected a decrease in the 39KD outer membrane protein and a slight increase in the ratio of saturated : unsaturated + cyclopropane fatty acids in phospholipids. The observed changes obviously had little effect in altering the serum resistance of *P. mirabilis* VI. In cell wall fragments isolated from spheroplasts, no alterations in proteins could be observed, but we did detect an apparent increase in outer membrane phospholipids. Whether this contributed to the increase in serum sensitivity of the spheroplasts is difficult to assess. It has been shown that the spheroplast L-form of *P. mirabilis* 19 induced by 50 μg/ml cefoxitin contained only 39% of the peptidoglycan of the bacterial form (E. Ruttkowski and H.H. Martin, unpublished data). Therefore, it may be that the spheroplasts contain areas completely devoid of outer membrane material, which we could not detect by the isolation methods used, and that such areas are more accessible for complement. This would coincide with electron micrographs showing partial breakdown of the cell wall and exposure of the cytoplasmic membrane in penicillin-induced spheroplast L-forms of *Escherichia coli* K12 [4].

Acknowledgements. This work was supported by the Deutsche Forschungsgemeinschaft. We thank Heide-Marie Rostek and Gesche Bauer for excellent technical assistance.

Summary

A comparison of the sensitivity of the bacterial form of the serum resistant strain *Proteus mirabilis* VI with its filament form induced by cefuroxime, its spheroplast L-form induced by cefoxitin and its stable protoplast L-form, which grows in the absence of antibiotics, to the action of normal human serum was made in a photometric assay for bactericidal activity. The bacterial form was relatively resistant to the action of normal serum; only 10% killing was achieved with the highest concentration of serum tested (75%). The sensitivity of the filament form was not significantly different. In contrast, up to 90% killing of the spheroplast L-form was observed. The protoplast L-form was most sensitive, in which case 100% killing was achieved with 30% normal human serum. All forms were much less susceptible to the bactericidal action of guinea pig serum, although the same comparative tendencies in sensitivity among the various forms were also observed. The results indicate that relatively extensive cell wall defects are required for conversion of this serum resistant strain of *P. mirabilis* to serum sensitivity.

References

1. Bub F, Bieker P, Martin HH, Nixdorff K (1980) Immunological characterization of two major proteins isolated from the outer membrane of *Proteus mirabilis*. Infect Immun 27: 315–321
2. Dutcher BS, Reynard AM, Beck ME, Cunningham KK (1977) Potentiation of antibiotic bactericidal activity by normal human serum. Antimicrob Agents Chemoth 13: 820–826
3. Gmeiner J, Martin HH (1976) Phospholipid and lipopolysaccharide in *Proteus mirabilis* and its stable protoplast L-form. Difference in content and fatty acid composition. Eur J Biochem 67: 487–494
4. Gumpert J, Schuhmann E, Taubeneck U (1971) L-Formen von *Escherichia coli* K12 (λ). II. Elektronenoptische Untersuchungen. Z Allg Microbiol 11: 283–300
5. Lorian V, Atkinson B (1979) Effect of serum and blood on *Enterobacteriaceae* grown in the presence of subminimal inhibitory concentrations of ampicillin and mecillinam. Rev Infect Diseases 1: 797–806
6. Martin HH (1964) Composition of the mucopolymer in cell walls of the unstable and stable L-form of *Proteus mirabilis*. J Gen Microbiol 36: 441–450
7. Martin HH (1983) Protoplasts and spheroplasts of gram-negative bacteria (with special emphasis on *Proteus mirabilis*). In: Potrykus I, Harms CT, Hinnin A, Hütter R, King PJ, Shillito RD (eds) Protoplasts 1983 (lecture proceedings), Experimentia supplementum vol. 46. Birkhäuser Verlag, Basel, pp 213–225
8. Martin HH, Staboulis D, Schilf W (1981) Penicillinbindeproteine als spezifische Wirkorte der β-Laktam-Antibiotika und als Faktoren der Antibiotikaresistenz. Immun Infekt 3: 99–105
9. Muschel LH, Jackson JE (1966) The reactivity of serum against protoplasts and spheroplasts. J Immunol 97: 46–51
10. Muschel LH, Treffers HP (1956) Quantitative studies on the bactericidal actions of serum and complement. I. A rapid photometric growth assay for bactericidal activity. J Immunol 76: 1–10
11. Siegmund-Schultze N, Nixdorff K (1985) Modulation of the interaction of bacteria and their membrane components with host defense systems by antibiotics. In: Pulverer G, Jeljaszewicz J (eds) Chemotherapy and immunity. Zentralbl Bakteriol Mikrobiol Hygiene I. Abteilung, supplement 13. Gustav Fischer Verlag, Stuttgart New York, pp 231–242
12. Taylor PW (1983) Bactericidal and bacteriolytic activity of serum against gram-negative bacteria. Microbiol Rev 47: 46–83
13. Traub WH, Sherris JC (1970) Studies on the interaction between serum bactericidal activity and antibiotics *in vitro*. Chemotherapy 15: 70–83

Discussion

Kroll: Can you comment on one assumption? You have a graded set of outermost layers with different amounts of phospholipids, like Salmonella core mutants have with increasing defects of their LPS moieties increasing amounts of phospholipids and are increasingly serum sensitive. What do you think about a correlation between fluidity of the outermost layer and serum sensitivity?

Siegmund-Schultze: It is indeed a possibility that changes such as increase of phospholipids in the outer membrane may affect the sensitivity. However, since the increase in phospholipids in the spheroplast-L-form was relatively small, we think that surface areas devoid of outer membrane are also a possible explanation for increased sensitivity. However at this point we do not know which possibility is relevant.

Verhoef: Are you sure that the complement system of the commercially available guinea pig serum was completely intact?

Siegmund-Schultze: I don't know the answer to this question, as we have not performed complement titration of guinea pig serum.

Vosbeck: Is the term bactericidal activity of your strain dependent on the bacterial growth rate, and if so, are the differences in bactericidal activity seen with different sera and complement sources due to intrinsic differences of the complement sources or to differences in bacterial growth rates?

Siegmund-Schultze: Bactericidal sensitivity can vary with growth rate. However, we always used logarithmic phase cultures in the assay system, and we also tested the growth rate of these cultures in heat-inactivated human and guinea pig serum and in medium under bactericidal assay conditions. In these cases, we observed no significant difference in growth rate.

Stephens: Did you look at possible differences in killing by classical versus alternative pathway in your system? This might suggest differences in LPS structure although killing was equal.

Siegmund-Schultze: We have not specifically determined which pathways of complement activation are operating in our system. We assume that several mechanisms might be, such as C1-activation without antibody, alternative pathway or even classical pathway with antibodies. All these mechanisms are possible.
In earlier studies we tested these same forms with an antibody-dependent bactericidal system with diluted guinea pig serum as complement source, in which the alternative pathway probably was not operating. Also in this system, bacteria and filaments were resistant, whereas spheroplasts and protoplasts were sensitive. That is, the results are essentially the same as with normal human serum.

Opferkuch: Is anything known about the resistance mechanism in Proteus mirabilis?

Siegmund-Schultze: Outside of the usual explanations for resistance in other gram-negative bacteria such as E. coli or Salmonella, I don't know any mechanism peculiar to Proteus.

Imipenem Renders Serum resistant *Enterobacter aerogenes* and *Escherichia coli* to Serum sensitive Bacteria

C. WIEMER, B. KUBENS, and W. OPFERKUCH

Medizinische Mikrobiologie und Immunologie, Ruhr-Universität Bochum, D-4630 Bochum, F.R.G.

Introduction

A main virulence factor of gram-negative bacteria is the resistance to the bactericidal activity of serum. The ineffectiveness of this host defense mechanism is due to a resistance to the attack of complement. According to the literature, certain β-lactam antibiotics influence the resistance of bacteria against the attack of complement. Several reports are listed in Table 1. Since the method of determination and the definition of serum-resistance differ from one report to another, the results of these experiments are very controversial and the findings are difficult to compare. The aim of this investigation was to study the influence of imipenem and other new β-lactam antibiotics on the serum resistance of enterobacteriaceae. In this study, serum resistance was defined merely on the basis of growth curves. A strain is determined as "serum-resistant" if there is growth or no decrease of viable counts in 20% NHS, and as "serum-sensitive" when no further growth takes place after incubation in 5% NHS.

Table 1. Influence of β-Lactam Antibiotics on the Serum Resistance of Entero-Bacteriaceae

Strain	Antibiotic	Test Serum	Serum Resistance	Ref.
E. coli	Ampicillin	NHS	↓	7
E. coli	Cyclacillin	HRS	↓	3
E. coli	Ampicillin	NHS	↓	1
Enterobacteriaceae	Ampicillin/ Mecillinam	NHS	↑	4
E. coli Klebs. pneumoniae	Ampicillin/ Cephalothin	NHS	∅	2
E. coli	Mecillinam	NHS	↓	6

NHS – normal human serum, HRS – hyperimmune rabbit serum, ∅ – no influence, ↑ – increase, ↓ slight decrease

The Influence of Antibiotics on the Host-Parasite Relationship II
Edited by D. Adam, H. Hahn, W. Opferkuch

Material and Methods

Bacteria

All strains used were originally isolated from blood cultures, except for *E. coli* WF 52, which is a urine isolate and was kindly provided by Prof. A. Glynn, London. The strains of *E. coli, Enterobacter cloacae, Klebsiella pneumoniae, Serratia marcescens, Proteus mirabilis,* and *Proteus morganii* grew after incubation in 20% NHS, whereas the strains of *Serratia marcescens* only tolerated a maximum of 8% NHS. The serogroups of the two *E. coli* strains were: BK 615: 09a; WF 52: 026:B6:H11.

Serum, Antibiotics

Blood was obtained from healthy volunteers, allowed to clot for two hours at 4 °C, and centrifuged at 3000 rpm at 4 °C; serum was stored in aliquots at −20 °C and thawed shortly before use. For inactivation serum was heated at 56 °C for 30 min. Complement activity was tested by determination of CH50, and ranged between 25 and 36 units.

The antibiotics were obtained by Merck, Sharp & Dohme, Munich, F.R.G. (imipenem, cefoxitin), Hoechst AG, Frankfurt, F.R.G. (cefotaxime, cefuroxime, cefodizime-HR 221), Leo Pharmaceuticals, Bellerup, Denmark (mecillinam) and Takeda-Grünenthal, Stolberg, F.R.G. (cefsulodin).

Serum Bactericidal Assay

Bacteria were inoculated from a nutrient agar slant into 0.9% sodium chloride up to a concentration of 0.5 McFarland standards, then transferred into a nutrient broth containing ¼ MIC of the different antibiotics and incubated at 37 °C for 18 h. 100 μl of this overnight culture were added to 900 μl of a medium containing the same concentration of the antibiotic and incubated in a shaking water bath for 3 h in order to obtain a log-phase culture. Thereafter, the cultures were standardized to a concentration of 0.5 McFarland standards, and the final test mixture contained 0.8 ml of this bacterial suspension, 0.2 ml freshly thawed serum and ¼ MIC of the antibiotic. The control mixtures contained either heat inactivated serum with or without antibiotics or active serum without antibiotic. The pH of the test mixtures ranged between 7.3 and 7.5. The bacteria were incubated for another 90 minutes in a shaking water bath, and samples were taken every 30 minutes to estimate the number of viable counts.

Results and Discussion

The MIC of the used antibiotics for *E. coli* (BK 615, WF 52) and *Enterobacter aerogenes* (BK 319) are listed in Table 2. A slight decrease in serum resistance from 20% to 15% NHS (serum resistance is given as the maximal serum concentration in

Table 2. MIC of six new β-Lactam Antibiotics for the Strains BK 615, WF 52, and BK 319

Antibiotic	MIC (μg/ml) BK 615	WF 52	BK 319
Cefsulodin	64	64	2048
Cefoxitin	4	2	2048
Cefuroxime	4	4	512
Cefotaxime	0.032	0.016	128
Cefodizime	0.125	0.125	128
Mecillinam	0.25	0.25	256

% (v/v), tolerated before and after treatment with the antibiotic) could be observed if ½ MIC of cefoxitin, cefodizime (HR 221), cefsulodin and cefuroxime was added. As shown in Fig. 1, the killing rate of NHS in the presence of ½ MIC cefuroxime is still less than 90%. Comparable results were obtained after treatment with cefoxitin, cefodizime and cefsulodin (data not shown). Serum resistance of the investigated strains was not influenced by addition of ½ MIC mecillinam (Fig. 2) and ½ MIC cefotaxime (data not shown).

All strains used to investigate the influence of imipenem on serum resistance are listed in Table 3. The MIC and the serum resistance with or without ¼ MIC imipenem are also given. As can be seen from these data, serum resistance of the

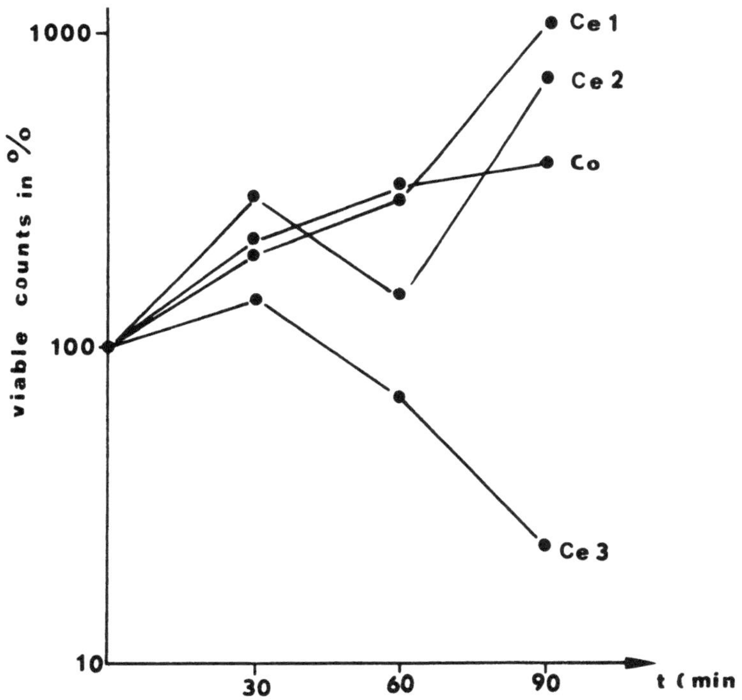

Fig. 1. Influence of ½ MIC cefuroxime on the serum resistance of BK 319. Ce 1, 20% heat-inactivated serum; Ce 2, 15% active serum; Ce 3, 20% active serum; Co, 20% active serum without antibiotic

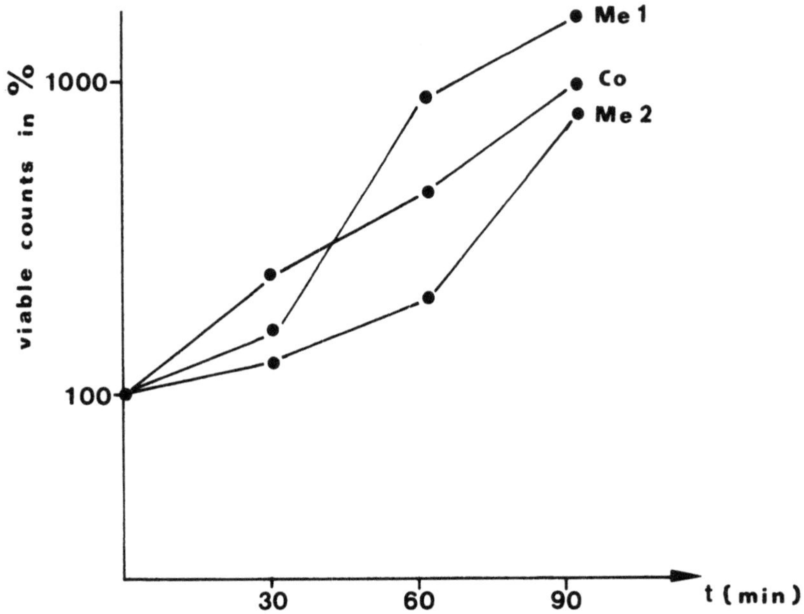

Fig. 2. Influence of ½ MIC mecillinam on the serum resistance of BK 319. Me 1, 20% heat-inactivated serum; Me 2, 20% active serum; Co, 20% aktive serum without antibiotic

Table 3. Mic of Imipenem for twelve Strains of Enterobacteriaceae; Serum Resistance as the maximal Serum Concentration in % (v/v) tolerated by the Bacteria before and after treatment with ¼ MIC Imipenem

Strains	MIC µ/ml	before treatment	after treatment
WF 52 Escherichia coli	0.125	20	15
BK 615 Escherichia coli	0.25	20	3
BK 28 Enterobacter aerogenes	0.5	20	15
BK 319 Enterobacter aerogenes	0.5	20	3
BK 383 Enterobacter cloacae	0.25	20	20
BK 967 Enterobacter cloacae	0.5	20	20
BK 177 Klebsiella pneumoniae	0.25	20	20
BK 1171 Klebsiella pneumoniae	0.25	20	20
BK 1435 Proteus mirabilis	4	20	15
BK 74 Proteus morganii	2	20	20
BK 31 Serratia marcescens	1	8	8
BK 1292 Serratia marcescens	0.5	8	8

strains WF 52, BK 615, BK 28, BK 319 and BK 1435 was altered. The decrease in serum resistance was especially impressive for the strains BK 615 and BK 319. The other seven strains did not show any alterations. Treatment of BK 28, BK 1435 and WF 52 with ½ MIC imipenem in 20% NHS caused a killing rate of about 90%, while the untreated cells grew in their normal rate. Under the same conditions more than 99,9% of the inoculum of BK 615 and BK 319 was killed. Even a serum concentration of only 5% NHS led to a reduction of viable counts by more than 90% of the inoculum, whereas heat inactivated serum showed no influence (Fig. 3). This phenomenon of "induced serum sensitivity" (from 20% to 3% NHS) proved to be dose dependent as shown in Fig. 4. The decrease of viable counts is two logs of growth smaller if a concentration of ⅛ MIC instead of ¼ MIC imipenem is added. After transfer into antibiotic free medium, the bacteria recovered their original resistance which indicates that the induced sensitivity is reversible.

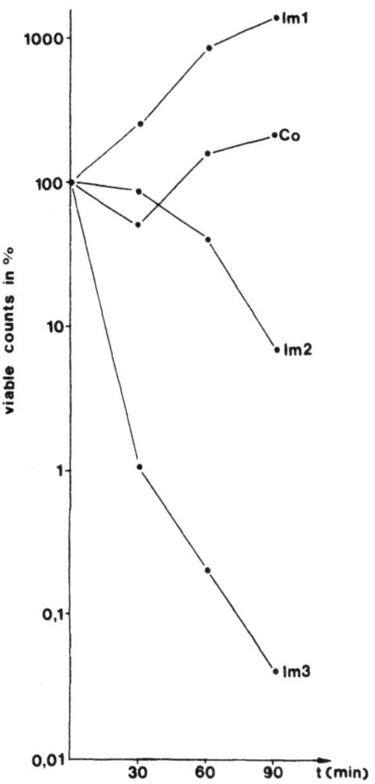

Fig. 3. Influence of different serum concentrations on the imipenem-treated serum-resistant bacteria of BK 319. Im 1, 20% heat-inactivated serum; Im 2, 5% active serum; Im 3, 20% active serum; Co, 20% aktive serum without antibiotic

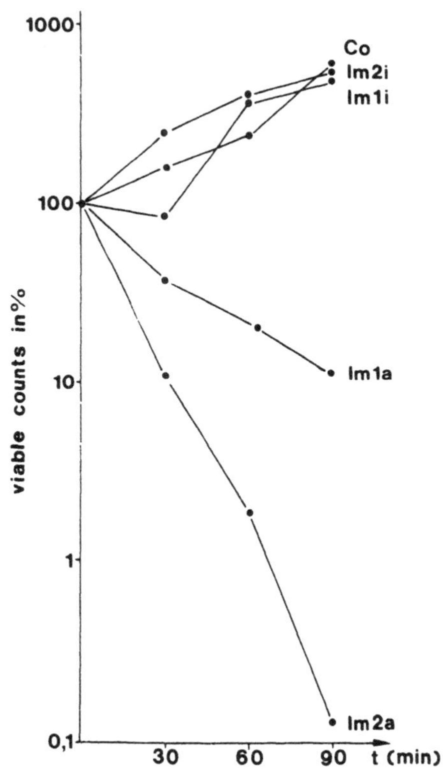

Fig. 4. Influence of different concentrations of imipenem on the serum resistance of BK 319. Im 1i, ⅛ MIC + heat-inactivated serum; Im 1a, ⅛ MIC + active serum; Im 2i, ¼ MIC + heat-inactivated serum; Im 2a, ¼ MIC + active serum; Co, 20% active serum without autibiotic. Serum concentration: 20%

Table 4. Comparison of Serum Resistance in % (v/v) after treatment with Mecillinam and Imipenem

Antibiotic	Serum resistance in % BK 615	WF 52	BK 319
Control	20	20	20
Mecillinam	20	20	20
Imipenem	3	15	3

A comparison of the serum resistance after treatment with imipenem and mecillinam is shown in Table 4. While imipenem caused a decrease in serum resistance, mecillinam had no influence at all. Taylor et al. [6] found a decrease in serum resistance in *E. coli* after treatment with ½ MIC or lower concentrations of mecillinam, whereas the strains used for this study were not killed under similar conditions. As both mecillinam and imipenem are round-cell forming β-lactam antibiotics and their main biochemical target is the penicillin-binding protein (PBP) 2, these results are of special interest. From our data it can be concluded that PBP 2 cannot be responsible for the observed phenomenon.

To be sure that there was a real change in serum resistance and not an additional effect of the antibiotic itself, it is very important to reduce the serum concentration. When the experiments were carried out under the same conditions using filament-forming antibiotics no killing could be observed if the serum concentration was reduced. Imipenem was the only antibiotic which influenced serum resistance of the strains BK 615 and BK 319 to such an extent that even 3% NHS (instead of more than 20%) was able to kill the bacteria.

To answer the question how complement kills the imipenem-treated cells, by the classical or by the alternative pathway, experiments were carried out with serum preincubated with bacteria to absorb antibodies or MG^{++} EGTA containing serum: no killing effect could be observed in the cases of absorbed serum or EGTA containing serum (Fig. 5). It was assumed that the killing works via the antibody dependent classical pathway. In order to find a serum protein responsible for the described effect of induced serum sensitivity serum proteins were fractionated by ammonium sulfate precipitation and gel chromatography. But up to now it has not been possible to restore the killing potency of the serum by adding fractions to absorbed serum. Further studies are necessary to identify the responsible factor.

Summary

Serum resistance of five strains of Enterobacteriaceae was reduced after growth in subinhibitory concentrations of imipenem. Incubation in 20% normal human serum (NHS) in the presence of imipenem caused a decrease of cfu of two logs of growth, compared with the control. Two of these strains (one *E. coli* and one *Enterobacter*

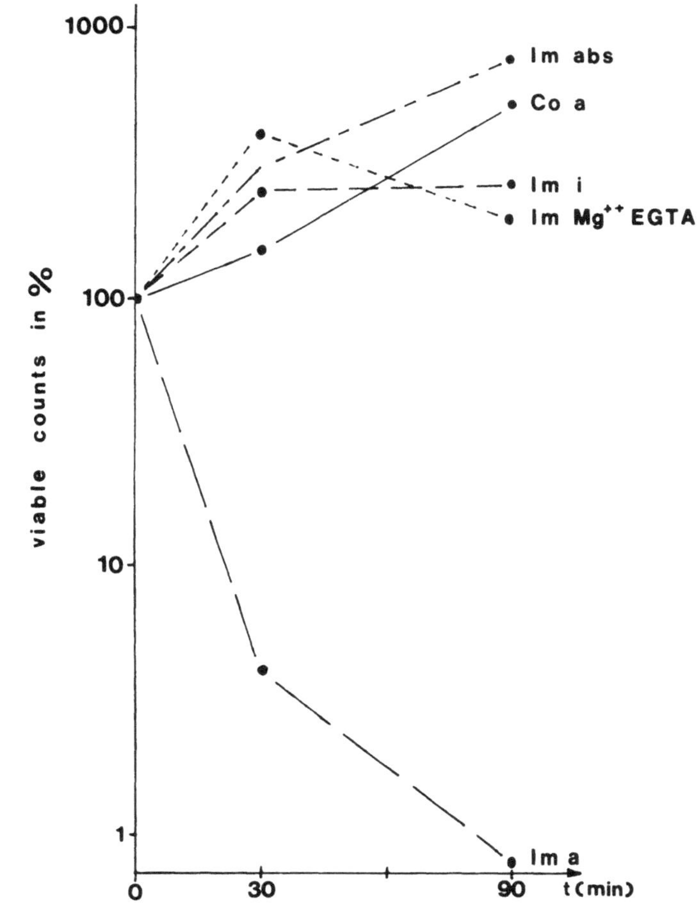

Fig. 5. Growth curves of imipenem-treated bacteria of BK 319 in absorbed or Mg^{++} EGTA-containing serum. abs, absorbed serum; a, active serum; i, heat-inactivated serum; Co, 20% active serum without antibiotic

aerogenes) showed this bactericidal effect of NHS in the presence of imipenem even when the serum concentration was reduced to 5% NHS. This phenomenon of "induced serum sensitivity" is dose-dependent and reversible.

References

1. Dutcher BS, Reynard AM, Beck ME, Cunningham RK (1978) Potentiation of antibiotic bactericidal activity by normal human serum. Antimicrob Agents Chemother 13:820–826
2. Fierer J, Finley F (1979) Lethal effect of complement and lysozyme on polymyxin-treated, serum-resistant gram-negative bacilli. J Infect Dis 140: 581–589
3. Friedman H, Warren GH (1976) Antibody-mediated bacteriolysis: Enhanced killing of cyclacillin-treated bacteria. Proc Soc Exp Biol Med 153: 301–304
4. Lorian V, Atkinson BA (1979) Effect of serum and blood on Enterobacteriaceae grown in the presence of subminimal inhibitory concentrations of ampicillin and mecillinam. Rev Infect Dis 1: 797–

5. Mayer MM (1964) Complement and complement fixation. In: Kabat EA, Mayer MM (eds), Experimental Immunochemistry, 2nd ed. C.C. Thomas, Springfield, pp. 133–240
6. Taylor PW, Gaunt H, Unger FM (1981) Effect of subinhibitory concentrations of mecillinam on the serum susceptibility of Escherichia coli strains. Antimicrob Ag Chemother 19: 786–788
7. Traub WH, Sherris JC (1970) Studies on the interaction between serum bactericidal activity and antibiotics in vitro. Chemotherapy 15: 70–83

Discussion

Bauernfeind: Were there any differences in the concentration of thienamycin in experiments with active or inactive serum or without serum at the end of the incubation time?

Wiemer: We did not test the thienamycin concentration after the test time.

Influence of subinhibitory concentrations of Clindamycin on the phagocytosis of Staphylococcus aureus

ETÈL VERINGA and J. VERHOEF

Laboratory of Microbiology, State University of Utrecht, Catharijnesingel 59, 3511 GG Utrecht, The Netherlands

Introduction

The process of phagocytosis by Polymorphonuclear leukocytes (PMN) and monocytes (MN), the major host defense mechanism against invading microorganisms, has been shown to be influenced by various antibiotics [1]. One way of interaction between antibiotics and phagocytic cells is that phagocytes may react differently with bacteria when the bacterial surface has been altered by antibiotics. It has been shown that exposure to some antibiotics at concentrations lower than the minimal inhibitory concentration (MIC), not only induce morphological changes in certain bacteria [7], but also a higher susceptibility of these bacteria to phagocytosis [2, 3, 4, 8, 9]. As opsonization of *S. aureus* is necessary for phagocytosis it may be possible that the enhanced uptake by phagocytes of these antibiotic treated bacteria that was observed takes place due to enhanced opsonization [2, 5]. Because opsonization is determined by the binding of antibodies and C3b to the bacterial surface, the enhanced uptake of microorganisms by PMN after exposure to subinhibitory concentrations of antibiotics may be due to changes in the structure of antibody binding or complement activating sites.

Therefore, the aim of our experiments was to study the influence of subinhibitory concentrations of clindamycin on opsonization and phagocytosis of *S. aureus*.

Materials and Methods

Bacteria

S. aureus Ev (a clinical isolate) and *S. aureus* M, an encapsulated strain, were grown overnight at 37 °C in Mueller-Hinton broth (Difco Laboratories) in the presence or absence of one-quarter or one-half of the minimal inhibitory concentration (MIC) of Clindamycin (The Upjohn Co., Ede, The Netherlands), washed three times in phosphate-buffered saline (PBS), and adjusted photometrically to yield 2.5×10^8 CFU/ml. For phagocytosis studies, the bacteria were radiolabeled by adding 20 μCi/ml of [^{3}H]-thymidine (Amersham International plc, Amersham, UK) to the Mueller-Hinton broth.

The Influence of Antibiotics on the
Host-Parasite Relationship II
Edited by D. Adam, H. Hahn, W. Opferkuch

PMNs

Heparinized venous blood (10U of heparin per ml of blood) was obtained from healthy volunteers. Fivehundred ml blood was mixed with 150 ml of 6% dextran. After sedimentation of the erythrocytes for 45 min at room temperature, the leucocyte rich plasma was removed, centrifuged for 10 min at 160 g and resuspended in Hanks' balanced salt solution with 0.1% gelatine (Gel-HBSS). Every 7.5 ml of this leucocyte rich suspension was layered onto 3 ml Ficoll-Paque (Farmacia, Fine Chemicals AB, Uppsala, Sweden), and centrifuged for 35 min at 160 g. After removal of the mononuclear cell layer residual erythrocytes in the sediment were lysed by adding 6 ml of a NH_4Cl solution (0.87% in sterile water). The PMN were washed three times (160 g) in Gel-HBSS, resuspended and diluted to a final concentration of 1×10^7 PMN/ml.

Contamination with mononuclear cells never exceeded 2%. More than 95% of the cells were viable as shown by trypan blue exclusion.

Serum sources

Normal human pooled serum (hps) was obtained from ten healthy donors. Antiserum, raised in rabbits, was heated (30 min 56 °C) in order to block the complement system. Serum of a patient with agammaglobulinaemia with less than 5 mg/100 ml IgG and no detectable amounts of IgM, IgA, IgE and IgD was further depleted of IgG by adsorption with sepharose CL4B-proteine A (0.5 g for 10 ml serum). All sera were stored at −70 °C until use.

Phagocytosis assay

The uptake of radiolabeled bacteria by PMNs was measured by the method of Verhoef et al [6]. For opsonization 2.5×10^8 radioactive labeled bacteria/ml were incubated in different concentrations of serum during different periods of time. At indicated times, opsonization was stopped by adding 2.5 ml ice-cold phosphatebuffered saline (PBS). The opsonized bacteria were centrifuged for 15 min at 1600 g, and suspended in 1.0 ml of Gel.-HBSS. A 0.2 ml amount of opsonized staphylococci (5×10^7/ml) was incubated with 0.2 ml of PMN (5×10^6/ml) in a 37 °C shaking water bath. After 2, 6 and 12 min respectively phagocytosis was stopped by adding 2.5 ml ice-cold PBS. The percentage of staphylococci taken up by the PMN after 2, 6 and 12 min incubation was calculated from the uptake of radio-activity by the phagocytes and the total added radio-activity, determined in a separate vial. Non-phagocyte-associated radioactivity was separated from PMN-associated radioactivity by three cycles of differential centrifugation (160 g for 5 min). Radioactivity was determined by liquid scintillation counting (Philips).

Results

Radiolabeled bacteria were opsonized in 10% normal human pooled serum (hps). The uptake rates of antibiotic treated bacteria did not differ significantly from that of the untreated control bacteria. However when 1% hps was used, *S. aureus* Ev grown in the presence of subinhibitory concentrations of clindamycin was taken up more readily by PMNs than were control bacteria. To further investigate the opsonic kinetics, bacteria were preincubated in different serum concentrations for various periods of time. The results are summarized in Fig. 1 and Table 1. Using 1% and 5% hps as opsonic source, antibiotic-treated *S. aureus* Ev was taken up faster than the control *S. aureus* Ev. The difference in uptake rates between the control and the

Table 1. Effect of subinhibitory concentrations of clindamycin on phagocytosis of *S. aureus* Ev by PMN

		Uptake %		
		hps (1%)	antiserum (1%)	agammaserum (10%)
S. aureus (control)				
opsonization time	1 min	5	30	—
	5	25	30	35
	10	35	30	65
S. aureus (¼ MIC clindamycin)				
opsonization time	1 min	25	40	—
	5	50	55	60
	10	65	60	70
S. aureus (¼ MIC clindamycin)				
opsonization time	1 min	60	50	—
	5	90	55	65
	10	85	55	80

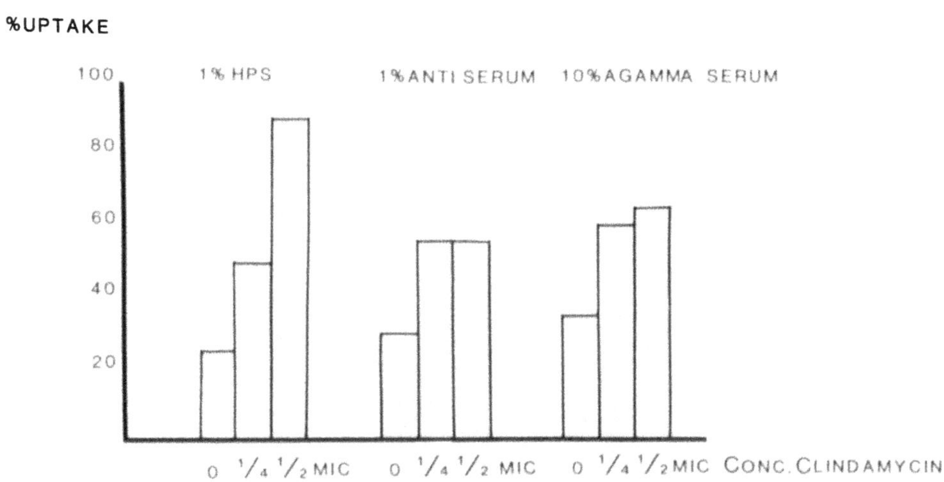

Fig. 1. Effect of sub MIC of clindamycin on phagocytosis of *S. aureus* by human PMN. Clindamycin treated and untreated (=control) bacteria were opsonized during 5 min

clindamycin treated bacteria were most obvious when 1% hps was used (compared with 5% hps). These results clearly reflect the opsonization kinetics. In hps opsonization is primarily mediated by complement, only amplified by antibodies. When only low concentrations of complement (and antibodies) are present (as in 1% hps), or when the opsonization time is short, (e.g. 1 min), clindamycin treated *S. aureus* Ev may bind complement and antibodies more readily than the control bacteria. Raising the serum concentration or extending the opsonization time however, enables the control bacteria to activate the same amount of complement as the clindamycin treated bacteria do.

Because normal human pooled serum contains both complement and antibodies as opsonins, we are not able to really distinguish between opsonization mediated by the classic and the alternative complement pathway and by antibodies. Therefore bacteria were opsonized in heated antiserum and in agammaglobulinaemic serum.

Using 10% agammaglobulinaemic serum, *S. aureus* Ev grown in the presence of subinhibitory concentrations of clindamycin was phagocytized more readily than *S. aureus* Ev grown in the absence of clindamycin. When 5% agammaglobulinaemic serum was used, the uptake rates of both clindamycin treated and untreated bacteria did not exceed 20% when the opsonization time was 1 min or 5 min. Opsonization with 5% agammaserum during 10 min resulted in an uptake rate of 20% for the untreated *S. aureus* Ev, and of 40% for the clindamycin treated bacteria. Thus the complement dependent phagocytosis appears to be time and concentration dependent.

To study the antibody dependent phagocytosis of *S. aureus* Ev, heated antiserum was used as opsonic source. When 10% and 5% heated antiserum was used for opsonization the uptake rates of antibiotic treated bacteria did not differ from that of the untreated control. Using 1% heated immuneserum, *S. aureus* Ev grown in the presence of subinhibitory concentrations of clindamycin was phagocytized faster than the control bacteria.

In all experiments where clindamycin treated bacteria were taken up more readily than control bacteria, clindamycin at ½ MIC was more effective than at ¼ MIC.

The uptake by PMN of *S. aureus* M, an encapsulated strain, was not enhanced after exposure to subinhibitory concentrations of clindamycin. Using hps, antiserum as well as agammaserum as opsonic source, no difference was observed in uptake by PMN of clindamycin treated and untreated bacteria.

Discussion

Subinhibitory concentrations of clindamycin induce morphological changes in *S. aureus*. In addition, *S. aureus* grown in subinhibitory concentrations of clindamycin, appear to be taken up more readily by PMNs compared with *S. aureus* grown in the absence of clindamycin.

This enhanced uptake was seen after opsonization with hps, antiserum and serum from a patient with agammaglobulinaemia. The mechanism of action is likely to be an increased susceptibility of clindamycin treated bacteria to opsonization. Therefore it is possible that PMNs can recognize the drug treated bacteria more easily as a phagocytizable particle than untreated bacteria.

Since clindamycin inhibits protein synthesis in *S. aureus*, subinhibitory concentrations possibly change the structure of antibody binding sites and complement activating sites. This could be an explanation for the observed enhanced immunoglobulin- and complement dependent phagocytosis of *S. aureus* Ev. In the *S. aureus* M strain no enhancement of phagocytosis by PMN was observed after exposure to subinhibitory concentrations of clindamycin. It is possible that the polysaccharide capsule covers the altered protein surface.

This findings may explain why certain drugs may be effective when they are given in a low concentration. Especially at the site of infection (e.g. abscesses), where it commonly occurs that the concentration of the antibiotic is lower than the MIC, certain antibiotics still may be effective.

Studies on the clinical relevance need to be extended. Also, more work is needed with other protein-synthesis inhibitors.

Summary

Staphylococcus aureus Ev and the encapsulated *S. aureus* M strain were grown in the presence of one-half and one-quarter of the minimal inhibitory concentration of clindamycin. Complement- as well as antibody dependent phagocytosis of the antibiotic treated *S. aureus* Ev by human polymorphonuclear leucocytes was significantly enhanced, compared with that of the untreated control.

No enhancement of complement- and antibody dependent phagocytosis was observed after exposure of *S. aureus* M to subinhibitory concentrations of clindamycin.

References

1. Milatovic D (1983) Antibiotics and phagocytosis. Eur J Clin Microbiol 2: 414–425
2. Gemmel CG et al (1981) Potentiation of opsonization and phagocytosis of Streptococcus pyogenes following growth in the presence of clindamycin. J Clin Invest 67: 1249–1256
3. Milatovic D (1982) Effect of subinhibitory concentrations on the phagocytosis of Staphylococcus aureus. Eur J Clin Microbiol 1: 97–101
4. Milatovic D (1984) Influence of subinhibitory concentrations of antibiotics on opsonization and phagocytosis of Pseudomonas aeruginosa by human polymorphonuclear leucocytes. Eur J Clin Microbiol 3: 288–293
5. Milatovic D et al (1983) Clindamycin enhances opsonization of Staphylococcus aureus. Antimicrob Agents Chemother 24: 413–417
6. Verhoef J et al (1977) Kinetics of staphylococcal opsonization, attachment, ingestion and killing by human polymorphonuclear leucocytes: a quantitative assay using [^{3}H]-thymidine labeled bacteria. J Immunol Methods 14: 303–311
7. Lorian V (1975) Some effects of subinhibitory concentrations of antibiotics on bacteria. Bull of New York Acad Med 51: 1046–1053
8. Friedman H et al (1974) Enhanced susceptibility of penicillinresistant staphylococci to phagocytosis after in vitro incubation with low dosis of nafcillin. Proc Soc Exp Biol Med 146: 707–711
9. McDonald PJ et al (1981) Postantibiotic leucocyte enhancement. Increased susceptibility of bacteria pretreated with antibiotics to activity of leucocytes. Rev Inf Dis 3: 38–44

Discussion

Mandell: Did you see any evidence of a dose response effect as far as clindamycin and C3b deposition are concerned?

Verhoef: Yes, a dose response was observed but only between ¼–½ of the MIC. However, the longer the opsonization time and the more serum was used, the differences between ¼ and ½ MIC were not so pronounced.

Pawelzik: You determined uptake of the bacteria by the phagocytes by measuring the ^{3}H cpm associated with the phagocytes. But association is not identical with phagocytosis. How do you distinguish between those bacteria that were only adherent, that means that were outside of the cells, and those that were already internalized by the phagocytes?

Verhoef: We used lysostaphin to lyse adherent non-ingested Staphylococci at 4 °C. Also by E.M. studies you get an impression that after e.g. 30 min incubation of staphylococci with PMN few bacteria are attached and many are ingested.

Roszkowski: I cannot agree with your statement that chemiluminescence can occur only in opsonized systems. Undoubtedly in the presence of opsonins, chemiluminescence is quick and high. However, it is possible to stimulate granulocytes without participation of opsonin. In this case we can observe slow but significant chemiluminescence.

Verhoef: I must confess, there are so many confusing data in the literature about the need for opsonins or not. It depends on the test system and the stimulus whether a chemiluminescence response is observed. We also were able to induce CL in the absence of opsonins when "surface" phagocytosis was tested. However, the field of opsonic recognition remains intriguing and an important subject to study.

Büscher: Do you have any information about which type of reactive oxygen species is liberated by PMN in response to stimulation by bacteria? Which method do you use to measure superoxide anion?

Verhoef: We have measured superoxide by means of the reduction of ferricytochromes. We measured H_2O_2 and indirectly $^{\cdot}OH$ (with scavengers). During stimulation of PMN by microorganisms we feel that at least these products are produced.

Büscher: Is the chemiluminescence response of PMN in your system induced by antibody, complement or by bacteria themselves?

Verhoef: We feel that in order to get a chemiluminescence response opsonized (by Fab, and/or complement) bacteria are needed.

The Effect of Antibiotics on the Killing of *E. Coli* by Macrophages of the Cell Line P 388 D1

M. PAWELZIK, C. WIEMER, K.H. BÜSCHER, and W. OPFERKUCH

Medizinische Mikrobiologie und Immunologie, Ruhr-Universität Bochum, D-4630 Bochum, F.R.G.

Introduction

Treatment of bacterial infections with antibiotics is only succesful if host defence mechanisms, especially phagocytosis and intracellular killing, are effective. Recent reports have provided evidence that subinhibitory concentrations of β-lactam antibiotics can change bacterial morphology and surface composition as well as the adherence properties of the bacteria to phagocytes (Leying et al., in press, 1985, and Leying, this volume; [1, 5, 8]). The question now arises whether these changes are of any influence on the phagocytosis and killing of the bacteria by these cells. In the present study the influence of β-lactam antibiotics on the attachment and killing of the bacteria by macrophages was investigated. The quantitation of bacteria in this study was done by determining bacterial cell masses.

Material and Methods

Antibiotics

The antibiotics used in this study were cefuroxime (Hoechst), moxalactam (Lilly) and aztreonam (Squibb), all of which induce filament formation, and imipenem (MSD), which induces round cell formation.

Bacteria

E. coli strains D 539 (MS^+/MR^+) and D 509 (MS^+/MR^-) were tested. Fresh Mueller-Hinton broth containing 10 μCi ^{14}C acetate/ml was inoculated with 1/100 of a labelled overnight culture (10 μCi ^{14}C acetate/ml). After 3 hrs growth with or without antibiotics, the bacteria were harvested by centrifugation (2 000 × g 10') and washed once with HEPES buffer. For opsonization the control culture was adjusted to a concentration of 5×10^8 bacteria/ml. The final concentration for the experiments was 2×10^8 bacteria/ml. The cpm of these suspensions were determined and the antibiotic treated cultures were adjusted to the same value as the control.

The Influence of Antibiotics on the Host-Parasite Relationship II
Edited by D. Adam, H. Hahn, W. Opferkuch

Opsonization

For antibody dependent opsonization bacteria were incubated at room temperature for 30 min with 0.5% heat inactivated antiserum. Opsonization was stopped with the three-fold volume of cold HEPES-buffer and the bacteria were sedimented by centrifugation. The control culture was adjusted to a final concentration of 2×10^8 bacteria/ml. The cpm of this suspension were determined again and the antibiotic treated cultures were adjusted to the same value.

Macrophages

The cell line P 388D1 was used. The Macrophages were maintained in culture bottles in DMEM + 10% FCS at 37 °C and 8% CO_2. Preparation of the macrophages for the experiments was done according to the method described by Vosbeck et al. [10]. One ml of fresh medium containing 3×10^5 macrophages was filled in each well of a Costar multiwell dish. After 20 hrs the confluent monolayers were washed once with HEPES-buffer and 0.5 ml of this buffer was added to each well.

Antisera

Antisera were raised by immunization of rabbits with bacteria that had not been treated with antibiotics.

Determination of attachment and killing

A modification of the method described by Roberts and Ford [6] was used. 0.5 ml of ^{14}C radiolabelled bacteria was added to each monolayer to give a final volume of 1 ml. After 15', 30' or 45' at 37 °C, the monolayers were washed four times with HEPES-buffer. To release intracellular bacteria the macrophages were lysed 8 min with 1 ml of distilled water by vigorous pipetting. Viable bacteria in the lysate were labelled by adding 1 ml double concentrated Mueller-Hinton broth containing 20 μCi ^{3}H-thymidine to the lysate. After further incubation for 1 hr at 37 °C the bacterial suspension was transferred into 2 ml ice cold 20% TCA + 0.1 M Na-pyrophosphate. The rest of the suspension was taken up with 0.5 ml of distilled water and 0.5 ml 20% TCA + 0.1 M Na-pyrophosphate was added to each simple to give a final volume of 5 ml. To remove excess ^{3}H-thymidine, 1 ml of the suspension was filtrated through a 0.45 μm nitrocellulose filter (Millipore). To diminish the quench effect the filter was dissolved by adding 10% methanol (v/v) to the scintillation liquid (KL 372, Zinsser). ^{3}H and ^{14}C counts were determined in a liquid scintillation spectrometer (LKB), and corrected to disintegrations per minute (dpm).
The measurement of ^{14}C dpm gives the total amount of the bacterial mass associated with the macrophages. Determination of ^{14}C dpm of killed bacterial mass was done according to the following formulae:

bacteria without macrophages:

$$\frac{^{3}\text{H dpm}}{^{14}\text{C dpm}} = \text{a}$$

bacteria with macrophages:

$$\frac{^{3}\text{H dpm}}{^{14}\text{C dpm}} = \text{b}$$

Viability in cultures without macrophages was determined as 100%. That means:
a = 100% viability
b = x% viability from the control culture without macrophages

$$= \text{x}\% = \frac{\text{b x } 100\%}{\text{a}}$$

The percentage of dead bacteria or the "killing rate" is obtained by simple subtraction:
100% live bacteria − x% live bacteria = % dead bacteria

The number of ^{14}C dpm of the dead bacteria is obtained as follows:

$$\frac{\%\text{ bacteria that were killed}}{100\%} \times {}^{14}\text{C dpm of associated bacteria} = {}^{14}\text{C dpm of killed bacteria}$$

Results

On Fig. 1–7 the influence of different β-lactam antibiotics on the association and killing of two strains of E. coli by macrophages is shown.

Treatment of strain D 539 (MS^+/MR^+) with cefuroxime, moxalactam or aztreonam (Fig. 1–3) causes an increase in adherence of non-opsonized bacteria ($p < 0{,}01$), while the killing of these bacteria is less efficient as compared with the control. No killing of cefuroxime treated bacteria could be observed even after 45' incubation ($p < 0{,}01$ at 45'). With moxalactam (Fig. 2) the first killing of non-opsonized filaments occurred after 30 min of incubation ($p < 0{,}01$ at 30'). With aztreonam, the absolute dpm value of killed filaments was the same as that of the control. For opsonized filaments the killing rate, that is the percentage of adherent bacterial mass that is killed is similar for drug treated and untreated bacteria. The absolute dpm value of killed filaments is lower because less bacterial mass attaches to the macrophages. In contrast to the experiments with antibiotics that induce filament formation, pretreatment of strain D 539 with imipenem (Fig. 4) seems to sensitize the bacteria to killing by macrophages, though the first dead bacteria could be observed after 30 min of incubation ($p < 0{,}2$). After 15' the killing rate of

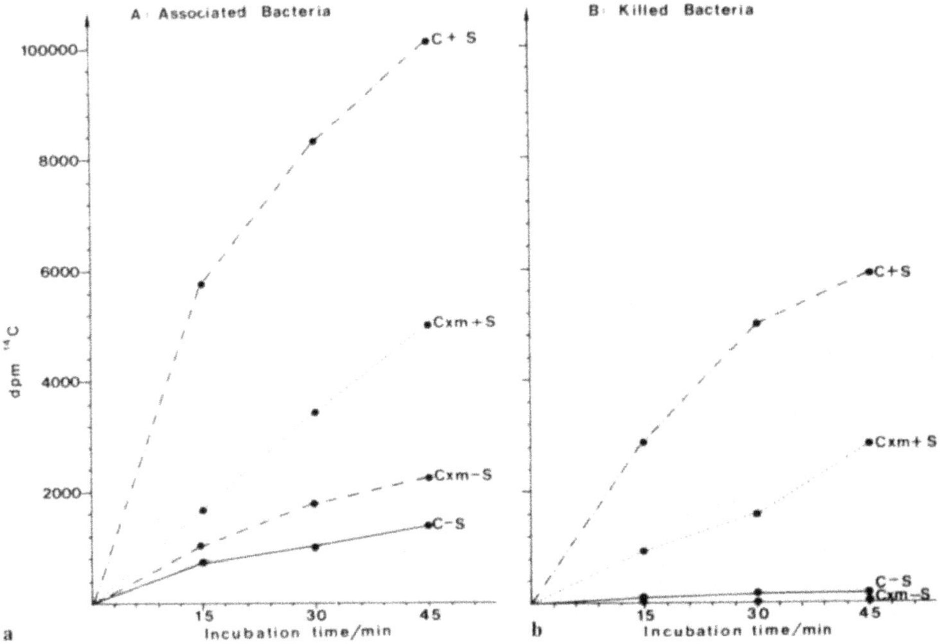

Fig. 1a, b. Influence of ⅛ MIC of cefuroxime on the association (**a**) of strain D 539 to macrophages and on the killing (**b**) of the bacteria by these cells.
Symbols: •—• control without antiserum; •–·–• control, opsonized with heat inactivated antiserum; •——• drugtreated bacteria without antiserum; •····• drug treated bacteria opsonized with heat inactivated antiserum

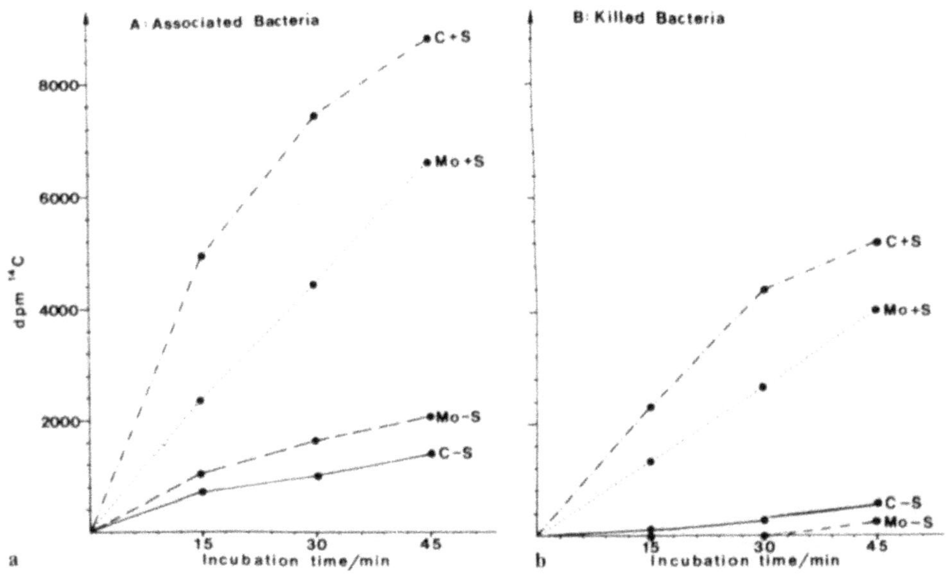

Fig. 2a, b. Influence of 1/16 MIC of moxalactam on the association (**a**) of strain D 539 to macrophages and on the killing (**b**) of the bacteria by these cells. Symbols are as described in Fig. 1

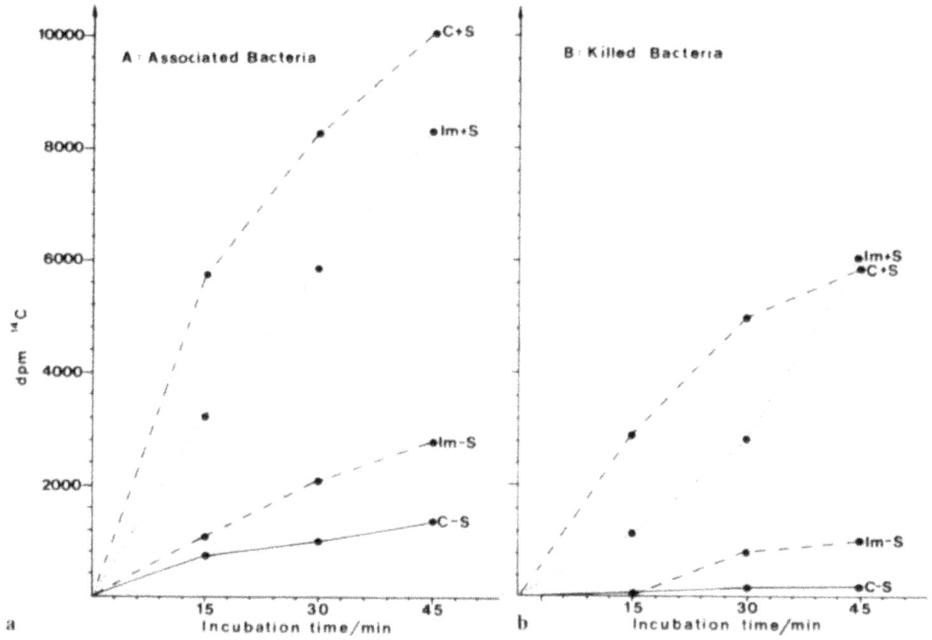

Fig. 3a, b. Influence of ⅛ MIC of aztreonam on the association (**a**) of strain D 539 to macrophages and on the killing (**b**) of the bacteria by these cells. Symbols are as described in Fig. 1

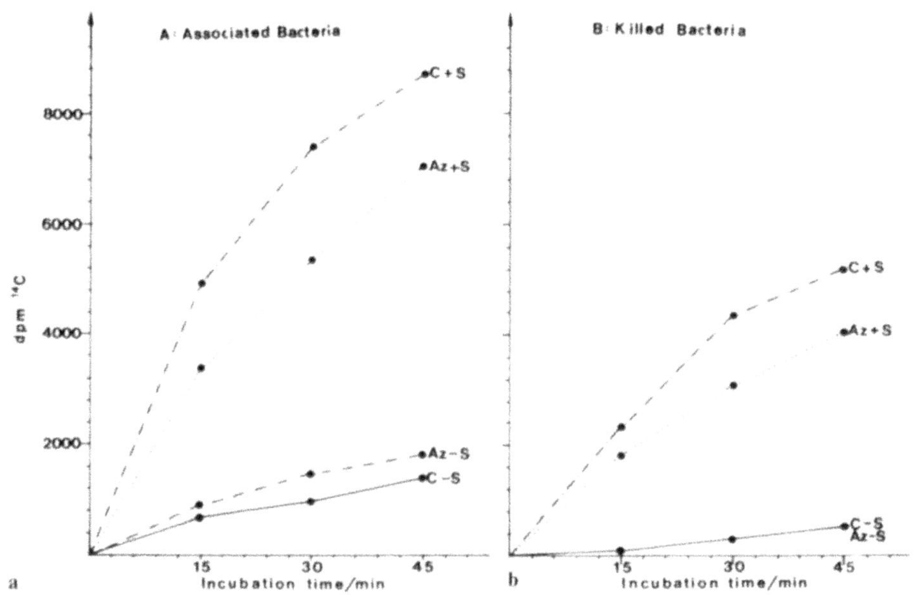

Fig. 4a, b. Influence of ⅛ MIC of imipenem on the association (**a**) of strain D 539 to macrophages and on the killing (**b**) of the bacteria by these cells. Symbols are as describes in Fig. 1

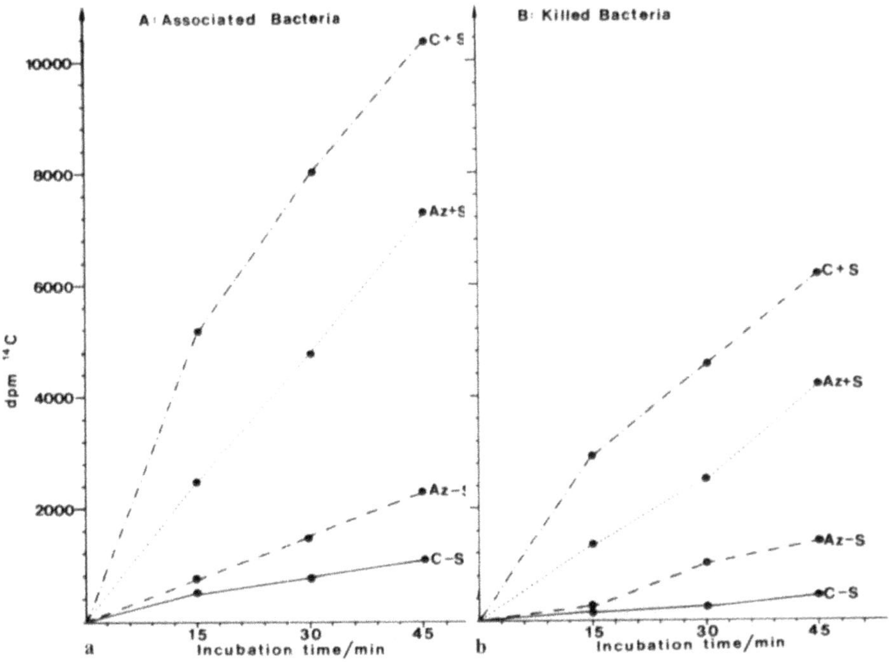

Fig. 5a, b

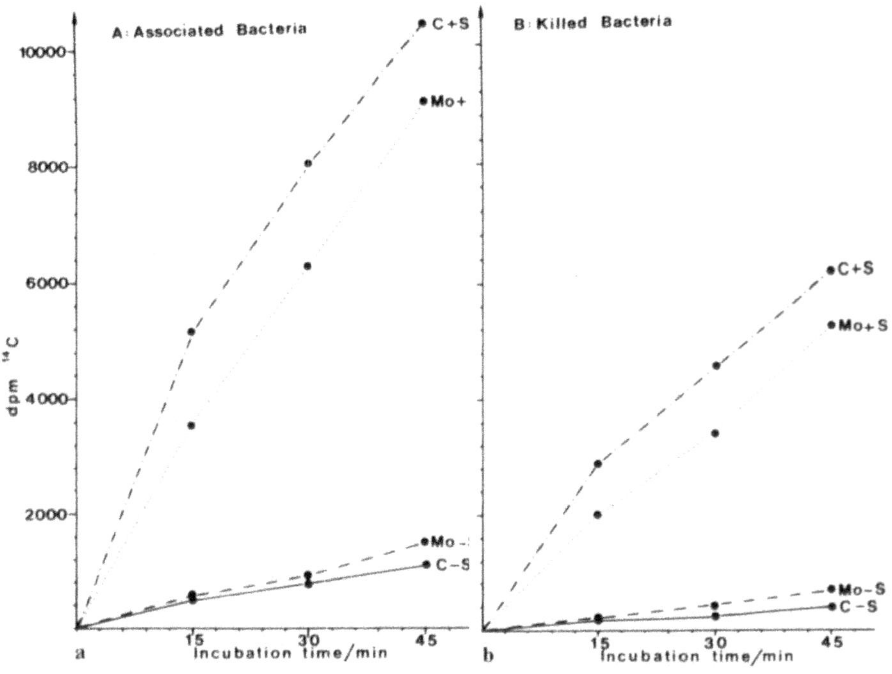

Fig. 6a, b

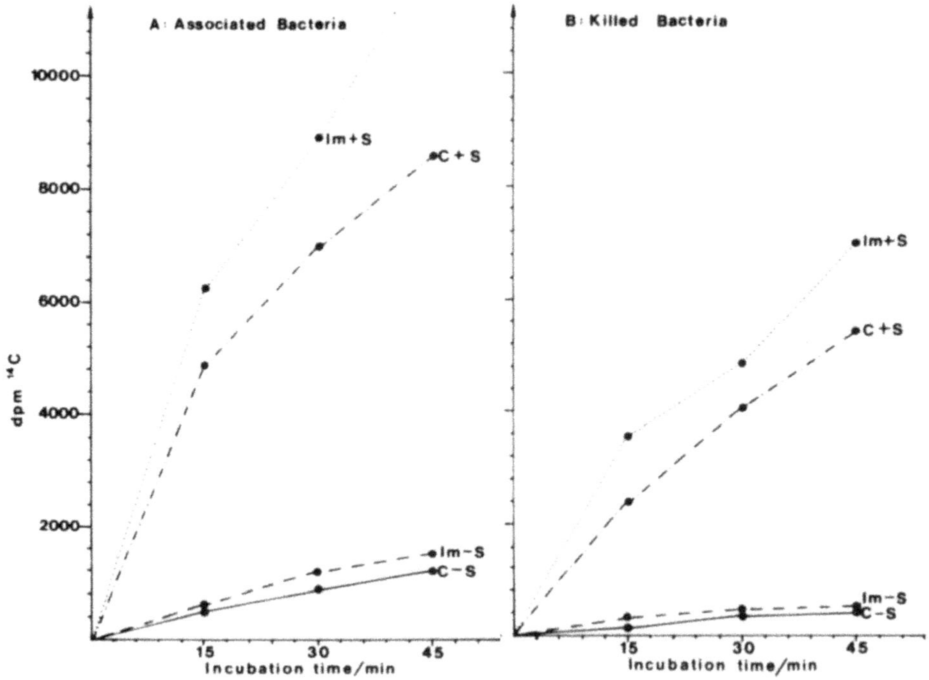

Fig. 7a, b. Influence of ⅛ MIC of imipenem on the association (**a**) of strain D 509 to macrophages and on the killing (**b**) of these bacteria by these cells. Symbols are as described in Fig. 1

opsonized drug treated bacteria is lower than that of the control ($p < 0{,}01$) but during the following incubation time the killing rate increases so that after 45' dead bacteria reach a dpm value similar to that of the control. Strain D 509 expressing MS adhesins behaves only slightly differently. Aztreonam treatment (Fig. 5) renders non-opsonized bacteria more susceptible to the bactericidal mechanisms of the macrophages ($p < 0{,}05$). Treatment with moxalactam and imipenem (Figs. 6 and 7) had no significant effect on the killing of non-opsonized bacteria, though the mean value of killed antibiotic treated bacteria seems to be different from that of untreated ones.

Fig. 5a, b. Influence of ⅛ MIC of aztreonam on the association (**a**) of strain D 509 to macrophages and on the killing (**b**) of the bacteria by these cells. Symbols are as described in Fig. 1

Fig. 6a, b. Influence of 1/12 MIC of moxalactam on the association (**a**) of strain D 509 to macrophages and on the killing (**b**) of the bacteria by these cells. Symbols are as described in Fig. 1

Discussion

A comparison of β-lactam treated bacteria and untreated controls with respect to phagocytosis and killing by phagocytes is problematic because these antibiotics cause morphological alteration of these cells, such as filament or round cell formation.

In general, two methods are used to quantitate and compare the killing of bacteria by phagocytes:

1. Determination of the number of cfu remaining after exposure of the bacteria to phagocytes;
2. determination of the bacterial mass killed by the phagocytes.

When comparison is performed on the basis of cfu, the difference between one filament and one normal bacterium is not taken into consideration. When bacterial suspensions are adjusted to the same cfu at the beginning of the experiment the mass of antibiotic treated bacteria offered to the phagocytes is much greater than that of untreated bacteria, because the actual mass of a filament is much greater than that of a normal bacterium [2]. Since we did not know whether these differences in masses have any influence on the killing capacity of the phagocytes, we decided to compare attachment and killing of antibiotic treated and untreated bacteria when equal masses of bacteria were offered to the phagocytes. Measurement of the mass is possible by

a) determination of the weight of the bacteria;
b) determination of the optical density of a bacterial suspension;
c) labelling the bacteria with radioactive isotopes and measuring the cpm.

The advantage of labelling the bacteria is that attachment and killing can be determined in the same sample. With a dual label assay it is possible to label the total mass of all bacteria with ^{14}C, while viable bacteria that are not killed by the phagocytes are detected by labelling with ^{3}H-thymidine.

Results concerning the adherence of E. coli treated with sub MIC's of β-lactam antibiotics are contradictory. While Vosbeck et al. [10] could not detect any significant influence, other authors found a decrease in adherence after treatment with ampicillin or penicillin [7, 9, 4]. We found that antibiotic pretreatment caused a slight increase of attachment of non-opsonized bacteria. After opsonization with specific antibodies, adherence of the filaments is lower than that of the opsonized control. One possible explanation for this is that antibodies that are raised against antibiotic untreated bacteria might not bind to the filaments very well, since β-lactam antibiotics are able to induce changes in the surface composition of bacteria (Leying, this volume). The killing of non-opsonized filaments of strain D 539 was less efficient compared with the control. With strain D 509, a significant increase of killed bacterial mass could be observed in the case of aztreonam while with moxalactam no significant difference between the killing of antibiotic treated and untreated bacteria could be detected. Similar results were published by Nishida et al. [3]. No effect on the phagocytosis and killing of non opsonized Pseudomonas aeruginosa by rabbit PMN's could be observed when the bacteria were pretreated with β-lactam antibiotics (Sulbenicillin, Carbenicillin). In contrast to the results obtained with non-opsonized bacteria, the percentage of killed opsonized bacteria that had been pretreated with antibiotics was not significantly different from that of untreated bacteria. These data are in good correlation with the results obtained by Lorian and Atkinson [2] who found similar PMN bactericidal activity against filaments and normal bacteria.

Summary

The influence of subinhibitory concentrations of different β-lactam antibiotics (cefuroxime, imipenem, moxalactam, aztreonam) on the attachment of E. coli to mouse macrophages (cell line P388D1) and the killing of the bacteria by these cells was investigated. ^{14}C radiolabelled bacteria were grown in the absence of presence of antibiotics for 3 hrs. Adherent bacteria were determined after 15', 30', or 45' by measuring the ^{14}C dpm associated with the macrophages. Viability of the bacteria was measured by their uptake of ^{3}H-thymidine. Experiments were performed with opsonized and non-opsonized bacteria. In general, non-opsonized antibiotic-induced filaments adhere slightly better than the non-opsonized control. In contrast, adherence of opsonized filaments to macrophages is lower than that of opsonized control. With strain D 539 (MS^+/MR^+) the killing of adherent non-opsonized bacteria treated with cefuroxime, moxalactam or aztreonam was decreased, whereas imipenem seems to sensitize the bacteria to killing by phagocytes. With strain D 509 (MS^+/MR^-) the killing of adherent aztreonam treated bacteria was slightly increased, whereas no significant differences could be observed with moxalactam and imipenem. The percentage of opsonized filaments killed by the phagocytes is the same as that of antibiotic untreated bacteria. The absolute ^{14}C dpm value of killed filaments is lower because less bacterial mass attaches to the macrophages.

Acknowledgement: This work was supported by a grant of the DFG Op 12/13.

References

1. Lorian V (1980) Effects of subminimum inhibitory concentrations of antibiotics on bacteria. In: Lorian V (ed) Antibiotics in laboratory medicine. Williams and Wilkins, Baltimore, p 342–408
2. Lorian V, Atkinson BA (1984) Bactericidal effect of polymorphonuclear neutrophils on antibiotic induced filaments of gram negative bacilli. J Infect Dis 149: 719–727
3. Nishida M, Mine Y, Nonoyama S, Yokota Y (1976) Effect of antibiotics on the phagocytosis and killing of Pseudomonas aeruginosa by rabbit polymorphonuclear leukocytes. Chemotherapy 22: 203–210
4. Ofek I, Beachey EH, Eisenstein BJ, Alkan MC, Sharon N (1979) Suppression of bacterial adherence by subminimal inhibitory concentrations of β-lactam and aminoglycoside antibiotics. Rev Infect Dis 1: 832–837
5. Opferkuch W, Büscher KH, Karch H, Leying H, Pawelzik M, Schumann U, Wiemer C (1985) The effect of sublethal concentrations of antibiotics on the host-parasite relationship. Zbl Bakt Suppl 13: 165–177
6. Roberts PJ, Ford JM (1982) A new combined assay of phagocytosis and intracellular killing of Escherichia coli by polymorphonuclear leukocytes. J Immunol Methods 49: 193–207

7. Sandberg T, Stenqvist K, Svanborg-Eden C (1979) Effects of subminimal inhibitory concentrations of ampicillin, chloramphenicol and nitrofurantoin on the attachment of Escherichia coli to human uroepithelial cells in vitro. Rev Infect Dis 1: 838–844
8. Spratt BG (1975) Distinct penicillin binding proteins involved in the division, elongation, and shape of Escherichia coli K 12. Proc Nat Acad Sci USA 72: 2999–3003
9. Svanborg-Eden C, Sandberg T, Stenqvist K, Ahlstedt S (1979) Effects of subinhibitory amounts of ampicillin, amoxycillin and mecillinam on the adhesion of E. coli bacteria to human urinary tract epithelial cells: a preliminary study. Infection 7 Suppl 5: 452–455
10. Vosbeck K, Handschin H, Menge EB, Zak O (1979) Effects of subminimal inhibitory concentrations of antibiotics on adhesiveness of E. coli in vitro. Rev Infect Dis 1: 845–851

Discussion

Vosbeck: When doing association studies using radiolabelled bacteria, it is important to control for unspecific binding to plastic surfaces. Such binding to the plastic surface might, e. g., explain your observation that bacteria bind to a significant extent in the absence of serum, but are not killed. Did you do such plastic binding controls?

Pawelzik: No.

Vosbeck: During the rather long incubation period with ^{3}H-thymidine changes in incorporation due to differences in growth rates, postantibiotic effects, etc. might manifest themselves and lead to erroneous conclusions. I would therefore recommend to use ^{3}H-uracil of high specific activity to measure bacterial viability, which yield much higher TCA-insoluble counts and thus permits pulses of just 1 to 5 minutes.

Questions from *Adam*, *Daschner* and *Mandell* were concerned with the statistical significance of the presented effects. Statistical evaluations were performed and have been added to the manuscript.

Antibiotics and Polyelectrolytes Modulate Bacteriolysis and the Capacity of Bacteria to Trigger an Oxygen Burst in Neutrophils*

ISAAC GINSBURG[1], RUTH BORINSKI[1], MILU SADOVNIK[1], SARA SHAULI[1], J. WECKE[2], P. GIESBRECHT[2], and MEIR LAHAV[1]

[1] The Department of Oral Biology, The Hebrew University-Hadassah Faculty of Dental Medicine, Jerusalem, Israel
[2] The Robert Koch Institute, Berlin, FRG

Introduction

The invasions of tissues by pathogenic microorganisms is followed by a sequence of events which culminate in phagocytosis and the intracellular killing of the ingested agents, by "professional" phagocytes [19]. It is also expected that the rich arsenal of hydrolases present in neutrophils and macrophages, including the muralytic enzyme lysozyme is adequate to degrade the complex architecture of the bacterial cells. Surprisingly, however, most pathogenic bacteria are extremely resistant to lysozyme action [14,21] and the fate of phagocytosed bacteria *in vivo* is not fully known [7, 8, 16, 23]. The sequelae of the lack of bacterial degradation by leukocytes may be the "storage" of peptidoglycan-polysaccharide or peptidoglycan-lipopolysaccharide complexes within macrophages leading to the generation of granulomas and to the initiation of prolonged immune responses. This is pivotal to the initiation of immunopathological reactions [7, 8, 16, 23]. We have recently proposed [10, 11, 12, 13, 15, 29] that the biodegradation of certain microorganisms can be mediated through the activation, by cationic agents and phospholipases, of the bacterial own autolytic wall enzymes (suicidal phenomenon) which leads to the breakdown of the rigid cell walls. On the other hand, a variety of sulfonated anionic polyelectrolytes [11–13, 15] likely to be present in inflamed issues, may inhibit the biodegradation of the walls by the autolytic enzymes. One of the ways by which antibiotics, of the beta lactam series, could be involved in the killing and biodegradation of bacteria by leukocytes is through their capacity to change the cell walls via the release of lipoteichoic acid and the reduction of peptidoglycan cross linking [17, 18, 24, 25].

It has been recently demonstrated, however, that the same anionic polyelectrolytes which inhibited autolysis in staphylococci [11, 12] also markedly interfered the penicillin-induced lytic phenomenon [28]. Thus, a delicate balance between activators and inhibitors of the autolytic wall enzymes may determine whether microbial cell wall components may persist in tissues and serve as phlogistic agents [see 11, 13]. Since antibiotics may also modify the bacterial surfaces and also

* This study was supported in part by a research grant obtained from Dr. S.M. Robbins, of Cleveland, Ohio, from a grant from the Israeli Academy of Sciences and Humanitis, and from a research grant from Deutsche Forschungsgemeinschaft

The Influence of Antibiotics on the
Host-Parasite Relationship II
Edited by D. Adam, H. Hahn, W. Opferkuch

interfere with the biosynthesis of certain virulence factors [1,3] it is conceivable that such modified cell surfaces may also result in the induction of an altered "oxygen burst" in phagocytic cells. Such a burst usually takes place during membrane perturbation by opsonized bacteria during phagocytosis [19,22]. The purpose of the present communication is a): to further describe the role played by beta lactam antibiotics in the biodegradation of staphylococci and micrococci by leukocytes and some of their factors *in vitro* and *in vivo*, and b) to describe the altered chemiluminescence patterns which occur when antibiotic-grown streptococci are employed to stimulate human neutrophils [9].

Material and Methods

Bacterial strains

Staphylococcus aureus (strain Peltzer), *Micrococcus lysodeikticus* and group A streptococci type 3 (M-protein negative) were cultivated in brain heart infusion broth (Difco) either in the absence or presence of added penicillin G and ^{14}C-N-acetylglucosamine 0.2–0.5 μC/ml (specific activity 30–50 mci/mmol) New England Nuclear Boston, Mass., USA. Both the labeled and unlabeled cells were harvested from different phases of growth. The cells were washed several times in saline and resuspended either in O.lM acetate buffer pH 5.0, saline buffered with 0.01M phosphate pH 7.4, 0.1M phosphate buffer pH 6.5 or in Hanks balanced salt solution (HBSS) containing 1 mM HEPES buffer pH 7.33. The bacterial suspension were adjusted to differenct cell densities (see below).

Lysis of staphylococci

^{14}C-NAGA staphylococci, which had been cultivated either in regular brain heart infusion broth or in broth containing sub-inhibitory concentrations of penicillin G (0.01–0.02 μg/ml) were washed in saline and suspended in 0.1M acetate buffer pH 5.0 to a cell density of 0.5 OD at 550 nm. The labeled cells were exposed for 2–4 hours at 37 °C to

1. buffer alone,
2. crystalline egg with lysozyme (50 μg/ml),
3. leukocyte cationic proteins (LCP) (50 μg/ml). Following incubation the reaction mixtures were centrifuged at 2 000 g for 15 minutes, and the solubilized radioactivity was determined in the supernatant fluid as described [11].

Lysis of Micrococcus lysodeikticus *(ML) by lysozyme*

ML was inoculated into brain heart infusion broth in the absence or presence of penicillin G (0.03–0.5 μg/ml). The cultures were harvested after 16 hrs of incubation at 37 °C. The cells were washed in saline and resuspended in 0.1M phosphate buffer pH 6.5 and adjusted to an O.D. of 0.7 at 540 nm. Egg white lysozyme (5 μg/ml) was added to all tubes, and the decrease in absorbance at 540 nm was monitored at 3 minute intervals at 37 °C. Results were expressed as percent decrease in turbidity at each point of measurement.

Determination of luminol-dependent chemiluminescence (LDCL)

Leukocytes were isolated from heparinized human blood after sedimentation with dextran as described in detail [9]. In some experiments we have further separated pure neutrophils on a Ficoll-Hypaque. The leukocytes were suspended in HBSS containing 1 mM HEPES buffer pH 7.33 and 10% fetal calf serum. Luminol dependent chemiluminescence (LDCL) was measured in an LKB Wallac 1250 Luminometer as described [9]. Group A streptococci which had been cultivated in BHI broth in the absence or presence of penicillin G (increasing concentrations) were opsonized either with 1) nuclear histone (Type II-A Sigma Chemical Co.) 1 mg/ml, 2) dextran sulfate (M.W. 500,000) 1 mg/ml. The streptococci were washed with saline to remove unbound ligands and employed to stimulate LDCL leukocytes [9]. Results were expressed as peak LDCL in millivolts after 5 min of incubation.

Induction of arthritis

Both regular staphylococci and bacteria which had been cultivated in the presence of different concentrations of penicillin G (0.1–1 μg/ml) were harvested from the logarithmic phase of growth, washed in saline and 50 μl of suspensions containing approximately 2.0 OD/ml were injected into the knee joints of male rats (Hebrew University strain). The animals were sacrificed after various time intervals. The synovial fluids and membranes were removed, fixed in glutaraldehyde and processed for transmission electronmicroscopy as described in detail [8, 12, 13, 29].

Results

Effect of penicillin G on lysis of staphylococci

Figure 1 shows that the cell walls of penicillin-grown bacteria are degraded to a larger extent than bacteria grown in regular medium following treatment either by lysozyme or by LCP. Similar results were obtained when the bacteria were treated with crystalline ribonuclease (also a cationic protein (not shown)). All these agents had been previously shown to lyse staphylococci, presumably by activating autolysis [11].

These findings are in accord with previous studies describing the higher susceptibility to lysozyme action of beta lactam-grown staphylococci [2, 4, 10, 26]. The Figure also shows that penicillin-grown staphylococci autolysed in buffer to a larger extent than control cells; however, this effect proved to be dependent on the growth medium. If we use peptone water instead of brain heart infusion broth, penicillin was even capable of inhibiting the activity of autolyticy wall enzymes [31].

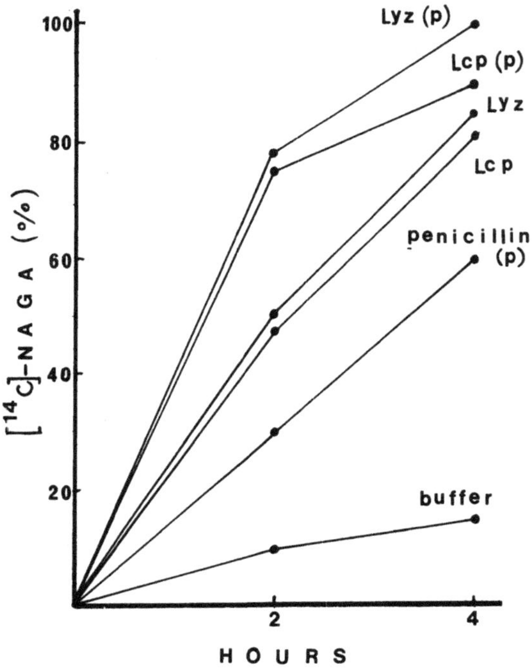

Fig. 1. *Degradation of penicillin-grown staphylococci by various agents.* ^{14}C-NAGA-staphylococci, which had been cultivated in the absence or presence of penicillin G (P) 0.01–0.02 µg/ml, were treated at 37°C with buffer alone, leukocyte cationic proteins (LCP) (50 µg/ml) or with egg white lysozyme (LYZ; 50 µg/ml). Note that bacteria pregrown in brain heart infusion broth containing penicillin were lysed to a greater extent than controls

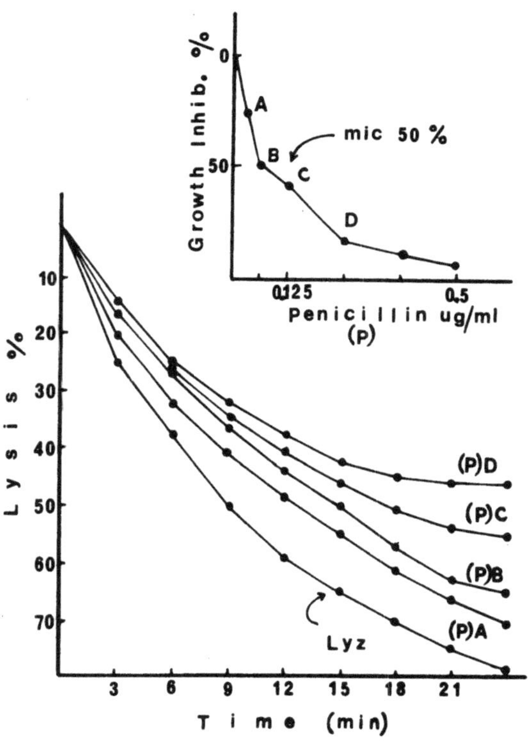

Fig. 2. *Effect of penicillin G on lysis of M. lysodeikticus (ML) by lysozyme (LYZ).* ML was cultivated in the absence or presence of increasing concentrations of penicillin G (P) (Upper curve) and expressed as percent inhibition of growth by turbidity. The concentrations of penicillin employed were: A = 0.03 µg/ml, B = 0.06 µg/ml, C = 0.125 µg/ml, D = 0.25 µg/ml. The cells were then treated with lysozyme (5 µg/ml) and the percent decrease in turbidity (percent lysis) was monitored for 23 min. Note that as the concentration of P used to grow the cells increased, the susceptibility to lysis by LYZ decreased

Lysis of penicillin-grown M. lysodeikticus *by lysozyme*

M.Lysodeikticus (ML) is the known most susceptible microorganism to lysis by lysozyme [21]. It was, therefore, interesting to know, whether or not this microorganisms will demonstrate altered susceptibility to lysis when grown in the presence of subinhibitory concentrations of penicillin. Fig. 2 shows that as the concentration of penicillin employed in the growth medium increased, the susceptibility of the bacteria to lysis by lysozyme decreased. Microscopic examination, however, revealed that the ML which grew in the presence of increasing concentrations of penicillin appeared swollen and in large clumps [20, 30].

In other studies we have found that ML, which had been cultivated in the presence of polyanethole sulfonate, and which also grew in large clumps also became highly refractory to lysozyme action as well. Similarly, ML, which had been treated *in vitro* with a variety of polycationic agents (histone, polyarginine, polyhistidine) which strongly agglutinated the cells, also became highly resistant to lysis by lysozyme [11, 12].

Degradation of penicillin-grown staphylococci in vivo

We have demonstrated that staphylococci which had been injected into the knee joints of rats, underwent a very significant cytoplasmic degradation (plasmolysis) within phagocytic cells, however, surprisingly their cell walls appeared more or less intact [12]. We postulated that this might have been caused, in part, by the accumulation in the inflammatory site of large amounts of anionic polyelectrolytes, which interfered with autolysis [11]. We have, therefore, examined the fate of penicillin-grown staphylococci in inflammatory sites. Sections taken from knee joints of rats, injected 7 days previously with penicillin-grown staphylococci, revealed the presence of more or less intact cell walls within macrophages, which however, appeared to be somewhat thinner than cell walls, from control cells (Fig. 3). It is suggested, therefore, that penicillin treatment failed to induce a complete biodegradation of staphylococcal cell walls within phagocytic cells.

Chemiluminescence induced by antibiotic-grown bacteria

Since beta lactam antibiotics are known to affect cell walls in bacteria, we examined the possibility that the cell surfaces of such antibiotic-grown cells might have also been altered. This phenomenon could perhaps be demonstrated by chemiluminescence techniques. Fig. 4 indeed shows that group A streptococci or staphylococci (not shown) which had been cultivated in the presence of increasing concentrations of penicillin G, induced much less chemiluminescence generation by leukocytes than controls, when stimulated by the ligands dextran sulfate or histone [9]. This phenomenon occurred despite the fact that we have employed bacterial suspensions of the same optical densities and protein content. It appears that penicillin-grown cells bind less polyelectrolytes than control cells, suggesting an alteration in their surface properties.

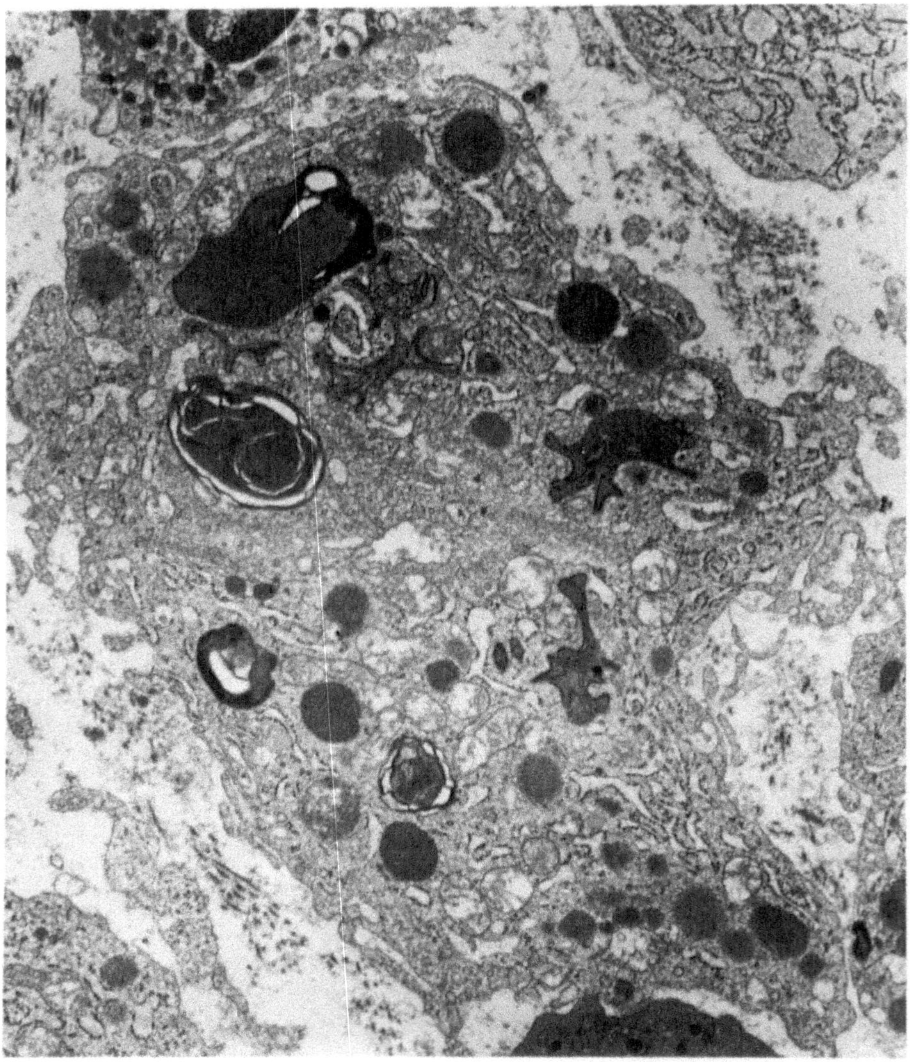

Fig. 3. Electron microscopy of a macrophage in the synovial tissue of a rat 5 days following injection of staphylococci, which had been cultivated 4 hours in the presence of sub-MIG concentrations of penicillin G (0.01 μg/ml). Note the survival of incompletely degraded cell walls within large macrophage which only appeared somewhat thinner than cell walls from control bacteria. Magnification bar = 1 μm

Discussion

The data presented (Fig. 1) show that staphylococci, which had been cultivated in brain heart infusion broth in the presence of subinhibitory concentrations of penicillin G, became much more susceptible to cell wall degradation *in vitro* follow-

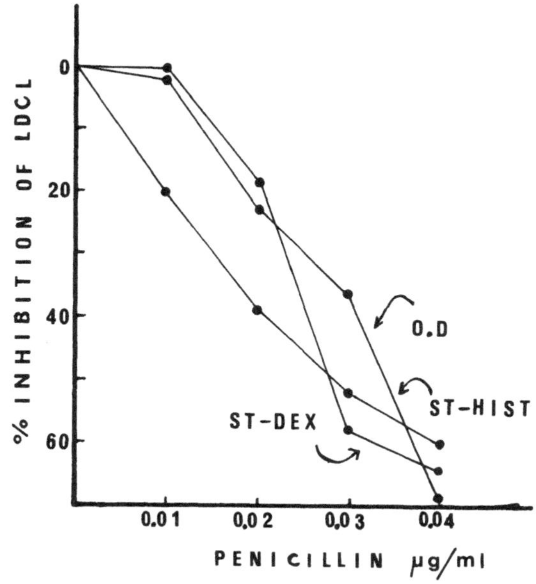

Fig. 4. *Luminol-dependent chemiluminescence (LDCL) induced by penicillin-grown streptococci.* Human blood neutrophils (3 × 10^6) were stimulated with streptococci (ST) which had been cultivated in the presence of increasing concentrations of penicillin G. The streptococci were "opsonized" either with histone (HIST) or with dextran sulfate (DEX). Note parallelism between decrease in the bacterial growth (O.D.) and the decrease in the LDCL responses

ing treatment with agents, previously shown by us to act as activators of the autolytic wall enzymes in the microbial cells [11–13, 28, 29]. However, the different results with different growth media remain to be considered, indicating the possibility that also substances of the growth medium may be involved in these effects.

Furthermore, the reduced degree of peptidoglycan cross linking, which is present in penicillin-grown cells [21] and also the possibility that penicilling treatment removed regulators of the autolytic wall enzymes [17, 25] both should facilitate the degradation of the bacterial cell walls. "Weaker" cell walls of bacteria should also be more susceptible to elimination by the lysosomal apparatus of phagocytic cells *in vivo*. However, penicillin-pregrown staphylococci, which had been injected intraarticularly persisted as scarcely degraded cell walls, within macrophages (Fig. 3). It is essentially similar to the findings on the persistence of penicillin treated streptococci in liver granulomas [16]. Although the reason for this phenomenon is still not fully known, it may be due to interference in the activity of endogenous autolysins, rather than to the acquisition of resistance of the PPG cell wall to degradation by lysozyme and by other hydrolases present in leukocytes. This assumption is also based on our findings that a variety of anionic polyelectrolytes markedly inhibited autolysis in staphylococci *in vitro* [11, 12]. Furthermore, staphylococci which had been cultivated in the presence of such polyanions also became highly resistant to degradation by the endogenous autolysins.

It is thus tempting to speculate that one reason for the resistance of staphylococcal cell wall to the degrading capacity of the lysosomes might be the accumulation of anionic agents in the inflammatory site, resulting from their pinocytosis by macrophages which, in turn, affects either the action of the bacterial own autolytic wall enzymes, or the degradability of the bacterial cell wall [12, 13]. Such anionic agents failed, however, to inhibit the degradation of cytoplasmic and membrane structures, which have been shown to undergo fast degradation within the phagolysosomes (Fig. 3) [7, 8]. The other reason may be due to the fact that proteinases being capable of destroying autolytic wall enzymes of penicillin pregrown cells staphylococci [6] could also in phagocytes hamper bacterial autolysins.

The findings, however, that *M. lysodeikticus,* which is the known most susceptible microorganism to lysozyme became highly resistant to lysis by this enzyme if pregrown with subinhibitory concentrations of penicillin is intriguing. Further scrutiny of this phenomenon revealed, however, that a large proportion of the micrococci, which escaped lysis by lysozyme, appeared in large clumps. Such clumps are probably caused by the induction, by penicillin, of pseudo multicellular forms due to the interference of this antibiotic with the separation of the daughter cells [20, 28]. The large clumps interfered with the penetration of lysozyme deep into the aggregates, thus the cells were spared. It is of interest that a similar formation of large pseudomulticellular staphylococci also took place, when staphylococci were cultivated in the presence of liquoid, presumably because this anionic agent inhibited the action of certain autolytic wall enzymes, responsible for cell separation [12, 28]. However, resistance to wall disintegration following growth in the presence of lytic concentrations or of large amounts of penicillin was also demonstrated [5, 27, 31]. This inhibiting effect of penicillin on the bacterial wall release may explain the resistance of penicillin-grown bacteria to degradation as well.

Another way in which antibiotics may alter host and parasite interrelationships is either through their direct action of leukocyte functions (chemotaxis, phagocytosis, intracellular killing, etc.) [3, 30] or through changes that antibiotics induce in the outer surfaces of bacteria. Although beta lactam antibiotics, at therapeutic concentrations, have not been found to alter leukocyte functions to an appreciable extent [30]; these agents may nevertheless modify the bacterial surfaces. This may alter the binding of ligands, which are associated with opsonization. Our findings (Fig. 4) indeed show that streptococci, which had been cultivated in the presence of subinhibitory concentrations of penicillin G or cephalosporins (not shown) induced much less LDCL in human blood leukocytes, as compared with controls. This was mainly observed when histone or dextran sulfate were employed to "opsonize" the bacteria [9]. Less conspicuous was the reduction of LDCL by antibody-opsonized bacteria, but still significantly lower than the LDCL, which had been induced by regular streptococci. Preliminary observations have also indicated that chloramphenicol-grown bacteria were also capable of binding much less polyelectrolytes than control cells.

Although the significance of our findings and their possible relation to the bactericidal effects of oxygen radicals, which are generated following stimulation of leukocytes by ligands, is still not known, one may postulate that penicillin grown bacteria may generate the production of lesser amounts of oxygen radicals, some of which have been shown to be deleterious to cells and tissues [19, 22]. On the other hand, a diminished production of oxygen radicals may hamper the bactericidal activity of leukocytes. Whatever the consequence of penicillin effect on the oxygen metabolism of leukocytes be, the employment of CL techniques may be important to establish subtle changes that take place in the bacterial surface properties under the effects of antibiotics. Penicillin-grown bacteria are known to contain much less lipoteichoic acid [1, 18] which alter their adherence to ephithelial surfaces [1]. Such altered virulence may be an important factor in the host and parasite interrelationships.

Studies on the biological properties of solubilized peptidoglycans derived from antibiotic-grown cells, will be presented elsewhere.

Summary

The biodegradation of bacteria by leukocytes in infectious and inflammatory foci many involve direct enzymatic cleavage by lysosomal hydrolases as well as the activation by leukocyte and tissue-derived cationic proteins and phospholipases of the bacterial own autolytic wall enzymes. Antibiotics of the lactame series may facilitate autolysis by removing regulators of the autolytic wall enzymes (i.e. acid phospholipids) thus leading to degradation of the bacterial cell walls from within. However, a variety of sulfated polyelectrolytes (dextran sulfate, heparin, polyanethole sulfonate) markedly inhibited penicillin G lysis of Staph. aureus presumably due to their capacity to inhibit autolytic wall enzymes. Staphylococci grown in the presence of subinhibitory concentrations of penicillin G became much more susceptible to lysis by leucocyte extracts and by lysozyme as compared to controls. This process was also markedly inhibited by anionic polyelectrolytes, which presumably interfered with the activity of the bacterial autolysines. Both intact staphylococci and penicillin-grown cells which had been injected intraarticularly into the knee joints of rats underwent massive plasmolysis within macrophages, but no apparent degradation of their cell walls was evident by electron microscopy, suggesting that the accumulation of inflammatory exudates, rich in anionic polyelectrolytes, might have been responsible for this effect. Staphylococci and group A streptococci, which had been cultivated in the presence of subinhibitory concentrations of penicillin G or cephalosporins induced a much lesser generation of chemiluminescence and superoxide responses in human PMNs when a variety of cationic and anionic agents were employed as ligands upon such bacteria. The results suggest that the antibiotics modulated bacterial surfaces. The survial of undegraded microbial cell wall components in tissue, even under penicillin treatment, may explain the perpetuation and propagation of certain post infectious sequellae. Bacteria treated with antibiotics may also generate lesser amounts of toxic oxygen radicals, which may contribute to longer survival of viable microorganisms in the tissues of the host.

The phlogistic role played by degradation products of bacteria will also be discussed.

References

1. Alkan ML, Beachey EH (1978) Excretion of lipoteichoic acid by group A streptococci. Influence of penicillin on excretion and loss of ability to adhere to human oral mucosal cells. J Clin Invest 61: 671–677
2. Efrati C, Sacks T, Ne'eman N, Lahav M, Ginsburg I (1976) The effect of leukocyte hydrolases on bacteria. VIII. The combined effect of leukocyte extracts, lysozyme, enzyme "cocktails" and penicillin on the lysis of *Staphylococcus aureus* and group A streptococci *in vitro*. Inflammation 1: 371–407
3. Eickenberg HU, Hahn H, Opferkuch W (1982) The influence of antibiotics on the host-parasite relationship. Springer, Berlin New York

4. Friedman H, Warren GH (1976) Antibody-mediated bacteriolysis: enhanced killing of cyclocillin treated bacteria. Proc Soc Exp Biol. Med. 153: 301–304
5. Giesbrecht P, Morioka H, Krüger D, Kersten T, Wecke J (1983) Restoration of penicillin-damaged cell walls by de novo murein synthesis and successive murein degradation in staphylococci, revealing a hitherto unknown mechanism of penicillin action: blockage of autolytic wall process by penicillin. In: Hakenbeck R, Höltje JV, Labischinski H (eds) The target of penicillin – The murein sacculus of bacterial cell walls. Architecture and growth. De Gruyter, Berlin New York, pp 243–248
6. Giesbrecht P, Wecke J, Reinicke B (1976) On the morphogenesis of the cell wall of staphylococci. Int Rev Cytol 44: 225–318
7. Ginsburg I (1972) Mechanisms of cell and tissue injury induced by group A streptococci: Relation to poststreptococcal sequelae. J Infect Dis 126: 294–340; 319–456
8. Ginsburg I (1979) The role of lysosomal factors of leukocytes in the biodegradation and storage of microbial constituents in infectious granulomas. In: Dingle JT, Jacques PJ, Shaw H (eds) Lysosomes in biology and applied therapeutics, Vol 6. North Holland Publications Amsterdam, pp 327–406
9. Ginsburg I, Borinski R, Lahav M, Gillert KE, Falkenberg S, Winkler M, Müller S (1982) Bacteria and zymosan opsonized with histone dextran sulfate and poly anethole sulfonate trigger intense chemiluminescence in human blood leukocytes and platelets and in mouse macrophages: Modulation by metabolic inhibitors in relation to the leukocyte-bacteria interactions in inflammatory sites. Inflammation 6: 357–377
10. Ginsburg I, Lahav M, Bergner-Rabinowitz S, Ferne M (1982) Effect of antibiotics on the lysis of staphylococci and streptococci by leukocyte factors, on production of cellular and extracellular factors by streptococci and on the solubilization of cell sensitizing agents from Gram negative rods. In: Eichenberg HU, Hahn H, Opferkuch W (eds) Influence of antibiotics on the host-parasite relationship, Springer Verlag Berlin 219–227
11. Ginsburg I, Lahav M, Giesbrecht P (1982) Effect of leukocyte hydrolases on bacteria XVI. Activation by leukocyte factors and cationic substances of autolytic wall enzymes in *Staphylococcus aureus*: Modulation by anionic polyelectrolytes in relation to the survival of bacteria in inflammatory exudates. Inflammation 6: 269–284
12. Ginsburg I, Lahav M (1983) Cationic polyelectrolytes activate autolytic wall enzymes in *Staphylococcus aureus*: Modulation by anionic polyelectrolytes in relation to the survival of bacterial constituents in tissues. In: Hakenbeck R, Holtje JV, Labischinski H (eds) Target of penicillin, Walter de Gruyter & Co Berlin, New York pp 341–346
13. Ginsburg I, Lahav M (1983) Now are bacterial cells degraded by leukocytes *in vivo*? An enigma. Clin Immunol Newsletter 4: 147–153
14. Hakenbeck R, Holtje JV, Labischinski H (eds) (1983) The target of penicillin, The murein sacculus of bacterial cell walls architecture and growth, Walter de Gruyter & Co Berlin, New York
15. Ginsburg I, Sela MN, Ne'eman N, Lahav M (1983) The role of leukocytes, serum factors and of cationic polyelectrolytes in the lysis and biodegradation of *Staphylococcus aureus*: relation to the pathogenesis of staphylococal infections. In: Easmon C (ed) Staphylococci and staphylococcus infections, Academic Press, New York, pp 325–355
16. Heymer B, Spanel R, Haferkamp O (1982) Experimental models of arthritis. In: Current topics in pathology 71: 123–152, Springer Verlag, Berlin
17. Holtje JV, Tomasz A (1975) Lipoteichoic acid: A specific inhibitor of autolysin activity in pneumococci. Proc Mat Acad Sci USA 72: 1690–1694
18. Kessler RE, Van de Rijn I, McCarty M (1981) Release of lipoteichoic acid by group A streptococci. In: Shockman GD, Wicken AJ (eds) Chemistry and biological activities of bacterial surface amphiphiles, Academic Press, New York, pp 239–246
19. Klebanoff S, Clark RA (1978) The neutrophil, function and clinical disorders. Elsvier, New York
20. Lorian V, Atkinson B (1984) Bactericidal effect of polymorphonuclear neutrophils on antibiotic-induced filaments of Gram negative bacilli. J Infect Dis 719–727
21. Rogers HL, Perkins HR, Ward JB (1980) Microbial cell walls and membranes. Chapman and Hall, London p 437

22. Sbarra AJ, Strauss RR (eds) (1980) The reticuloendothelial system Volume 2 Biochemistry and metabolism, Plenum Press, New York
23. Schwab JH, Ohanian SH (1967) Degradation of streptococcal cell wall antigens *in vivo*. J Bact 94: 1346–1352
24. Sela MN, Lahav M, Ginsburg I (1977) Effect of leukocyte hydrolases on bacteria IX. The release of lipoteichoic acid from group A streptococci and from *Strep.mutans* by leukocyte extracts and by lysozyme: Relation to issue damage in inflammatory sites. Inflammation 2: 151–164
25. Tomasz A (1979) The mechanism of irreversible antimicrobial effects of penicillins: How beta lactam antibiotics kill and lyse bacteria. Ann Rev Microbiol 33: 113–137
26. Warren GH, Gray J (1967) Influence of nafcillin on the enzymatic lysis of *Staphylococcus aureus*. Canad J Microbiol 13: 321–328
27. Wecke J, Giesbrecht P (1981) Electron microscopic studies on the "paradoxical" reaction of staphylococci during treatment with antibiotics. In: Jeljaszewicz J (ed) Staphylococci and staphylococcal infections, Zbl Bakt Suppl 10, Gustav Fischer Verlag, Stuttgart New York pp 461–467
28. Wecke J, Lahav M, Ginsburg I, Kwa E, Giesbrecht P (1983) Cell wall degradation of antibiotic-treated staphylococci under phagocyte-specific conditions: Distinction between penicillin-induced lytic effects and wall alterations caused by anticoagulants. In: Hakenbeck R, Holtje J, Labischinski H (eds) Target of penicillin, Walter de Gruyter Berlin New York, pp 329–334
29. Wecke J, Lahav M, Ginsburg I, Giesbrecht P (1982) Cell wall degradation of *Staphylococcus aureus* by lysozyme. Arch Microbiol 131: 116–123
30. Yourtree EL, Root RK (1982) Antibiotic-neutrophil interactions in microbial killing. In: Gallin JI, Fauci AS (eds) Advances in host defence mechanisms, Vol I, Raven Press, New York, pp 187–209
31. Reinicke B, Blümel P, Labischinski H, Giesbrecht P (1985) Neither an enhancement of autolytic wall degradation nor an inhibition of the incorporation of cell wall material are pre-requisites for penicillin-induced bacteriolysis in staphylococci. Arch Microbiol (in press)

Discussion

Martin: It is evident that the lysis of bacterial cell wall mureins, complex lysozyme-resistant structures, in phagocytized bacteria depends on the continued activity of bacterial murein hydrolases ("autolytic" enzymes). If these enzymes are destroyed in the killed bacteria, for instance by the phagocyte's proteases, enzymatic degradation of the murein will no longer be possible. A modulation of murein hydrolase activity (stimulation or inhibition) by effectors such as penicillin or phagocyte's polyanionic compounds or others is, of course, a possibility. But such postulated mechanisms must still be regarded as hypothetical as long as there is no biochemical evidence for a defined mechanism of interaction between the murein hydrolases and the presumed effector molecules.

Ginsburg: Since both penicillin and liquoid induced cell clumping due perhaps to interference with cell reparation and change is the structure of the cell wall, it may be possible to check resistance to lysozyme by breaking up the clumps by mechanical means and then test their susceptibility to lysozyme. Liquoid may also act as a distinct anionic polyelectrolyte. It may cover the bacterial cells and prevent the interaction of the cationic lysozyme with its targets.

Verhoef: Treatment of the staphylococci with your agents changes the intraphagosome pH, could that be the reason why degradation is ineffective? You showed that superoxide was produced, what about chemiluminescence?

Ginsburg: There is a controversy over the question of the intraphagosomal pH. Some claim the pH is rising in the phagosome, but some claim it becomes acid in order to facilitate the activity of the numerous acid hydrolases which have a pH optimum of around 5.0.

Chemiluminescence of Macrophages induced by Non-Opsonized *E.coli* treated with Subinhibitory Concentrations of β-lactam Antibiotics

K.-H. Büscher[1], H. Leying[1], M. Pawelzik[1], V. Klimetzek[2], and W. Opferkuch[1]

[1] Institute of Med. Microbiol. & Immunol., Ruhr-Universität, D-4630 Bochum, FRG
[2] Institute of Immunol. & Oncol., Bayer AG, D-5600 Wuppertal, FRG

Introduction

The colonization of mucosal surfaces by bacteria is dependent on the expression of various bacterial adhesins [2]. Furthermore, these adhesins might be involved in the interaction of non opsonized bacteria and phagocytes, leading to adherence, internalization and activation of phagocytes [3, 4, 5, 10, 12]. During this process macrophages may produce and liberate reactive oxygen species ($^1O_2$, O_2^-, H_2O_2, OH•), which can be measured by luminol-dependent chemiluminescence (LdCl) [1]. The chemiluminescence reaction of phagocytes in response to bacteria is influenced by the surface properties of the stimulating bacteria, e. g. the type and number of adhesins expressed on the bacterial surface [3, 4, 5, 10].

As the composition of the outer membrane of E.coli is modified after growth in presence of subinhibitory concentrations of β-lactam antibiotics (Leying et al., this volume), it was of interest to study the effect of such alterations on the interaction with phagocytes.

In order to exclude any influence of the morphology of the bacteria (filaments, round cells) on the activation of phagocytes, outer membrane vesicles prepared from bacteria grown either in presence or absence of subinhibitory concentrations of antibiotics were used to stimulate the macrophages.

Material and Methods

Bacteria

E.coli Wf96, serogroup 07:Kl:H6, was originally obtained from St. Mary's Hospital, London. E.coli strain D 509, serogroup 086, was obtained from Dundee University, England. Some characteristics of these strains are shown in Table 1. The minimal inhibitory concentration of the antibiotics was determined according to DIN 58940 in Mueller-Hinton-Broth (Merck, Darmstadt, FRG).

The Influence of Antibiotics on the
Host-Parasite Relationship II
Edited by D. Adam, H. Hahn, W. Opferkuch

Table 1. Surface characteristics of the strains of E. coli used

strain	serotype	hemagglutination	fimbriae	capsule
Wf 96	07:K1:H6	MS, MR	++	+
D 509	086	MS	++	–

Reagents

The following antibiotics were used: lamoxactam (Lilly), mezlocillin (Bayer), piperacillin (Cyanamid), cephaloridine (Lilly), imipenem (Merck, Sharp & Dohme) and aztreonam (Squibb).

Preparation of outer membrane vesicles

Outer membrane vesicles were prepared from bacteria grown for three hours in Mueller-Hinton-Broth either containing subinhibitory concentrations of antibiotics or not according to the method described by Kroll et al. [9].

Animals

10–16 weeks old B6D2 male mice were used in this study.

Macrophages

Peritoneal exudate cells were harvested from mice which had been injected i.p. with Propionibacterium acnes (Corynebacterium parvum, 0.7 mg/mouse, Wellcome, Beckenham, UK) 14 days before. Purified macrophages (Mø) were isolated according to [8].

Chemiluminescence assay

Measurement of the luminol-dependent chemiluminescence was done according to [11]. In short: 1×10^6 purified Mø in 0.5 ml Dulbecco's-PBS supplemented with 1% v/v amino acids (Biochrom, Berlin, FRG), 1% v/v non essential amino acids (Biochrom), 0.5 mM l-glutamine (Biochrom), 1 mg/ml bovine serum albumin (Sigma Chemicals, Munich, FRG) and 2.5 mg/ml glucose were incubated for 10 minutes at 37°C in the presence of 5×10^{-5} M luminol before the addition of 60 µg outer membrane material. The resulting chemiluminescence response was registrated continuously by use of a six-channel luminometer (LB 9505, Berthold, Wildbad, FRG).

Table 3. Luminol-dependent chemiluminescence response of P. acnes activated macrophages to different outer membrane vesicles (OM) from E. coli D 509.

stimulus[1]	counts/15 min × 10^6
OM D509	2.0±0.4
OM D509-piperacillin[2]	8.8±1.4
OM D509-lamoxactam[2]	14.3±0.3
OM D509-mezlocillin[2]	64.1±5.4
OM D509-aztreonam[3]	8.4±0.9
OM D509-cephaloridine[3]	1.4±0.4
OM D509-imipenem[3]	3.4±0.6

[1] 60 μg OM/10^6 Mø
[2] 4/12 MIC
[3] 5/12 MIC

Table 3. Luminol-dependent chemiluminescence response of P. acnes activated macrophages to different outer membrane vesicles (OM) from E. coli D 509.

stimulus[1]	counts/15 min × 10^6
OM D509	2.0±0.4
OM D509-piperacillin[2]	8.8±1.4
OM D509-lamoxactam[2]	14.3±0.3
OM D509-mezlocillin[2]	64.1±5.4
OM D509-aztreonam[3]	8.4±0.9
OM D509-cephaloridine[3]	1.4±0.4
OM D509-imipenem[3]	3.4±0.6

[1] 60 μg OM/10^6 Mø
[2] 4/12 MIC
[3] 5/12 MIC

Results

Outer membrane vesicles from E. coli Wf 96 were capable to induce a significant chemiluminescence response of Mø. This chemiluminescence was markedly increased, if bacteria had been pretreated with antibiotics (Table 2). In this case the best stimulation was found, if outer membrane vesicles from bacteria grown in presence of imipenem were used. On the other hand, if bacteria of strain D 509 were used, imipenem caused only a slight increase in regard to control vesicles (Table 3). With this particular strain no stimulatory effect of cephaloridine could be found (Table 3); the highest stimulation was obtained by pretreatment of bacteria with mezlocillin.

The amount of chemiluminescence was dependent on the amount of outer membrane vesicles used for stimulation (Fig. 1). The differences in chemiluminescence induced by outer membrane vesicles from bacteria treated with antibiotics or not could not be overcome by increasing the amount of stimulus, because different plateaus of chemiluminescence were reached with the respective stimuli.

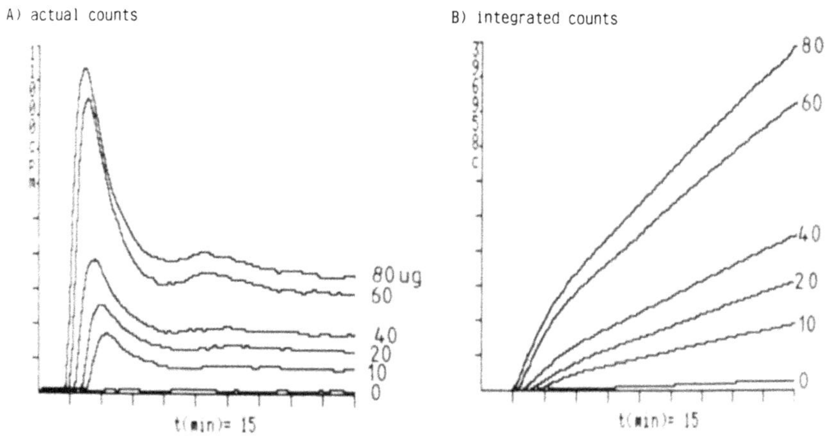

Fig. 1. Kinetics of the luminol-dependent chemiluminescence response of *P. acnes* activated macrophages in response to different amounts of outer membrane vesicles from *E.coli* D 509 treated with ⅓ MIC mezlocillin

The enhancing effect of the antibiotics on the stimulating properties of outer membrane vesicles was dependent on the concentration of antibiotic bacteria had been grown in. In case of E. coli D 509 there was an increase in stimulation of Mø with increasing concentrations of lamoxactam and aztreonam for example, whereas imipenem seems to have a stimulatory effect only at lower concentrations.

As E. coli Wf 96 expresses different types of fimbriae [7] and shows mannose-sensitive as well as mannose-resistant hemagglutination, it was interesting to know, whether treatment of bacteria with different antibiotics has a selective influence on one type of adhesin. In order to answer this question experiments were performed by use of outer membrane vesicles pretreated with 50 mM mannose. The comparison of mannose-treated and untreated vesicles from E.coli Wf 96 showed that about 30% of the chemiluminescence reaction was due to mannose-resistant adhesins. The increased stimulatory effect of outer membranes from antibiotic pretreated cells was mainly due to the participating mannose-sensitive adhesins, whereas mannose-resistant adhesins were not influenced or even slightly depressed in case of lamoxactam. Similar results were obtained with outer membranes from bacteria of strain D 509, which also induces some chemiluminescence that could not be inhibited by mannose.

In order to determine the quality of radical production experiments were performed in presence of the superoxide anion scavenger superoxide dismutase (SOD) or the hydrogen peroxide scavenger catalase. No inhibition of LdCl induced by outer membrane vesicles of Wf 96 by SOD could be found and only a slight (15%) inhibition occurs in case of imipenem pretreated bacteria (Table 4). Catalase reveals an inhibition of LdCl induced by most outer membrane preparations except those from lamoxactam treated bacteria (Table 4).

Table 4. Influence of superoxide dismutase (SOD) and catalase on the luminol-dependent chemiluminescence of 1×10^6 macrophages induced by 60 μg of differnt outer membrane vesicles

stimulus	% inhibition by presence of	
	250 U SOD	1.125 U catalase
OM Wf 96	—	13.9
OM Wf 96-imipenem	15±9	17,0
OM Wf 96-aztreonam	—	29,2
OM Wf 96-lamoxactam	—	—

Discussion

Even non opsonized bacteria can interact with Mø via different adhesins. Mannose-sensitive [4, 10] as well as mannose-resistant [3, 5] adhesins of E. coli have been shown to participate in the biochemical activation of phagocytes. Our studies show that even purified outer membrane vesicles are capable to activate Mø. These outer membrane vesicles allow the study of modulations of the activating surface properties of E. coli by growth in presence of antibiotics with morphological alterations (filament our round-cell formation) excluded.

Growth of bacteria in presence of sub-MIC of various β-lactam antibiotics lead to a significant increase of the LdCl of P. acnes activated Mø (Table 2). Only cephaloridine and imipenem showed slight effects in case of E.coli D 509 (Table 3). The magnitude of the LdCl was dependent on the amount of outer membrane material used for stimulation (Fig. 1) and the concentration of antibiotic bacteria had been grown in. As the differences in LdCl induced by outer membranes from bacteria grown in presence or absence of antibiotics could not be overcome with increasing amounts of stimulus, it seems to be likely that treatment of bacteria with sub-MIC β-lactam antibiotics lead to alterations in the surface properties. As the LdCl is dependent on the density of stimulating epitopes [6], this might be an explanation for the effect of antibiotics.

Inhibition experiments with superoxide dismutase (Table 4) studies on the lucigenin-dependent chemiluminescence as well as measurement of superoxide anion by cytochrome C reduction (data not shown) showed that there was no liberation of superoxide anion by Mø in response to outer membrane vesicles, indicating that there was no stimulation of the O_2^- generating NAD(P)H-oxidase. Even pretreatment of bacteria with antibiotics could not lead to the activation of this enzyme.

Summary

Different bacterial adhesins might be involved in the interaction of bacteria and phagocytes. These adhesins can induce the production and liberation of reactive oxygen species by macrophages. This study shows the effect of treatment of bacteria with sub-MIC of different β-lactam antibiotics on the stimulating properties of outer membrane vesicles of two strains of E. coli. The following results were obtained:

1. Outer membrane vesicles prepared from two strains of E. coli induced a dose-dependent luminol-amplified chemiluminescence response of P. acnes activated macrophages.
2. In most cases treatment of bacteria with antibiotics lead to an augmented chemiluminescence.
3. The stimulatory effect of antibiotic treatment was dependent on the concentration of the antibiotic used.
4. Pretreatment of the outer membrane vesicles with mannose lead to a marked suppression of the resulting chemiluminescence response.
   Treatment with antibiotics lead to an increase of the mannose sensitive part of chemiluminescence only.
5. Inhibition experiments of luminol-dependent chemiluminescence with superoxide dismutase showed that all outer membrane vesicles of bacteria either treated with antibiotics or not were uneffective in the activation of the superoxide generating oxidase of macrophages.

Acknowledgement: This work was supported by a grant of the DFG Op 12/13.

References

1. Allen RC, Loose LD (1976) Phagocytic activity of a luminol-dependent chemiluminescence in rabbit alveolar and peritoneal macrophages. Biochem Biophys Res Commun 69: 245–252
2. Beachey EH (1981) Bacterial adherence: Adhesin-receptor interactions mediating the attachment of bacteria to mucosal surfaces. J Infect Diseases 143: 325–345
3. Björksten B, Wadström T (1982) Interaction of *Escherichia coli* with different fimbriae and polymorphnuclear leukocytes. Infect Immun 38: 298–305
4. Blumenstock E, Jann K (1982) Adhesion of piliated *Escherichia coli* strains to phagocytes: Differences between bacteria with mannose-sensitive pili and those with mannose-resistant pili. Infect Immun 35: 264–269
5. Büscher K-H, Klimetzek V, Opferkuch W (1984) Chemiluminescence studies on the adhesin-receptor interaction of different strains of *E.coli* and mouse macrophages. In: Kricka LJ, Stanley PE, Thorpe GHG, Whitehead TP (eds) Analytical applications of bioluminescence and chemiluminescence. Academic Press, Orlando, pp 327–330
6. Büscher K-H, Opferkuch W, Klimetzek V (1984) Effects of different opsonins on the production of oxygen radicals and chemiluminescence of macrophages. In: Bors W, Saran M, Tait D (eds) Oxygen radicals in chemistry and biology, de Gruyter, Berlin, pp 893–896
7. Karch H, Leying H, Büscher K-H, Kroll H-P, Opferkuch W (1985) Isolation and separation of physicochemically distinct fimbrial types on a single culture of *Escherichia coli* 07:Kl:H6. Infect Immun 47: 549–554
8. Klimetzek V, Remold HG (1980) The murine bone marrow macrophage, a sensitive indicator cell for murine migration inhibitory factor and a new method for their harvest. Cell Immunol 53: 257–266
9. Kroll H-P, Bhakdi S, Taylor PW (1983) Membrane changes induced by exposure of *Escherichia coli* to human serum. Infect Immun 42: 1055–1066
10. Mangan DF, Snyder JS (1979) Mannose-sensitive stimulation of human leukocyte chemiluminescence by *Escherichia coli*. Infect Immun 26: 1014–1019
11. Müller S, Falkenberg S, Fromtling RA, Fromtling AM, Klimetzek V (1981) Signals of chemiluminescence emitted by spleen cells and bone marrow macrophages after stimulation with mitogens and particulate substances. In: deLuca MA, McElroy WD (eds) Bioluminescence and chemiluminescence, Academic Press, New York, pp 721–727
12. Svanborg Eden C, Bjursten L-M, Hull R, Hull S, Magnusson K-E, Moldovano Z, Leffler H (1984) Influence of adhesins on the interaction of *Escherichia coli* with human phagocytes. Infect Immun 44: 672–680

Discussion

Stephens: Strain Wf 96 in addition to MS^+/MR^+ fimbriae has capsular polysaccharides. Are the results you obtained due to a decrease in capsular polysaccharide upon exposure to sub-MIC antibiotics? Possibly you are unmasking additional ligands on your OM preparations that recognize macrophage mannose receptors?

Büscher: Leying reported yesterday that the amount of capsular polysaccharide of E.coli Wf96 is reduced after growth in the presence of sub-MIC of β-lactam antibiotics. Though partial removal of capsular polysaccharides by heating did not enhance the chemiluminescence in the case of Wf 96, it might be that the decrease in K1-content might play a role in the augmentation of the chemiluminescence reaction.

Vosbeck: What were the size and the size-distribution of your outer membrane vesicles and were they right-side out or erected? Could some of the effects you have observed be explained by changes in these properties?

Büscher: Electronmicroscopic studies of the outer membrane vesicles showed that outer membrane vesicles from bacteria grown either in presence or absence of antibiotics have almost the same size. As fimbriae could be observed on the outside, they seemed to be right-side out.

Ginsburg: We do not understand what oxygen radicals are responsible for the CL-reaction. SOD will only block 30–40% of the CL-peak. SOD of Sigma is much inferior to the SOD of Grünenthal. CL-techniques are however very helpful to assess membrane perturbation and attachment of ligands to cell surfaces.

Büscher: I do agree with this.

Beachey: You stated that you looked at the vesicles by electron microscopy and found them to contain fimbriae on the outside. What was the effect of the antibiotics on the content of fimbriae on these vesicles?

Büscher: The amount of fimbriae on the outer membrane vesicles seemed to be not markedly altered, but at the present time we did not quantify the number of fimbriae exactly.

Beachey: What compound or compounds did you use to study the mannose sensitivity of the vesicles-macrophage-interaction? As you know, simple monosaccharides are easily washed after pretreatment of either bacteria or macrophages. You must also be aware that macrophages contain mannose receptors. How did you determine if the effect of your mannose pretreatment was the result of blocking the mannose-binding sites on the E. coli vesicles as opposed to the mannose binding sites on the macrophages? The mannose receptors of macrophages may be recognizing sugar residues of the E. coli LPS.

Büscher: For those experiments, alpha-D-mannose or alpha-methyl-mannoside were used. We could exlude a participation of the mannose receptor of macrophages, because the same concentration of mannose used in this study did not effect the chemiluminescence of zymosan, which acts via the mannose receptor.

Ghione: What kind of SOD did you use?

Büscher: We used superoxide dismutase from bovine liver for the inhibition experiments.

Ghione: Did you test the activity of Fe-dependent bacterial SOD?

Büscher: Our studies using whole bacteria showed that the amount of bacterial superoxide dismutase or catalase is too low to affect chemiluminescence under the conditions used.

Indirect Antibiotic Effects on Granulocyte Migration and Phagocytosis

HÅKAN GNARPE and JUDY BELSHEIM

Department of Clinical Bacteriology, Gävle Central Hospital, S-801, 17 Gävle, Sweden

Introduction

It is well known that patients with low numbers of neutrophil granulocytes or with different mal-functions of neutrophil granulocytes [6] do become infected more easily and that infections tend to be more serious in these patients than in individuals with normal granulocytes [6].

Antimicrobial agents may influence the functioning of the host defence mechanisms. In most cases, these effects have been inhibitory or depressing as described for tetracyclines [8, 10], fusidic acid [5] and rifampin [2]. Some of the cephalosporin antibiotics have been found to have no effect in low concentrations, however, to be inhibitory at high concentrations [11]. Only few reports have described a direct stimulatory effect of antibiotics on granulocyte function [9]. Antibiotics at subinhibitory concentrations have been found to indirectly stimulate phagocytosis of certain bacteria as described by Gemmell in a recent leading article (1984).

In this work, we describe the indirect antibiotic effects on granulocyte migration when using bacterial cultures exposed to subinhibitory concentrations of various beta-lactam antibiotics as chemoattractant. We also report the indirect effects on phagocytosis and the direct effects of the present thienamycin preparation on granulocyte migration.

Materials and Methods

Bacterial strains

A Pseudomonas aeruginosa strain and an Escherichia coli strain, originally isolated from wound cultures were used as the bacteria for production of chemotactic factors. All MIC determinations were done by 1:2 tube dilutions in freshly made trypticase soy broth (TSB). The highest dilution giving no turbidity on visual inspection after 18 hours at 37 °C was considered at the MIC for the respective drug. All bacterial inoculates were 10^5 cfu per ml.

The Influence of Antibiotics on the Host-Parasite Relationship II
Edited by D. Adam, H. Hahn, W. Opferkuch

Production of Bacterial Cell free supernates

E.coli and Ps.aeruginosa were grown in TSB without additions and in TSB with $\frac{1}{10}$ or $\frac{1}{20}$ MIC of different antibiotic preparations. All E.coli cultures were allowed to grow 4 hours in a shaking water bath at 37 °C. Ps. aeruginosa, having a longer generation time, was allowed 16 hours for growth. When the allowed time interval had elapsed, the bacterial suspensions were centrifuged at 3 000 × G for 15 min. to remove the majority of bacteria, then the supernatants were filtered through 0.22 micron sterile millipore filters to obtain sterile, bacteria-free supernatants. All supernatants were then frozen at −20 °C in aliquots for future use. Sterility of all preparations was checked by culture.

Antibiotics used

Antibiotics tested were: benzyl penicillin (Astra, Södertälje), ampicillin trihydrate (Glaxo Laboratories, Greenford), mecillinam hydrochloride (Leo Pharmaceuticals, Denmark), cefuroxime (Glaxo Laboratories, Greenford), sodium cefoxitin (Merck, Sharp & Dohme, Rahway, N.J.), cefotaxime (Hoesch AG, Frankfurt am Main), ceftazidime (Glaxo Laboratories, Greenford), cefotriaxone (F. Hoffmann La Roche Co. Ltd., Basle), thienamycin, Batch no. L−642, 957 −01D 07, and Batch no. L-638 596 −01D 419 (new) (Merck, Sharp & Dohme, Rahway, N.J.), piperacillin (Lederle, Gosport, England), Batch FP 65 −1; gentamicin (AB Ferrosan, Malmö), amikacin (Bristol Laboratories, Syracuse, N.Y.), doxycycline (Pfizer Europe, Belgium), and lymecycline (Chimica Industriale, Milan).

Leukocyte suspensions

Whole blood, drawn in heparinized vacutainer tubes from normal healthy volunteers, was sedimented by gravity after the addition of 3 ml dextran (Macrodex, Pharmacia, Uppsala). The resulting suspension was concentrated and washed three times with Krebs Ringers Phosphate Medium before a final adjustment to 2.0×10^7 cells per ml with 3% bovine serum albumin for chemotactic assays, or 2.0×10^5 PMNL per ml in 1% human serum albumin for phagocytosis assays.

Chemotaxis Assay

Chemotactic migration was assayed using the LMAT as described earlier (Belsheim, 1981). Chemotactic factors used were in each case TSB only, E.coli or Ps. aeruginosa grown in TSB or the culture supernatants when bacteria were grown in the presence of $\frac{1}{10}$ or $\frac{1}{20}$ of an MIC of each of the antibiotics.

Phagocytosis Assay

Phagocytosis was measured using the method described earlier [9] according to Hed [12]. Saccharomyces cervisiae yeast cells were used as phagocytic particle at a final concentration of 20:1 per PMNL. Yeast cells were first opsonized 30 min. at 37 °C in 25% serum, then diluted to the appropriate concentration in TSB or the cell free culture supernatants obtained from growth of E.coli in TSB, with and without subinhibitory concentrations of ceftazidime, piperacillin and thienamycin. Phagocytosis was stopped by immersion in an ice water bath after a 15 min period for phagocytosis and then read microscopically.

Results

Using a cell free culture filtrate of *Escherichia coli* incubated with 0.1 MIC of ten betalactam antibiotics, mecillinam, six different cephalosporins and thienamycin did cause a significant increase in production of chemotactic factor, reflected by an increase in the granulocyte migration (Table 1). The values at 0.05 MIC for piperacillin and thienamycin were not different from 0.1 MIC values, and are not given in the Table.

Using a strain of *Pseudomonas aeruginosa* for generation of chemotactic factor, the five investigated cephalosporin derivatives, but not mecillinam, caused an indirect increased granulocyte migration (Table 2).

A cell-free filtrate from an E. coli culture reduced the number of granulocyte-associated as well as the percentage of internalized yeast cells to less than half and less than a quarter of bacterial-factor free TSB, respectively. A cell free filtrate from the E. coli culture exposed to 0.1 MIC of piperacillin and of thienamycin did not cause any alteration of the number of internalized yeast cells as compared to antibiotic-free supernatant. However, at a concentration of 0.05 of MIC, there was a

Table 1. Influence of subinhibitory concentrations of different betalactam antibiotics on the release of chemotactic factor(s) from an E. coli culture, exposed for 0.1 MIC of the respective antibiotic. Values are expressed as % of migration stimulated by antibiotic-free bacterial culture filtrate. Values below 92 and above 108 indicate significant deviation

Antibiotic preparation	stimulated migration in % of drug-free culture filtrate
penicillin G	107
ampicillin	106
mecillinam	125
piperacillin	131
cefuroxime	120
cefoxitin	116
cefotriaxone	122
cefotaxime	118
ceftazidime	120
thienamycin	123

Table 2. Influence of subinhibitory concentrations of different betalactam antibiotics on the release of chemotactic factor(s) from a Ps. aeruginosa culture, exposed for 0.1 MIC of the respective antibiotic. Values are expressed as in Table 1; the same border values are valid

Antibiotic preparation	stimulated migration in % of drug-free culture filtrate
penicillin G	95
ampicillin	95
mecillinam	105
cefuroxime	109
cefoxitin	138
cefotriaxone	120
cefotaxime	115
ceftazidime	115

significant reduction in phagocytosis using piperacillin and a significant increase using thienamycin (Table 3). Ceftazidime, at a concentration of 0.1 of MIC, caused a slight increase in the total number of cell-associated yeast particles and a significant increase in the number of internalized yeast cells.

None of the above antibiotics had any detectable direct effects on the granulocyte migration at the low concentrations mentioned above.

When tested in the same way, neither lymecycline, doxycycline, gentamicin nor amikacin stimulated an increased release of chemotactic factors at subinhibitory concentration levels.

Table 3. Quantitative phagocytosis of opsonized yeast cells expressed as % of total amount of granulocyte – associated yeast cells in antibiotic-free, sterile TSB (trypticase soy broth). *Internalized* yeast cells are completely phagocytosed; PMNL-associated yeast cells are internalized + to PMNL attached yeast cells

Polymorphonuclear granulocytes exposed to	% internalized yeast cells	% PMNL – associated yeast cells
TSB (control)	58	100
TSB + cell-free filtrate from E. coli culture	12	41
TSB + filtrate from E. coli, exposed for 0.1 MIC of piperacillin	12	41
TSB + filtrate from E. coli, exposed for 0.1 MIC of thienamycin	12	41
TSB + filtrate from E. coli, exposed for 0.1 MIC of ceftazidime	37	52
TSB + filtrate from E. coli, exposed for 0.05 MIC of piperacillin	3	22
TSB + filtrate from E. coli, exposed for 0.05 MIC of thienamycin	23	63

When granulocytes were exposed for high concentrations of the new, stabilized thienamycin, no stimulation of the directed migration was found.

Discussion

The direct effects of most antibiotics on various parts of the host defence mechanisms are well known as studied *in vitro*. Less is known about the effects of low concentrations of antibiotics on the production or release of bacterial factors into the environment and the consequences of this for leukocyte function.

Antibiotics inhibiting the protein synthesis, for example the tetracyclines, have been shown to depress granulocyte functions at higher drug concentrations [4, 8]. The same has been shown for rifampicin, fusidic acid, high concentrations of cefamandole and cefoxitin as well as therapeutic concentrations of cephalothin [17]. Most often, these observations have been interpreted as giving negative side effects for the infected patients. However, there is no documentation to substantiate the effects of the observed *in vitro* stimulation or depression of granulocyte function for the clinical outcome of infection.

Betalactam antibiotics are poor penetrators of mammalian cell membranes [16] and are not accumulated intracellularly. This may explain the lack of reports on possible alterations of the unspecific host defences by low concentrations of betalactam antibiotics. At high concentrations, Grant et al. [11] found ampicillin depressing but cefaclor stimulating the PMNL function. Our earlier findings of a direct stimulation of the granulocyte migration by thienamycin were most likely due to instable impurities in the early preparation of the drug as we have not been able to reproduce these findings with the present preparation.

The effects of betalactam antibiotics on bacteria depend on the interaction between the penicillin binding proteins of a particular organism and the individual antibiotic. The consequences may be of relevance for the *in vivo* situation. Actively growing, gram-negative bacteria do continually release proteins, lipids and lipopolysaccharides [18]. These substances are stimulators of PMNL function due to the activation of the alternative path of complement activation and their effects on granulocyte membrane receptors. When the normal growth of bacteria is altered by subinhibitory concentrations of betalactam antibiotics, an increased leakage of bacterial cell contents seem to increase the quantitative amount of PMNL activation. The effects of subinhibitory concentrations of betalactams has been shown to differ depending on the strain of bacteria used. As shown in this report, Ps.aeruginosa and E.coli are affected differently by different antibiotics. This effect is probably due to the individual PBP composition of the gram-negative bacterial cells. As is also shown by our results, even smaller variations of the concentration level may change the net result.

Extrapolating the data, one may suppose that the effects of subinhibitory concentrations of betalactam antibiotics on a localized bacterial infection may result in an increased mobilization of PMNLs to the site. However, it is well known that growing bacteria elicit substances toxic for PMN [14, 1]. The *in vivo* situation may therefore not reflect the *in vitro* situation [15, 13]. There is no data to show that an increased inflammatory reaction, caused by an increased granulocyte migration, is beneficial for the patient. In fact, if the excess stimulation is of a large order, it may amplify the inflammatory response resulting in increased tissue damage. This may in part explain the general good effects of antibiotics, known to depress granulocyte functions in clinical infections when the antibiotic kills the invading bacteria and at the same time may depress the inflammatory response.

Summary

Most betalactam antibiotics, in contrast to for example the tetracyclines, have no direct effects on the various stages of the granulocyte function. When bacteria were exposed to subinhibitory concentrations of some betalactam antibiotics, the production of chemotactic factor(s) was altered, resulting in a significant increase of the granulocyte directed migration. Subinhibitory concentrations of ceftazidime, piperacillin and of thienamycin caused different, dose-dependant responses of the quantitative phagocytosis with opsonized yeast cells.

Possible clinical implications of these findings were discussed.

References

1. Anderson OK, Volden G, Revhaug A, Lundgren TI, Giercksky K-E (1984) Lysosomal hydrolases in Neutrophils, Mononuclear Cells and Serum during Escherichia coli Endotoxemia in Pigs. Scand J Infect Dis 16: 349
2. Bellashéne A, Forsgren AJ (1980) Effect of rifampin on the immune response in mice. Infect Imm 27: 15
3. Belsheim J (1981) A modified leukocyte migration under agarose technique (LMAT). Acta path mibrobiol scand Sect C 89: 167
4. Forsgren A, Schmeling D (1977) Effect of antibiotics on chemotaxis of human leukocytes. Antimicrob Agents Chemother 11: 580
5. Forsgren A, Bacnk G, Beckman H, Bellashéne A (1980) Antibiotic-host defence interactions in vitro and in vivo. Scand J Infect Dis 24: 195
6. Gallin JI (1981) Anormal phagocyte chemotaxis: pato-physiology, clinical manifestations and management of patients. Rev Infect Dis 3: 1196
7. Gemmell CG (1984) Potentiation of phagocytosis of pathogenic bacteria by exposure to low concentrations of antibiotics (Leading article) J Antimicrob Chemother 13: 407
8. Gnarpe H, Belsheim J (1981) Direct and Indirect effects of antibiotics on granulocyte activity. J Antimicrob Chemother 8, Suppl. C. 71
9. Gnarpe H, Belsheim J, Blomqvist C, Lundbäck A (1984) Stimulation of Granulocyte Functions in Vitro by Imipenem and the Renal Enzyme Inhibitor MK 0791. Antimicrob Agents Chemother 25: 179
10. Gnarpe H, Belsheim J (1985) Interaction of Antibiotics with Granulocytes and Lymphocytes. Zbl Bakt Suppl 13, 123
11. Grant M, Raeburn JA, Sutherland R, Harkness RA, Gormley JP, Kowolik MJ (1983) Effect of two antibiotics on human granulocyte activities. J Antimicrob Chemother 11: 543
12. Hed J (1977) The extinction of fluorescence by crystal violet and its use to differentiate between attached and ingested organisms in phagocytosis. FEMS Microbiol. Lett. 1: 357
13. Miller TE (1984) Host Defence Mechanisms, Antibiotics and Immunosuppressive Agents – A complex Relationship. (Editorial) Eur J Clin Microbiol 3: 285
14. O'Brien J, Friedlander A, Dreier T, Ezzell J, Leppla S (1985) Effects of Anthrax Toxin Components on Human Neutrophils. Infect Imm 47: 306

15. Oleske JM (1984) Effects of antimicrobials on host defence mechanisms (Leading article) J Antimicrob Chemother 13: 413
16. Prokesh RC, Hand WL (1982) Antibiotic entry into human polymorphonuclear leukocytes. Antimicrob Agents Chemother 21: 373
17. Welch WD, Davis D, Thrupp LD (1981) Effect of antimicrobial agents on human polymorphonuclear leukocyte microbicidal function. Antimicrob Agents Chemother 20: 15
18. Yoshida T, Cohen S, Bigazzi PE, Kuraswij T, Amsden A (1975) Inflammatory mediators in culture filtrates of Escherichia coli. Amer. J. Path. 81: 389

Discussion

Ghione: Years ago we carried out tests on the influence of antibiotics on PMN functions, sometimes with erratic results. A preeminent source of errors was due to the presence, in some antibiotic samples, of contaminants, chiefly protein of bacterial origin. Highly purified samples of antibiotics displayed a limited influence on PMN function. Will you comment on this point?

Gnarpe: All the antibiotic preparations used in this investigation were obtained directly from the manufacturer and were of a specific batch. We have made no attempts to check for possible impurities in these.

Gemmell: In your preliminary experiments you interacted the PMN with culture filtrates of drug-exposed E. coli in a chemotaxis assay lasting for 3 hours. Can you tell me whether a similar contact time was made between the PMN and culture filtrate before measurement of phagocytosis of yeast cells?

Gnarpe: The method we used for determination of the quantitative phagocytosis is based simply on adding the yeast cells to the granulocyte suspension. At the same time, we added the bacterial culture filtrate – with or without antibiotic pretreatment – to the PMN/yeast cell mixture. As follows from this, there was no preincubation.

Interactions between human Polymorphonuclear Neutrophils and Ampicillin treated *Escherichia coli*

LIONEL A. MANDELL

Division of Infectious Disease, McMaster University, Henderson General Hospital, 711 Concession Street, Hamilton, Ontario, Canada. L8V 1C3

Introduction

Recovery from acute bacterial infection depends upon the response of host defenses, particularly humoral components such as antibody and complement and phagocytic cells such as the polymorphonuclear neutrophil (PMN). In cases of more serious infection, antimicrobial drugs may be required as well.

A good deal has been learned about the mechanisms involved in opsonization, phagocytosis and killing of invading pathogens [1, 2, 3, 4] and about the mechanisms of action of antimicrobial drugs [5, 6]. Much less, however, is known about how antimicrobial drugs might affect the interactions between the PMNs and bacteria.

We are specifically interested in the effects of sub-inhibitory concentrations (sub-MICs) of antibiotics in view of the fact that sub-MICs of beta lactam antibiotics and gentamicin have been reported to prolong survival of animals infected with *Escherichia coli* or *Proteus mirabilis* beyond that of control animals [7].

This study was undertaken to evaluate the effects of sub-MIC ampicillin on the uptake and killing of a known strain of *E. coli* by human PMNs.

Materials and Methods

Microorganisms and Radioactive Labelling

The method of radioactive labelling and the techniques used for measurement of phagocytosis and killing were those of Verbrugh and Verhoef [8, 9]. A serum resistant strain of *E. coli* was used as the test organism. It was maintained on nutrient agar (Difco, Detroit, USA) at 4 °C. When required, several colonies were transferred to Mueller-Hinton broth (Difco, Detroit, USA) and after 5 hours incubation, 0.5 ml were subcultured into 5 ml of Mueller-Hinton broth containing 0.02 mCi thymidine-methyl-3H (specific activity 5 Ci/mmol., Amersham, Bucks, U.K.). The organisms were grown for 18 hours at 37 °C. The culture was then washed 3 times with phosphate buffered saline (PBS) (GIBCO, Grand Island, N.Y.), pH 7.4 and adjusted to a final concentration of 2.5×10^8 cfu/ml using a standard turbidimetric technique [10]. This was confirmed by pour plate colony counts. The minimal inhibitory concentration (MIC) of the organism for ampicillin was 6.0 μg/ml and was determined using a dilution test procedure [11].

The Influence of Antibiotics on the
Host-Parasite Relationship II
Edited by D. Adam, H. Hahn, W. Opferkuch

Serum

Normal human serum was obtained from the clotted blood of 7 healthy volunteers on no medication and was pooled and stored at −70 °C until required.

Leukocyte Preparation

30 ml of blood from healthy volunteers on no medication were drawn into heparinized 50 ml plastic syringes (10 units heparin/ml blood). 15 ml of 6% dextran T70 (Pharmacia Fine Chemicals AB, Uppsala, Sweden) were added and the red blood cells were allowed to settle for 1 hour.

The neutrophil rich supernatant was separated and centrifuged at 160 g for 5 minutes. The resulting cell pellet was washed 3 times in 0.87% ammonium chloride to lyse red cells. The neutrophils were resuspended in Hank's balanced salt solution (HBSS) containing 0.1% gelatin and total and differential cell counts were obtained. Following this, neutrophils were adjusted to a final concentration of 25×10^6 cells/ml.

Pretreatment of E. coli *with Ampicillin*

A stock solution of ampicillin (Ampicillin Trihydrate, 845 μg active substance per mg, Bristol Canada) was prepared using 0.1 M phosphate buffer, pH 6.0 as a diluent and stored at −70 °C.

The bacterial suspension was divided into 0.2 ml aliquots in plastic tubes (12×75 mm Falcon, Oxnard, California, USA) to which were added either 0.2 ml sterile PBS (control group) or 0.2 ml of a predetermined ampicillin concentration (final concentration ½, ⅛, 1/32 MIC). The antibiotic-bacterial mixtures were incubated at 37 °C for 75 minutes. At the end of this period, the bacteria were removed by centrifugation at 1600 g for 15 minutes. The antibiotic supernatant was then discarded leaving behind the bacterial pellet.

Opsonization

Sera were thawed and diluted to 10% in HBSS. To each of the bacterial pellets prepared as above 0.8 ml PBS (control) or 0.8 ml serum were added and incubated at 37 °C for 30 minutes. Following this, the tubes were centrifuged at 1600 g for 15 minutes and the supernatants discarded. When required, serum was inactivated by heating at 56 °C for 30 minutes.

Phagocytosis and Killing Assays

PMN-Associated Radioactivity

To determine PMN associated radioactivity, 0.2 ml of the PMN suspension were added to the pellets of opsonized bacteria in plastic tubes. The bacteria: PMN ratio was 10:1 for all experiments. The capped tubes were rotated at 150 rpm at a 30 degree angle at 37 °C. At each of three predetermined times (3, 10 and 20 minutes), appropriate tubes were removed, placed in an ice bath and 2.5 ml of ice cold PBS were added to each tube. The tubes were then centrifuged at 160 g for 5 minutes at 4 °C and the PMN pellets were washed twice with ice cold PBS to remove non-PMN associated bacteria. The PMNs were then solubilized into 6.0 ml scintillation liquid [toluene containing PPO, POPOP (Amersham) and 20% Triton X-100 (Amersham)] and counted in a liquid scintillation counter (Beckman LS3133T).

Total Radioactivity (PMN associated plus non-PMN associated bacteria)

0.2 ml of the PMN suspension were added to an opsonized bacterial pellet followed immediately by 2.5 ml ice cold PBS. The tubes were centrifuged at 2000 g for 15 minutes and the pellet was resuspended in 6.0 ml scintillation liquid and counts taken.

Colony Forming Units at Measured Times

To determine the colony forming units (CFU) which represented the viable PMN associated bacterial population after 3, 10 and 20 minutes incubation at 37 °C, 2.5 ml ice cold PBS were added to the designated tubes. These tubes were centrifuged at 160 g for 5 minutes at 4 °C and the PMN pellets washed twice with ice cold PBS. The PMNs were lysed by vigorous mixing in 2.0 ml sterile distilled water. Samples from the tubes containing lysed cells were serially diluted in PBS, and 0.2 ml samples then pour plated in nutrient agar. Colonies were counted after 48 hours incubation at 37 °C.

Total Colony Forming Units

The initial number of CFU added to the vials were determined by a turbidimetric method [10] and confirmed by pour plate counts.

The following formulas were used in calculating the results. The percent of the total bacterial population that was leukocyte associated at a given sampling time (% uptake) was calculated using formula 1:

1. % uptake at measure time =

$$\frac{\text{counts per minute in leukocyte pellet at measured time}}{\text{counts per minute in total bacterial pellet}} \times 100$$

2. % cfu in leukocyte pellet at measured time =

$$\frac{\text{cfu in leukocyte pellet at measured time}}{\text{cfu at time zero}} \times 100$$

The leukocyte-associated bacterial population that was viable at 3, 10 and 20 minutes incubation was calculated according to formulas 3:

3. % viable leukocyte-associated bacteria at measured time =

$$\frac{\text{\% cfu in leukocyte pellet at measured time}}{\text{\% uptake at measured time}} \times 100$$

where the numerator and denominator were derived from formulas 2 and 1 respectively.

At any measured time % killed = 100% – % viable.

Electron Microscopy

Bacteria pretreated with varying antibiotic concentrations were prefixed for one hour at 4 °C in equal amounts of culture medium and 2% glutaraldehyde in 0.2 M sodium cacodylate buffer, pH 7.3. Cells were then washed in 0.2 M sodium cacodylate and postfixed in 1% osmium tetroxide in 0.1 M sodium cacodylate. The cells were dehydrated using a graded series of alcohols then embedded in Spurr's resin. Ultrathin sections were stained and coated using uranium and lead respectively. The cells were examined using a Phillips 301 transmission electron microscope.

Statistical Analysis

Comparisons of means at specific times were done using two way analysis of variance. The Dunnett t test was applied to localize differences. Where appropriate, comparisons of multiple means for an overall effect were done using 3 way analysis of variance for repeated measures. In such cases, square root transformation of the data was carried out. Direct comparisons of two means were done using paired t tests. P values of less than 0.05 were considered statistically significant.

Results

Effects of Pretreatment with Sub-MIC Ampicillin on Bacterial Morphology

Incubation of bacteria with ½ MIC ampicillin for 75 minutes resulted in morphologic appearances consistent with previously reported results [12, 13]. At ½ MIC, there was obvious elongation of the organisms compared to control cells without evidence of septation (Fig. 1). At ⅛ MIC (Fig. 2) the cells were still longer than control cells with some constrictions evident indicating initiation of division. At 1/32 MIC, there was obvious septum formation and cell size was approximately equal to control cell size.

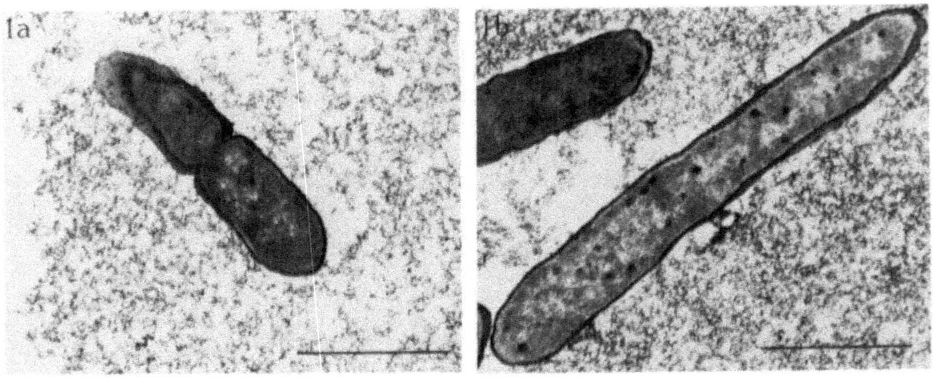

Fig. 1a, b. Transmission electron micrographs. **a**) is an untreated E. coli. **b**) is an E. coli exposed to ½ MIC ampicillin. Horizontal bar = micron

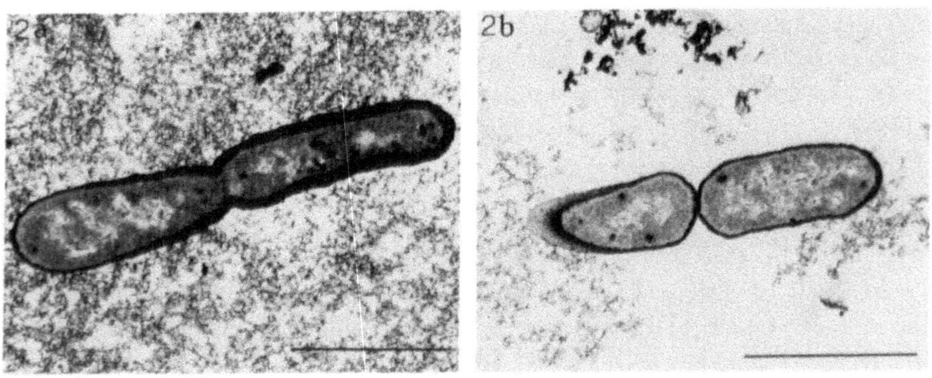

Fig. 2a, b. Are E. coli exposed to ⅛ and 1/32 MIC ampicillin respectively

Effects of Pretreatment with Sub-MIC Ampicillin on Phatocytosis of E. coli *by PMNs*

Mean results were calculated daily and an overall mean for each drug concentration group and control group at each time was determined. The overall means for 8 experiments with their standard errors are shown in Table 1. Fig. 3 shows a graph of the % uptake (phagocytosis) versus time for the control and drug treated cells. Analysis at each time showed a statistically significant difference at 3 minutes across the four groups ($P < 0.002$). This difference was lost at 10 ($P = 0.07$) and 20 minutes ($P = 0.25$). A separate analysis at the 3 minute point comparing uptake of the ½ MIC group with the control group also showed a statistically significant result ($P < 0.01$).

Figure 4 shows results for the log ampicillin concentration plotted against % uptake at 3 minutes. A linear relationship was found supporting a dose response relationship. Statistical analysis using the technique of linear contrast also provided strong evidence in favour of a dose response relationship ($P < 0.025$).

Table 1. *Effects of sub-MIC Ampicillin on Phagocytosis of E. coli*

	% uptake				
Time (minutes)	½ MIC	⅛ MIC	1/32 MIC	Control	P
3	58.75 (9.30)	47.62 (10.13)	36.88 (6.51)	34.50 (5.50)	.002
10	63.5 (8.52)	47.75 (4.68)	56.63 (9.87)	55.12 (9.01)	.07
20	61.37 (7.82)	62.25 (7.96)	65.25 (7.63)	62.12 (7.46)	.25

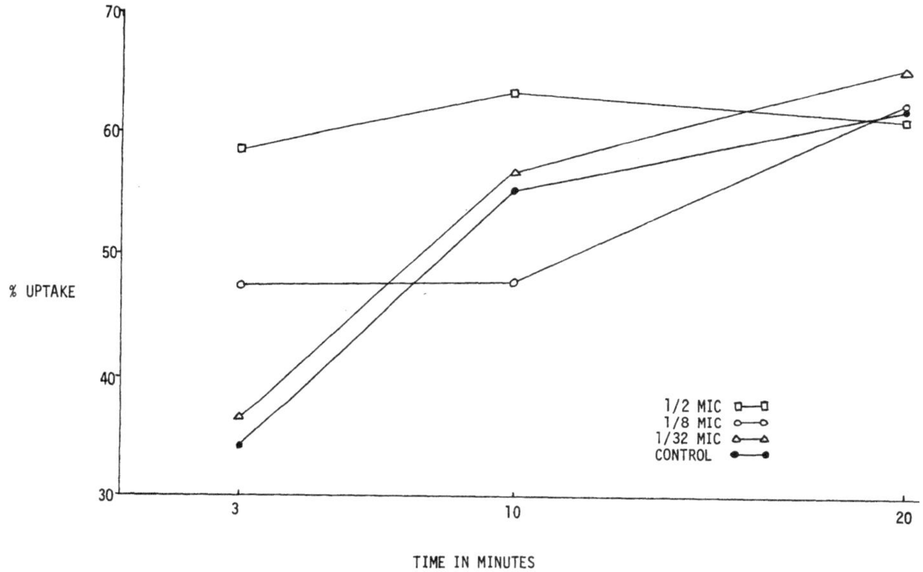

Fig. 3. Effects of sub-MIC ampicillin on phagocytosis of E. coli

Effects of Pretreatment with Sub-MIC Ampicillin on Killing of E. coli by PMNs

The overall means and their standard errors for killing of *E. coli* are shown in Table 2. Fig. 5 shows a graph of the % *E. coli* killed versus time for the control and drug treated cells. No significant differences were noted at 3 minutes ($P = 0.07$) or 10 minutes ($P = 0.19$). At 20 minutes, a clearly significant difference was seen ($P < 0.01$). Further analysis at 20 minutes comparing killing of ½ MIC treated organisms with the control cells was also significant ($P < 0.01$).

Combining and analyzing the data for concentration across all time points shows a statistically significant result ($P = 0.04$) in favour of a concentration effect.

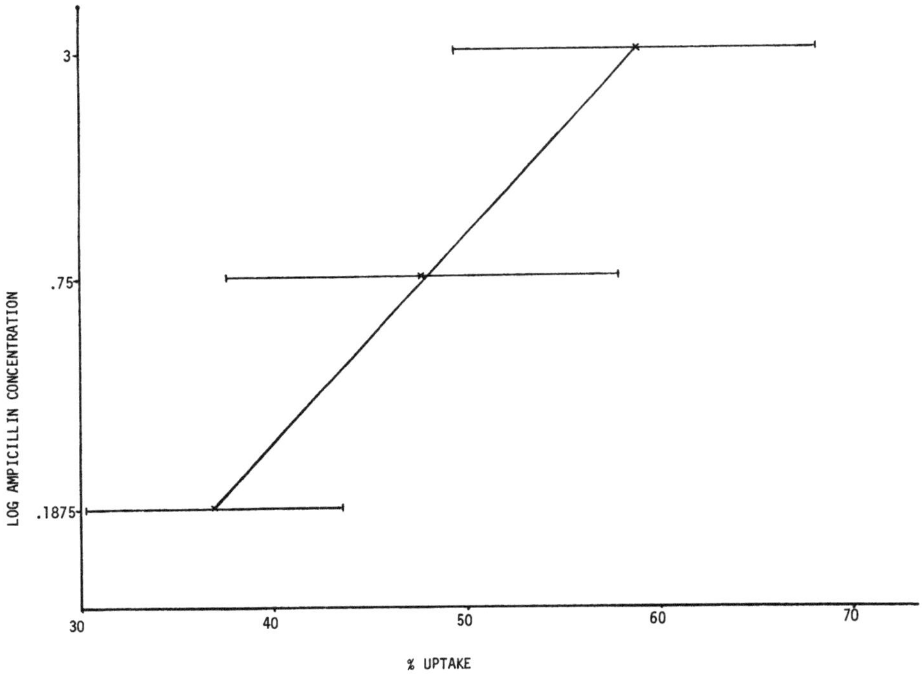

Fig. 4. Log ampicillin concentration versus % uptake of E. coli (3 minutes)

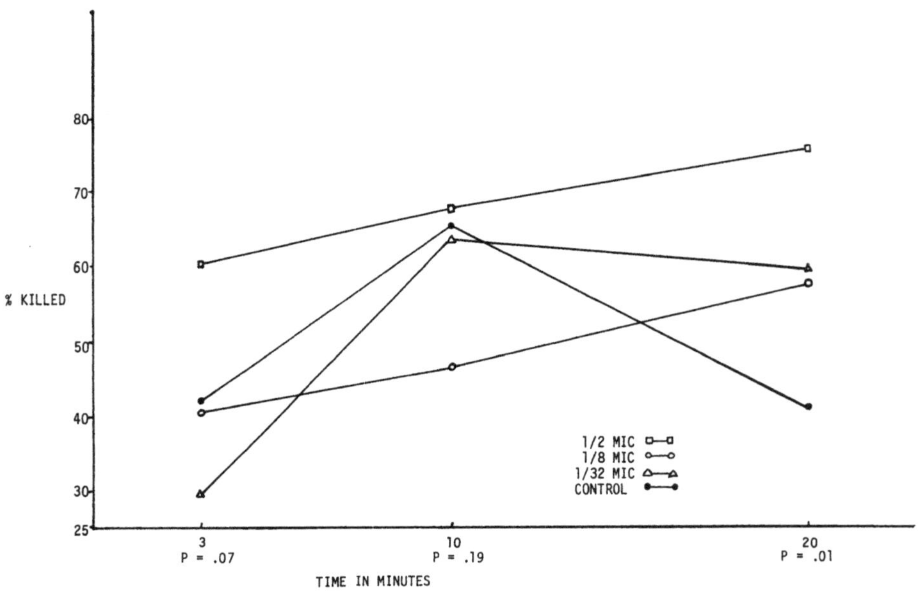

Fig. 5. Effects of sub-MIC ampicillin on killing of E. coli by PMNS

Table 2. *Effects of sub-MIC Ampicillin on Killing of E. coli by PMNS*

Time (minutes)	% Killed				
	½ MIC	⅛ MIC	1/32 MIC	Control	P
3	60.5 (6.08)	40.75 (10.72)	29.88 (5.84)	42.19 (11.54)	.07
10	67.62 (5.63)	46.75 (7.47)	63.62 (9.81)	65.12 (9.64)	.19
20	75.75 (5.18)	57.75 (10.89)	59.88 (11.13)	41.25 (7.85)	.01

(standard error)

Opsonin Requirements for Killing

Figure 6 shows the data in histogram form for control and drug treated groups (½ MIC) with % organisms killed measured at the 3 minute time only. Control and drug treated bacteria were exposed to either serum, heat inactivated serum or PBS prior to incubation with PMNs. The data show essentially no reduction in % killed of the drug treated organisms ($P = 0.25$) across the 3 groups but a significant reduction in % killed for the control bacteria was seen when PBS was used ($P < 0.05$). Direct comparisons of % killed for control and drug treated groups under each of the three PBS conditions showed a significant difference only when PBS was used ($P < 0.001$)

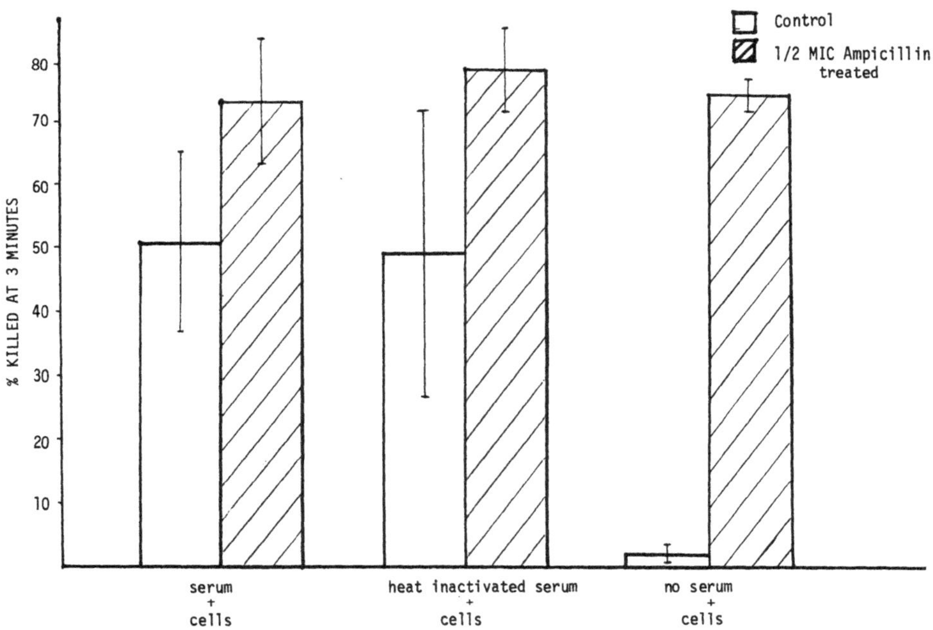

Fig. 6. Killing by PMNS of control and ampicillin treated E. coli under varying serum conditions

Discussion

Antimicrobial drugs may affect PMNs directly [14] or may alter humoral defense mechanisms, for example by inhibiting complement activation [15]. By their action on bacteria, antibiotics even in subinhibitory concentrations may also render the organisms more susceptible to the bactericidal activity of serum [16, 17] or to ingestion and killing by phagocytic cells [18, 19]. Our experimental design precluded any direct effects of ampicillin on either serum or PMNs since at no time did the drug come in contact with either of these. Any effects, therefore, would appear to be due solely to alteration of the organisms by the drug.

Following exposure to sub-MICs of beta lactam antibiotics, Gram negative bacilli such as *P. mirabilis* and *E. coli* become elongated with no evidence of septation [12, 13]. This effect may possibly be due to differential attachment by various antibiotics to penicillin binding proteins. These proteins can have different functions including septum formation, extension of the peripheral cell wall and maintenance of rod shape [20]. In the case of *E. coli* exposed to ampicillin at $\frac{1}{10}$ MIC, normal growth and cell division took place but at $\frac{1}{5}$ to 5 times MIC, cell division was completely inhibited resulting in production of long filaments [21].

Our transmission electron microscopy studies showed that elongated cells were clearly present at $\frac{1}{2}$ and $\frac{1}{8}$ MIC with septum formation barely perceptible at $\frac{1}{8}$ MIC but obvious at $\frac{1}{32}$ MIC.

As far as phagocytosis is concerned, our data strongly indicate increased uptake of drug treated bacteria compared with control cells at 3 minutes with clear evidence of a dose response relationship. A number of investigators have reported increased phagocytic uptake as well when organisms have been exposed to sub-MIC antibiotics.

A variety of bacteria have been used as the test cells and antibiotics studied have varied as well. Two reports with *E. coli* used doxycycline and clindamycin respectively as the antibacterial drugs [22, 23]. With doxycycline, it was found that preincubation of the organisms with low doses (2–10 μg/ml) resulted in no enhancement of uptake by PMNs but when higher doses were used (25–100 μg/ml) increased phagocytosis was found [22].

Subinhibitory clindamycin decreased adherence of *E. coli* to buccal epithelial cells and promoted uptake and killing by human PMNs [23]. Sub-MIC carbenicillin and sulbenicillin treatment of *Pseudomonas aeruginosa* resulted in increased uptake and killing of the bacteria by rabbit PMNs [18]. Gentamicin had no such effect on the organisms while polymyxin B and colistin produced increased uptake only at MIC [18].

In studies with mononuclear phagocytes, pretreatment of *Listeria monocytogenes* with sub-MIC ampicillin, chloramphenicol or tetracycline resulted in increased uptake as did nafcillin pretreatment of *S. aureus* [24, 25].

Effects of antibiotic pretreatment of bacteria on killing by PMNs have also been studied. As mentioned above, carbenicillin and sulbenicillin treated *Ps. aeruginosa* were more readily ingested and killed by PMNs than untreated organisms [18]. Other investigators have reported enhanced PMN killing of *Ps. aeruginosa* treated with various drugs including carbenicillin and ticarcillin [26], azlocillin and cefsulodin [27] and ceftriaxone [28].

Data that would more directly support our findings of enhanced killing of drug treated *E. coli* at 20 minutes are those reported by Lorian and Atkinson [27]. Exposure of various strains of *E. coli* to ampicillin (¼ to ½ MIC) resulted in filament formation. These were found to be much more susceptible to the bactericidal activity of PMNs than were the non-drug treated bacteria [27].

S. aureus is another organisms for which data exist although there are some conflicting results. Bassler [28] and Root [19] reported that ceftriaxone or Pen G and vancomycin respectively enhanced PMN bactericidal activity for *S. aureus* sensitive to the antibiotics. When the bacteria were pretreated with sub-MIC oxacillin, however, they were killed less readily by PMNs than the untreated bacteria [29]. Uptake, however, was unaffected. These conflicting results may be due to methodologic differences.

A study with more apparent in vivo relevance showed that certain strains of *P. mirabilis* made more sensitive to killing by serum and PMNs following treatment with sub-MIC carbenicillin were cleared more rapidly from the blood in experimental bacteremia [30].

Phagocytosis of particles (including bacteria) by PMNs depends upon the surface features of the particles and the presence on their surface of appropriate ligands [2]. In the case of encapsulated organisms, C3b and IgG are generally necessary, whereas with unencapsulated bacteria, the presence of specific serum ligands on their surface may not be required for phagocytosis to proceed [31].

There is in fact quite a heterogeneity in the requirements for immunoglobulin and complement for phagocytosis and killing by PMNs [32]. This applies not only to various genera of Gram negative aerobic bacilli and Gram positive microorganisms but to different strains of the same species as well [32]. In experiments with *S. aureus* it was found that complement itself or specifically the opsonically active C3b component may not be required for the increased killing of the penicillin treated organisms since identical results were obtained when heat inactivated serum was used or when PMN C3b receptors were blocked with Trypan blue [33]. Elimination of any opsonins, however, resulted in no PMN killing

Our data from the experiments conducted using serum, heat inactivated serum or no serum suggest that drug treated bacteria (½ MIC) are killed as well whether serum is present or not. The same is not true, however, for the control organisms since there was a dramatic reduction in killing in the absence of serum with this group.

It appears that antibiotic pretreatment minimizes opsonic requirements at least in the case of this particular *E. coli*. Since the organism does not appear to be encapsulated, the result is not entirely surprising. The action of the ampicillin on the organism could result in surface structural changes making the bacteria more readily ingestable by the PMNs.

The lack of a difference for the control group between serum and heat inactivated serum implies that complement is not necessary. The drop in killing without serum suggests a role for opsonically active immunoglobulins.

These results are supported by two of the studies referred to earlier [18, 28]. In Nishida's study, no serum was used in any of the incubation mixtures. Treatment of the organisms with sub-MIC carbenicillin and sulbenicillin resulted in greater uptake and killing by PMNs than the non-drug treated cells [18]. In Bassler's study, the results found depended on the organism tested [28]. *Ps. aeruginosa* and *S. aureus* exposed to sub-MIC ceftriaxone were involved. In the absence of serum, PMNs had no marked inhibitory effect on *S. aureus* at any ceftriaxone concentration studied but excellent activity against *Ps. aeruginosa* under such circomstances was found.

Summary

1. Exposure to sub-MIC ampicillin resulted in elongated bacilli and inhibition of septum formation.
2. Drug-treated organisms were ingested more readily than control cells at 3 minutes.
3. This early effect exhibited a linear log concentration versus uptake response supporting a dose response relationship.
4. Drug-treated organisms were killed more readily than control cells at 20 minutes.
5. Drug treatment minimized opsonic requirements.
6. Drug effects were due to direct interaction with bacteria, not to effects on serum or PMNs.

References

1. Fearon DT, Austen KF (1980) The alternative pathway of complement – a system for host resistance to microbial infection. N Engl J Med 303: 259–263
2. Horwitz MA (1982) Phagocytosis of microorganisms. Rev Infect Dis 4: 104–123
3. Spitznagel JK (1983) Microbial interactions with neutrophils. Rev Infect Dis 5(Suppl): 806–822
4. Root RK, Cohen MS (1981) The microbicidal mechanisms of human neutrophils and eosinophils. Rev Infect Dis 3: 565–598
5. Kucers A, Bennet NMcK (eds) (1979) The use of antibiotics. 3rd edn. William Heinemann Medical Books Ltd., London
6. Lietman PS (1985) Pharmacokinetics of antimicrobial agents. In: Mandell GL, Douglas RGJr, Bennett JE (eds) Principles and practice of infectious diseases. 2nd edn. John Wiley & Sons, New York, p. 164–243
7. Zak O, Kradolfer F (1979) Effects of subminimal inhibitory concentrations of antibiotics in experimental infections. Rev Infect Dis 1: 862–879
8. Verbrugh HA, Peters R, Peterson PK, Verhoef J (1978) Phagocytosis and intracellular killing of staphylococci by polymorphonuclear and mononuclear leukocytes. J Clin Pathol 31: 539–545
9. Verhoef J, Peterson PK, Sabath LD and Quie PG (1977b) Kinetics of staphylococcal opsonization, attachment, ingestion and killing by human polymorphonuclear leukocytes: a quantitative assay using (^{3}H) — thymidine labelled bacteria. J Immunol Meth 14: 303–311
10. McFarland J (1907) Nephelometer: an instrument for estimating the numbers of bacteria in suspensions used for calculating the opsonic index and for vaccines. JAMA 49: 1176–1178
11. Washington JA, Barry AL (1974) Dilution test procedures. In: Lenaette EH, Spaulding EH and Truant JP (eds) Manual of Clinical Microbiology, 2nd edn. American Society for Microbiology, Washington DC p. 410–417
12. Lorian V (1980) Effects of subminimum inhibitory concentrations of antibiotics on bacteria. In: Lorian V (ed) Antibiotics in laboratory medicine. Baltimore, William & Wilkins p. 342–408

13. Washington JA II (1979) The effects and significance of subminimal inhibitory concentrations of antibiotics. Rev Infect Dis 1: 781–786
14. Mandell LA (1982) Effects of antimicrobial and antineoplastic drugs on the phagocytic and microbicidal function of the polymorphonuclear leukocyte. Rev Infect Dis 4: 683–697
15. Alexander JW (1975) Antibiotic agents and the immune mechanisms of defense. Bull N Y Acad Med 51: 1039–1945
16. Traub WH, Sherris JC (1970) Studies on the interaction between serum bactericidal activity and antibiotics in vitro. Chemotherapy 15: 70–83
17. Friedman H, Warren GH (1976) Antibody-mediated bacteriolysis: enhanced killing of cyclacillin-treated bacteria (39533). Proc Soc Exp Biol Med 153: 301–304
18. Nishida M, Mine Y, Nonoyama S, Yokota Y (1976) Effect of antibiotics on the phagocytosis and killing of *Pseudomonas aeruginosa* by rabbit polymorphonuclear leukocytes. Chemotherapy 22: 203–210
19. Root RK, Isturiz R, Molavi A, Metcalf JA, Malech HL (1981) Interactions between antibiotics and human neutrophils in the killing of staphylococci. J Clin Invest 67: 247–259
20. Tipper Dj (1979) Mode of action of B-lactam antibiotics. Rev Infect Dis 1: 39–53
21. Rolinson GN, Macdonald AC, Wilson DA (1977) Bactericidal action of B-lactam antibiotics on *Escherichia coli* with particular reference to ampicillin and amoxycillin. J Antimicrob Chemother 3: 541–553
22. Melby K, Midtwedt T (1981) Effects of doxycycline on the phagocytosis of ^{33}P-labelled *Escherichia coli* by human polymorphonuclear cells. Chemotherapy 27: 264–269
23. Bassaris HP, Lianou PE, Papavassiliou JTh (1984) Interaction of subminimal inhibitory concentrations of clindamycin and *Escherichia coli:* effects on adhesion and polymorphonuclear leukocyte function. J Antimicrob Chemother 13: 361–367
24. Adam D, Schaffert W, Marget W (1974) Enhanced in vitro phagocytosis of *Listeria monocytogenes* by human monocytes in the presence of ampicillin, tetracycline and chloramphenicol. Infect Immun 9: 811–814
25. Friedman H, Warren GH (1974) Enhanced susceptibility of penicillin-resistant staphylococci to phagocytosis after in vitro incubation with low doses of nafcillin (38177) Proc Soc Exp Biol Med 146: 707–711
26. Petit J-C, Daguet G-L (1981) Enhanced killing of *Pseudomonas aeruginosa* by human polymorphonuclear leukocytes in presence of subinhibitory concentrations of carbenicillin and ticarcillin. Biomedicine 34: 29–33
27. Lorian V, Atkinson B (1984) Bactericidal effect of polymorphonuclear neutrophils on antibiotic-induced filaments of Gram-negative bacilli. J Infect Dis 149: 719–727
28. Bassler M, Blaschke H, Just M, Daschner FD (1982) Effect of ceftriaxone on *Pseudomonas aeruginosa* and *Staphylococcus aureus* in broth, serum and in combination with human polymorphonuclear leukocytes. Chemotherapy 28: 390–396
29. Lorian V, Atkinson B (1980) Killing of oxacillin-exposed staphylococci in human polymorphonuclear leukocytes. Antimicrob Agents Chemother 18: 807–813
30. Nishida M, Mine Y and Nonoyama S (1978) Relationship between the effect of carbenicillin on the phagocytosis and killing of *Proteus mirabilis* by polymorphonuclear leukocytes and therapeutic efficacy. J Antibiot (Tokyo) 31: 719–724
31. Horwitz MA, Silverstein SC (1980) Influence of the *Escherichia coli* capsule on complement fixation and on phagocytosis and killing by human phagocytes. J Clin Invest 65: 82–94
32. Leist-Welsh P, Bjornson AB (1979) Requirements for immunoglobulin and the classical and alternative complement pathways for phagocytosis and intracellular killing of multiple strains of Gram-negative aerobic bacilli. Infect Immun 26: 99–109
33. Yourtee EL, Metcalf JA, Root RK (1980) Augmented killing of penicillin pretreated *S. aureus* by human neutrophils: role of complement. Clin Res 28: 383A

Discussion

Daschner: Using a different experimental approach we have very similar or even identical results as you, that is: improvement of leukocyte killing by subinhibitory concentrations and "opsonizing" effect of sub-MIC concentrations of various antibiotics. Do you have any explanation for this "opsonizing" effect of antibiotic in a test system of leukocytes and antibiotic, but *without* serum?

Mandell: It is possible that the sub-MIC drugs are able to alter some of the surface structural characteristics of the organism. This may for example result in alteration in hydrophobicity thereby allowing more bacteria to be ingested. We also have to consider however that extracellular killing is taking place. In other words the treated organism converts the polymorph into a "secretory cell".

Verhoef: Have you done a control experiment at 4 °C to role out any aspecific "uptake" by unopsonized drug treated bacteria?

Mandell: No, we did not do this specific control. However as part of a general control for the assay system we did experiments at 4 °C and found that no phagocytosis took place.

Ginsburg: Did you try to employ cytochalasin B to prevent phagocytosis and thus look for extracellular killing?

Mandell: We have not yet done this experiment but we are planning to. Dr. Root in a study published in the Journal of Clinical Investigation a few years ago found that cytochalasin B treated PMNs were still able to kill staphylococci treated with sub-MIC penicillin.

Ginsburg: I would also like to mention that one should also consider the possibility that the PMN itself may be "opsonized". For example, PMN coated by very small concentrations of histone (a poly-cation) can now interact with non-opsonized streptococci able to induce chemiluminescence.

The Effect of Subinhibitory Concentrations of selected Antibiotics on Bacteria-Phagocyte Interaction

W. ROSZKOWSKI[1], P. CIBOROWSKI[2], H.L. KO[3], F. SCHUMACHER-PERDREAU[3], K. ROSZKOWSKI[1], J. JELJASZEWICZ[2], and G. PULVERER[3]

[1] Institute of Lung Diseases, Warsaw, Poland
[2] National Institute of Hygiene, Warsaw, Poland
[3] Institute of Hygiene, University of Cologne, Cologne, FRG

Introduction

Antibiotics, microorganisms and host defence mechanisms exhibit different interrelations which undoubtedly are important for the results of chemotherapy. Several effects of antimicrobial agents on host defence cells have already been identified [2]. It has been shown that some antibiotics are able to modulate immune responses as well as to change the susceptibility of bacteria to host defence mechanisms [1, 5]. Several papers have already been published where the effects of antibiotics in subinhibitory concentrations on different bacteria have been investigated. To our knowledge these studies have only been performed in experimental models using opsonin-mediated interactions between phagocytes and bacteria. The aim of our present paper is to demonstrate and to discuss the influence of antibiotics on opsonin-independent phagocytosis of staphylococci. The phagocytosis activity and bacterial killing degree of human neutrophils of different donors as well as the susceptibility of different staphylococcal strains have been taken into consideration, too. Preliminary data of studies concerned with the chemical structure of staphylococcal surface responsible for the opsonin-independent phagocytosis are presented as well.

Materials and Methods

Bacteria

Six staphylococcal strains from the collection of the Institute of Hygiene, University of Cologne, were used for the experiments [6]. *Staphylococcus aureus* strains T1, T 14 and SG 511-Jena as well as the coagulase-negative strains *Staphylococcus sciuri* S I, *Staphylococcus saprophyticus* S III were isolated from human infections, whereas *Staphylococcus simulans* S V was a human skin strain. The bacteria were grown overnight at 37 °C in Müller-Hinton broth (Difco Lab.), washed three times in saline and finally resuspended in RPM I 1640 (Gibco Lab.), or MEM (Gibco Lab.) medium.

The Influence of Antibiotics on the
Host-Parasite Relationship II
Edited by D. Adam, H. Hahn, W. Opferkuch

Antibiotics

Cefotaxime (Hoechst Co.) ciprofloxacin (Bayer Co.) and amikacin (Grünenthal Co.) were used in this study. Minimal inhibitory concentrations of these antibiotics were determined in liquid cultures, *Staphylococcus aureus* strain SG 511-Jena showed the following MICs: amikacin: 0.87 mcg/ml, cefotaxime: 1.56 mcg/ml, ciprofloxacin: 1.56 mcg/ml. $\frac{1}{10}$ of MIC was taken for the experiments.

Neutrophils

In our study we examined blood cells from 27 healthy volunteers of both sexes and of different age (range 19–51 years). Neutrophils were separated from heparinized blood by dextran sedimentation and the residual red cells were lysed by 30 sec treatment with 0.2% saline. This hypotonic shock was stopped by adding an equal volume of a 1.6% sodium chloride solution. The cells were washed three times and resuspended in RPMI 1640 or MEM (free of phenol red).

Bactericidal assay

The bacteria were suspended in RPMI medium and adjusted to a concentration of 10^6 bacteria/ml. For the assay we used a 1:1 ratio of bacteria to phagocytic cells. The mixture in plastic tubes was incubated at 37 °C for 2 hours under frequent gentle mixing. After 2 hours the suspension was sonicated for 15 sec and tenfold dilutions of the sonicate were made in saline. Aliquods (0.1 ml) of the dilutions were pipetted onto agar plates and the surface colony counts were performed after 24 hours incubation at 37 °C. The percent of killing was calculated according to the formula: $\frac{N_c - N_{ex}}{N_c} \times 100$ (N_c = number of viable bacteria in control tubes without phagocytic cells, N_{ex} = number of viable bacteria in tubes with phagocytes).

Luminol-dependent chemiluminescence

Luminol dissolved in dimethyl sulphoxide was used at a concentration of 10^{-5} M. The reaction mixture consisted of 0.5 ml suspension of neutrophils at a concentration of 10^6 cells/ml and of 0.1 ml bacteria suspension containing 10^8 staphylococci. Chemiluminescence measurements were carried out at 37 °C using a LKB Wallace Luminometer 1250 and recorded as mV.

Results

20 healthy volunteers were choosen as donors of peripheral neutrophilic granulocytes. The bactericidal activity of all these samples of neutrophils was examined using six different staphylococcal strains, three of them belonging to the species *S. aureus*,

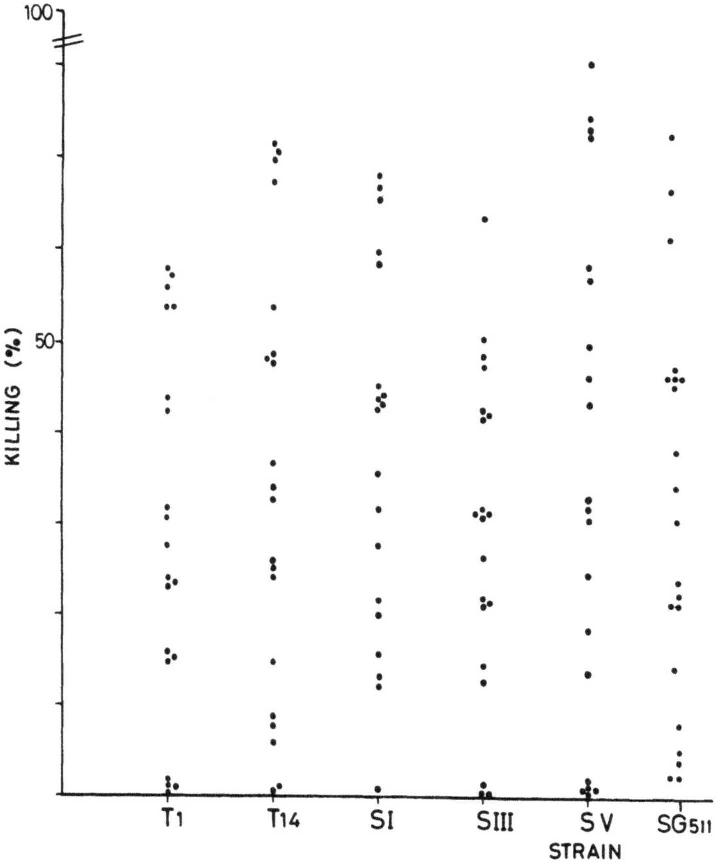

Fig. 1. Bactericidal activity of human neutrophils from 20 healthy donors directed against six different strains of staphylococci

the other strains being coagulase-negative. These staphylococcal strains have been selected for this study because of their differing cell walls structure [6] . As shown in Fig. 1, the bactericidal activity spectrum of human neutrophils as well as the differing susceptibility of staphylococci are widely ranged. Each human being has obviously his own individual pattern of antistaphylococcal activities of neutrophilic granulocytes. (Fig. 2).

Three of these healthy volunteers were examined in monthly intervals to see if their antibacterial activity of neutrophils remains more or less constant over a longer time. These donors were selected because of "high", "moderate" and "low" activities of neutrophils. Three corresponding staphylococcal strains were used as test organisms. As demonstrated in Fig. 3, the individual ability of staphylococcal killing by neutrophils persists over at least 3–5 months in a rather constant range. The phagocytic activity of granulocytes can be easily investigated with the help of the chemiluminescence assay. As expected, the data of this chemiluminescence

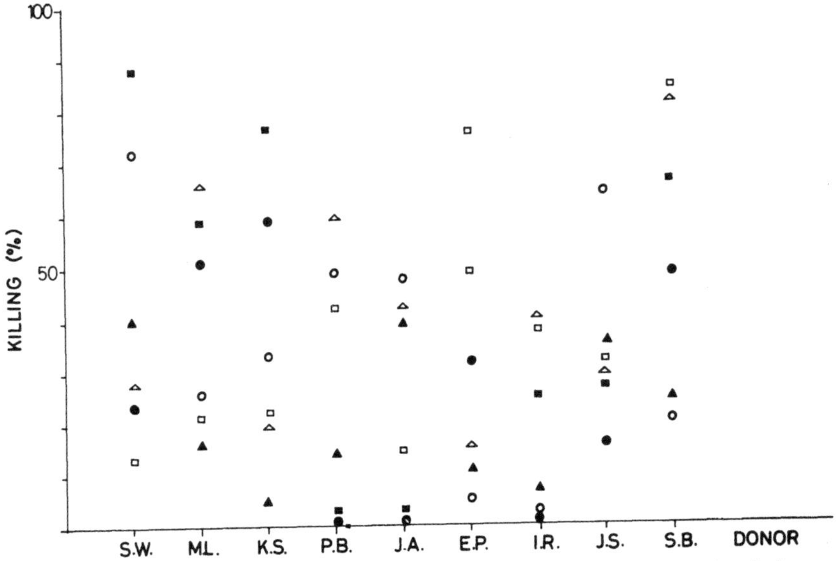

Fig. 2. Individual "pattern" of bactericidal activity directed against *Staphylococcus aureus* strain T 1 (•), T 14 (○), SG 511 (▲), *Staphylococcus sciuri* S I (△), *Staphylococcus saprophyticus* S III (□) and *Staphylococcus simulans* S V (■)

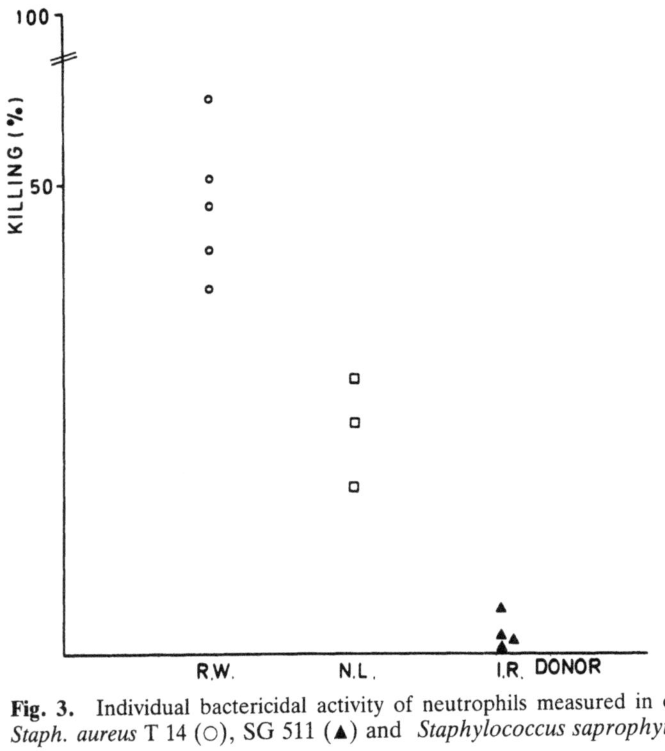

Fig. 3. Individual bactericidal activity of neutrophils measured in one month intervals against *Staph. aureus* T 14 (○), SG 511 (▲) and *Staphylococcus saprophyticus* (□)

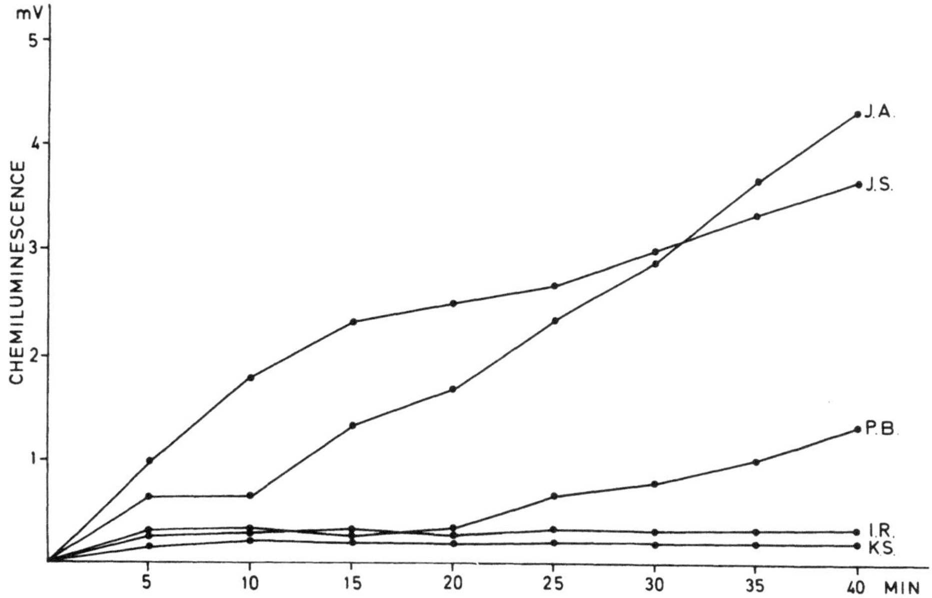

Fig. 4. Chemiluminescence of granulocytes obtained from different donors in the presence of *Staphylococcus aureus* SG 511

assay and the results of the bactericidal activity test are more or less congruent. Neutrophilic granulocytes with a low killing ability for a certain staphylococcal strain express only a low reactivity towards the same strain performing the chemiluminescence assay (Fig. 4).

S. aureus strain SG 511-Jena was exposed to subinhibitory concentrations (1/10 MIC) of cefotaxime, ciprofloxacin and amikacin. Thereafter phagocytosis was tested using "high" and "low" reacting neutrophils. Choosing "low" reacting neutrophilic granulocytes, *S. aureus* SG 511-Jena showed no increase in chemiluminescence when pretreated in 1/10 MIC of cefotaxime, ciprofloxacin or amikacin. On the other hand, *S. aureus* SG 511-Jena pretreated in subinhibitory concentrations of all three antibiotics changed significantly the results of the chemiluminescence assay when "high" reacting neutrophils were taken (Fig. 5–7).

As demonstrated in Fig. 8, *S. aureus* strain SG 511-Jena lost its granulocyte stimulating activity completely after exposure to 2 M urea over 30 min at 37 °C. Urea-treatment of staphylococcal cells is a rather gentle procedure, the viability of staphylococci is not affected. We collected the supernatants of urea-treated staphylococci, filtered them through millipore filters (0,2 mcm) and urea was thereafter removed by molecular sieving on a Sephadex G-25 column. The fraction with a maximal absorption at 215 nm was separated and lyophilyzed. When this fraction was added to urea treated staphylococci in a concentration of 5 mg to 10^9 bacteria, the granulocyte stimulating activity of this staphylococcal strain was completely restored (Fig. 8).

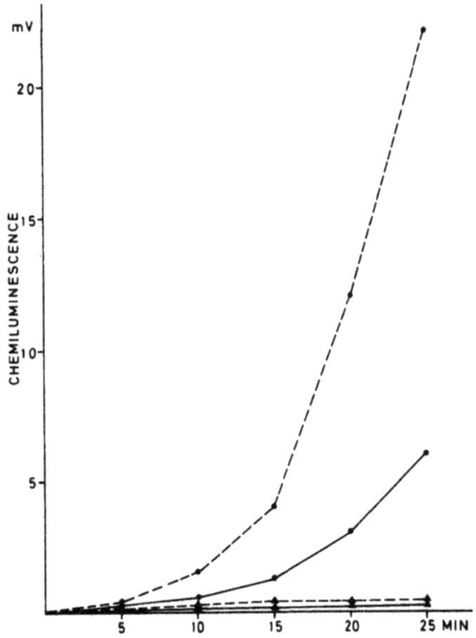

Fig. 5. Chemiluminescence induced by *Staph. aureus* SG 511 cultured alone (——) or in subinhibitory concentration of cefotaxime (——) of granulocytes with high (•) and low (△) bactericidal activity against this strain

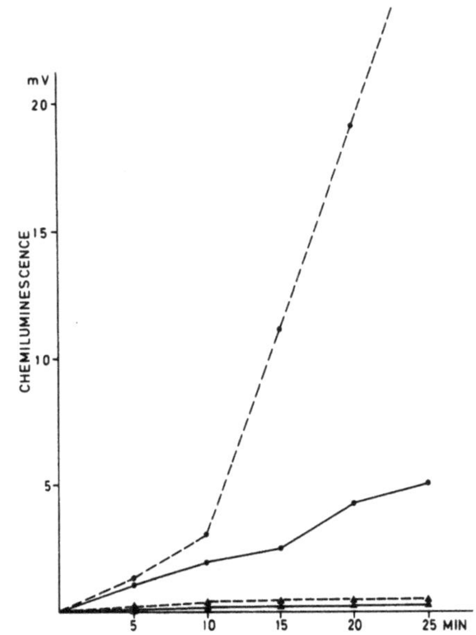

Fig. 6. Chemiluminescence induced by *Staph. aureus* SG 511 cultured alone (——) or in subinhibitory concentrations of ciprofloxacin (——) of granulocytes with high (•) and low (△) bactericidal activity against this strain

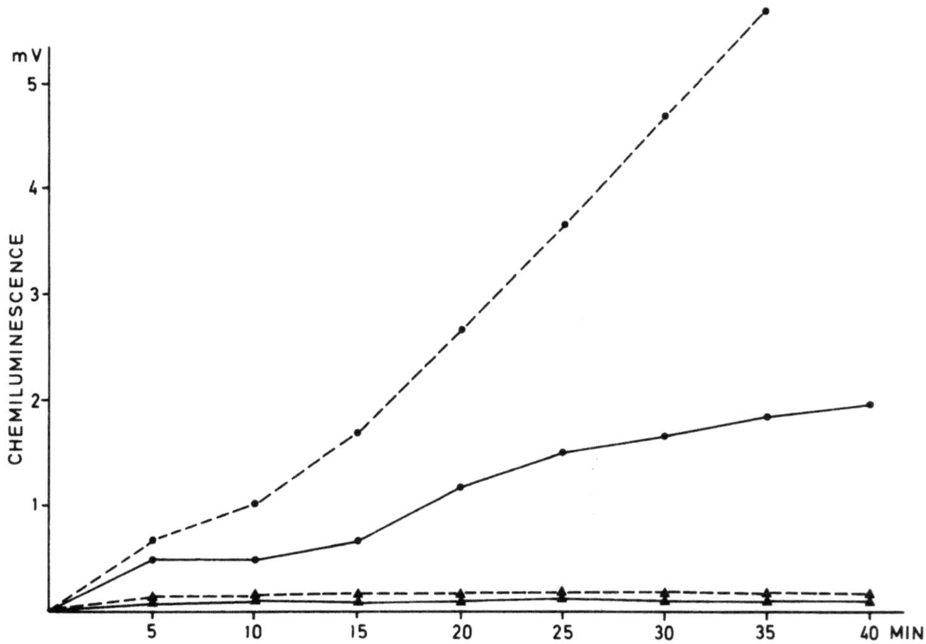

Fig. 7. Chemiluminescence induced by *Staph. aureus* strain SG 511 cultured alone (——) or in subinhibitory of amikacin (——) of granulocytes with high (•) and low (△) bactericidal activity against this strain

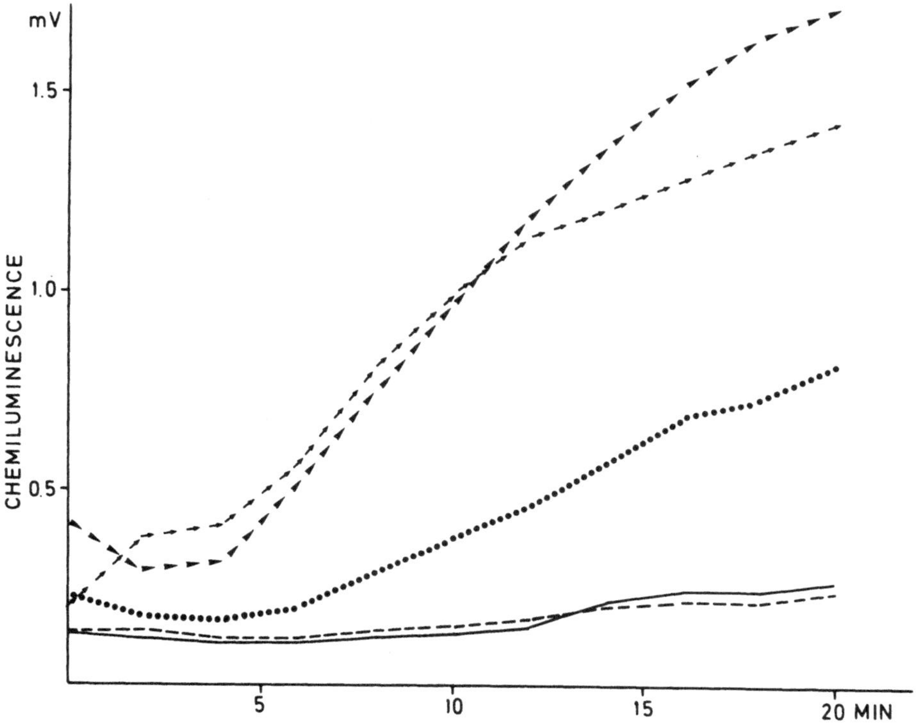

Fig. 8. Chemiluminescence of granulocytes induced by *Staphylococcus aureus* SG 511. (• • • •) control, (——) bacteria treated with 2 M urea, (▶ ▶ ▶) bacteria cultured in subinhibitory concentration of cefotaxime, (——) bacteria cultured in subinhibitory concentration of cefotaxime and treated with 2 M urea, (→ →) bacteria cultured with cefotaxime, treated with urea and incubated with fraction 215 nm

Discussion

Different extracellular and bacterial factors influence the phagocytosis and killing of staphylococci by human neutrophils [3,4]. The study of *Easmon* and *Desmond* [5] indicates that phagocytosis can be influenced when bacteria are pretreated with subinhibitory concentrations of different antibiotics. The authors showed that the exposure of streptococci, staphylococci and *E. coli* to antibiotics change the response of granulocytes to the bacteria when performing the chemiluminescence assay. However, in their experiments opsonization of microorganisms was applied.

In our present paper the direct interaction between staphylococci and phagocytic cells without participation of opsonins was investigated. We could demonstrate that the bactericidal ability as well as the chemiluminescence response for every given staphylococcal strain is highly depending on the individual donor. It was also noticed that each donor has his own pattern of reactivity to different strains of staphylococci. Such findings makes it rather difficult to establish conclusions on the influence of antibiotics on the bacterial susceptibility to neutrophilic killing. Therefore, when analysing the antibiotic effects on staphylococci one should always have in mind the individual activity pattern of neutrophils as well as the differing behaviour of staphylococcal strains used for the experiment.

Amikacin, ciprofloxacin and cefotaxime used in subinhibitory concentrations (1/10 MIC) facilitated the interaction between *Staphylococcus aureus* SG 511-Jena and neutrophils, as measured by a chemiluminescence assay. On the other hand it is known that the chemiluminescence of granulocytes reflects the membrane activation of these cells by bacteria. Assuming that this activation occurs through a contact with the antibiotic changed bacterial surface, in next experiments we used 2 M urea treatment for further modification of staphylococci. After urea-treatment staphylococcal cells lost their stimulatory effect on granulocytes. The activity was restored by a fraction present in supernatants of urea treated staphylococci. UV spectrum of this fraction showed a maximal absorption at 215 nm and no absorption at all at 280 nm, suggesting its rather non protein composition. We are now trying to achieve the biochemical characterization of this supernatant fraction.

Summary

The influence of cefotaxime, ciprofloxacin and amikacin on the opsonin-independent phagocytosis of staphylococci was examined. The anticiotics were used in subinhibitory concentration (1/10 MIC). Three *S. aureus* strains (T 1, T 14, SG 511-Jena) and three coagulase-negative staphylococci (*S. sciuri* S I, *S. saprophyticus* S III, *S. simulans* S V) were investigated as well as the phagocytosis activity and bacterial killing of human neutrophilic granulocytes from 20 healthy volunteers. A wide and constant activity range could be demonstrated depending on the donor of neutrophils as well as on the staphylococcal strain tested. The phagocytosis activity of "high" reacting neutrophils was markedly increased when antibiotic pretreated staphylococci were examined. "Low" reacting neutrophils showed no similar changes. Preliminary results on the biochemical characterization of the antibiotic induced modification of the staphylococcal surface were discussed.

References

1. Easmon CSF, Desmond AM (1982) The effect of subinhibitory antibiotic concentrations on the opsonization, uptake and killing of bacteria by human neutrophils. In: Eickenberg HU, Hahn H, Opferkuch W (eds) The influence of antibiotics on the host-parasite relationship, Springer-Verlag, Berlin, Heidelberg, New York, pp. 202–207
2. Milatovic D (1983) Antibiotics and phagocytosis. Review Eur J Microbiol 2: 414–425
3. Peterson PK, Verhoef J, Sabath LD, Quie PG (1976) Extracellular and bacterial factors influencing staphylococcal phagocytosis and killing by human polymorphonuclear leukocytes. Infect Immun 14: 496–501
4. Peterson PK, Verhoef J, Sabath LD, Quie PG (1977) Effect of Protein A on staphylococcal opsonization. Infect Immun 15: 760–764
5. Roszkowski W, Ko HL, Roszkowski K, Jeljasezwicz J, Pulverer G (1985) Effect of selected antibiotics on the cellular and humoral immune response in mice. In: Pulverer G, Jeljaszewicz J (eds) Chemotherapy and Immunity. Zbl Bakt Hyg Suppl 13, Gustav Fischer Verlag Stuttgart, New York, pp. 59–72
6. Schumacher-Perdreau F, Rotering H, Pulverer G (1983) Cell wall analysis and taxonomy of Staphylococci. Zbl Bakt Hyg I Abt 256: 25–36

Discussion

Limbert: Were the bacteria opsonized prior to the chemiluminescence assay?

Roszkowski: No, we did not use opsonins in our experiment.

Daschner: Did you study the phagocytosis of various patients or healthy volunteers at different day times?

Roszkowski: No, we did not perform such experiments. All blood samples for leukocyte isolation were taken in the morning. However, we examined the donors at monthly intervals, as presented by me. As you could see, the results of killing were more or less constant in time.

Haag: If you add antibiotics in ½ to 1/32 of the MIC to a homogeneous bacterial population, you impose a selection pressure resulting in selection of differing subpopulations which is also influenced by the growth medium. My question is, have you or has anybody else information on differences in phagocytosis rates due to these differing subpopulations of the same strain? Is the population which is not engulfed different from that which is engulfed?

Roszkowski: We have not performed such experiments. Obviously the possibility mentioned by you may exist.

Ginsburg: I presume that you have employed an LKB luminometer to determine chemiluminescence. Have you tested the non-responding PMNs for their capacity to produce superoxide by PMA? I would like to report that in my experience bacteria which grew in different media (brain heart, trypticase soy broth, Todd Hewitt, etc.) and washed trigger different CL signals. Presumably the bacteria adsorb from the medium ligands which may mask the bacterial surface.

The effect of subinhibitory concentrations of monobactams on bactericidal activity of macrophages

K. IIDA-TANAKA[1], T. TANAKA[1], S. IRINO[1], and A. NAGAYAMA[2]

[1] 1st Department of Internal Medicine, Kagawa Medical School, Ikenobe, Miki, Kagawa, 761–07, Japan
[2] Department of Microbiology, Saga Medical School, Nabeshima, Saga, 840–01, Japan

Introduction

It has been demonstrated that subminimum inhibitory concentrations (sub-MICs) of antibiotics have some effects on bacterial growth rate and morphology [8]. In other studies, it has been shown that sub-MICs of antibiotics alter adhesiveness of bacteria to mucosal surfaces [4] and enhance the bactericidal activity of human polymorphonuclear-leukocytes [1, 10, 11]. In addition to chemotherapy, host-defence reactions are important to host resistance against microbial infection. It is not fully understood how sub-MICs of antibiotics influence interactions between bacteria and the human host. We have been studying the effect of sub-MICs of antibiotics on such interactions, and found that monobactams, monocyclic β-lactam antibiotics (Aztreonam and AMA1080) potentiate the bactericidal activity of macrophages. Thus, monobactams may be expected to be useful agents for the clinical treatment of bacterial infections.

Materials and Methods

Bacterium

The bacterium used in this study, *Escherichia coli* S615 was originally isolated from clinical specimen.

Phagocyte

Mouse macrophage strain CA1SL1, which was established from C3H/HeN mice peritoneal exudate cells transformed with SV40 virus (A. Nagayama, manuscript in preparation). Cells were cultured in MEM medium supplemented with 10% fetal calf serum and 10% L cell-conditioned medium.

The Influence of Antibiotics on the
Host-Parasite Relationship II
Edited by D. Adam, H. Hahn, W. Opferkuch

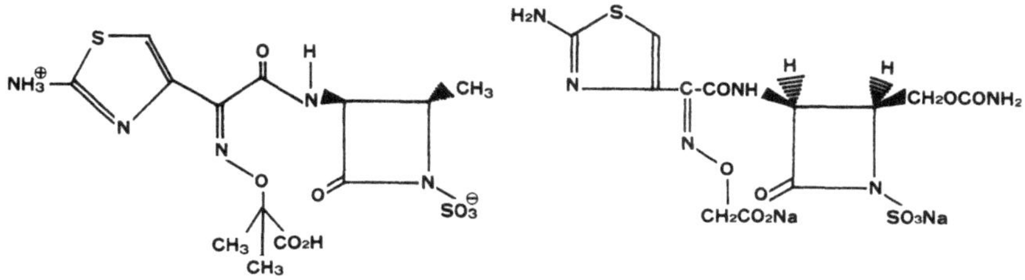

Aztreonam **AMA 1080**

Fig. 1. Structure of Aztreonam and AMA1080

Antibiotics

Antibiotics were kindly provided by each company, Aztreonam (AZT, Fig. 1) by Eisai Co., Tokyo, AMA1080 (AMA, Fig. 1) by Takeda Pharma. Co., Osaka, Cefoperazone (CPZ) by Toyama Chem. Co., Tokyo, Latamoxef (LMOX) by Shionogi Co., Osaka and AC-1370 by Mochida Pharma. Co., Tokyo, Japan. The minimum inhibitory concentration (MIC) was determined by serial two-fold dilutions method. Under these conditions, the MICs for these antibiotics except for AC-1370 to *E. coli* S615 were 0.1 μg/ml. That of AC-1370 was 6.25 μg/ml.

Phagocytosis assay

The phagocytosis of viable ^{14}C-leucine labeled *E. coli* by macrophages was assayed in the way described by Verhoef (14).

Bactericidal assay

Bactericidal assay was performed according to modified method of Zerigs' [15]. One ml of suspension containing 2×10^6 macrophages and 2×10^6 bacteria in Hanks Balanced Salt Solution (HBSS) supplemented with 5% newborn calf serum was incubated in a polypropylene test tube at 37 °C with shaking with or without antibiotics. Sample (0.1 ml) was removed at various intervals and added to 0.4 ml of sterilized water for osmotic disruption of the macrophages. After vigorous mixing with Vortex, additional dilutions were made with saline for plating.

Determination of intracellular killing

After the mixtures of *E. coli* and macrophages were incubated for 30 min in the presence or absence of antibiotics, macrophages were harvested, washed and suspended in HBSS. Then, they were incubated for 60 min at 37 °C and followed by bactericidal assay.

Pretreatment of macrophages

The monolayers of macrophages were treated with antibiotics for 16 hr at 37 °C in a 5%-CO_2 atmosphere. After washing, the macrophages were used for bactericidal assay.

Examination of viability of macrophages

To examine the viability of macrophages after phagocytosis, 0.2% trypanblue exclusion test was performed. The percentage of surviving macrophages was immediately determined under a light microscope.

Macrophage lysate

An equivalent number of macrophages to that used in the bactericidal assay was disrupted by freezing-thawing 3 times. Cell lysis was confirmed light-microscopically.

Electron microscopy

For scanning electron microscopy, the bacterial suspension was added to the monolayer of macrophages cultured on glass cover slips. After incubation, they were washed with phosphate buffered saline (PBS) and fixed with 2% glutaraldehyde in PBS. Then the specimens were dehydrated through an ethanol series, immersed in amylacetate and dried by the critical-point method. Platinum-palladium was evaporated on the surface of the specimens, and they were examined by a Hitachi S-550 scanning electron microscope operating at 20 kV. *E. coli* cells were fixed in a mixture of 0.25% glutaraldehyde and 5% acrolein in cacodylate buffer according to Murray [2]. After dehydration in a graded ethanol series, the specimens were embedded in Epon 812. Thin sections were examined with a Hitachi H600 transmission electron microscope.

Results

Bactericidal activity of macrophages in the presence of Antibiotics

Figure 2 shows that increased susceptibility of *E. coli* S615 to killing by macrophages was elicited by sub-MICs of AZT or AMA. Recovery of surviving bacteria after 1 hr incubation in the presence of monobactams (AZT or AMA) was significantly reduced when macrophages were present. The concentrations of AZT and AMA in which the effect was observed were between $\frac{1}{32}$ and 1 MIC and between $\frac{1}{16}$ and 1 MIC, respectively (Fig. 3). In addition to monobactams, three other β-lactam

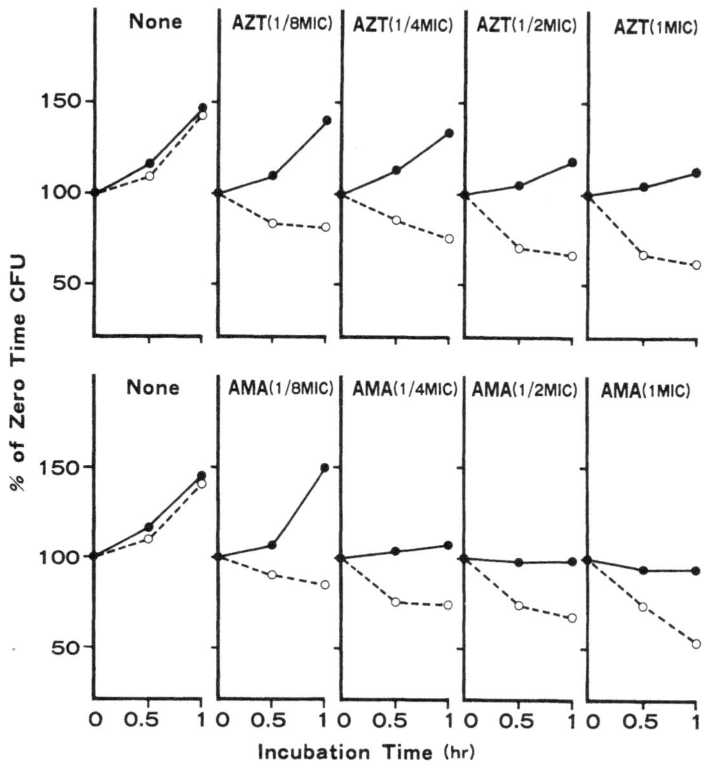

Fig. 2. Enhancement of bactericidal activity of macrophages; Symbol: —• , without macrophages; --•-, with macrophages

Table 1. Effect of β-lactam Antibiotics on Killing of *E. coli* S615 by Macrophages (MΦs)

Antibiotics	Conc. (MIC)	Untreated E. coli + MΦs	Antibiotic treated E. coli + MΦs	Antibiotic treated E. coli
AZT	½	434	41.2	88.8
	¼		75.0	107
AMA 1080	½	422	47.8	124
	¼		72.8	183
AC 1370	½	341	55.6	133
	¼		73.3	141
CPZ	½	303	58.6	29.3
	¼		82.1	78.0
LMOX	½	400	163	210
	¼		275	313

Results are expressed as percentages of the zero-time sample surviving after 2 hr

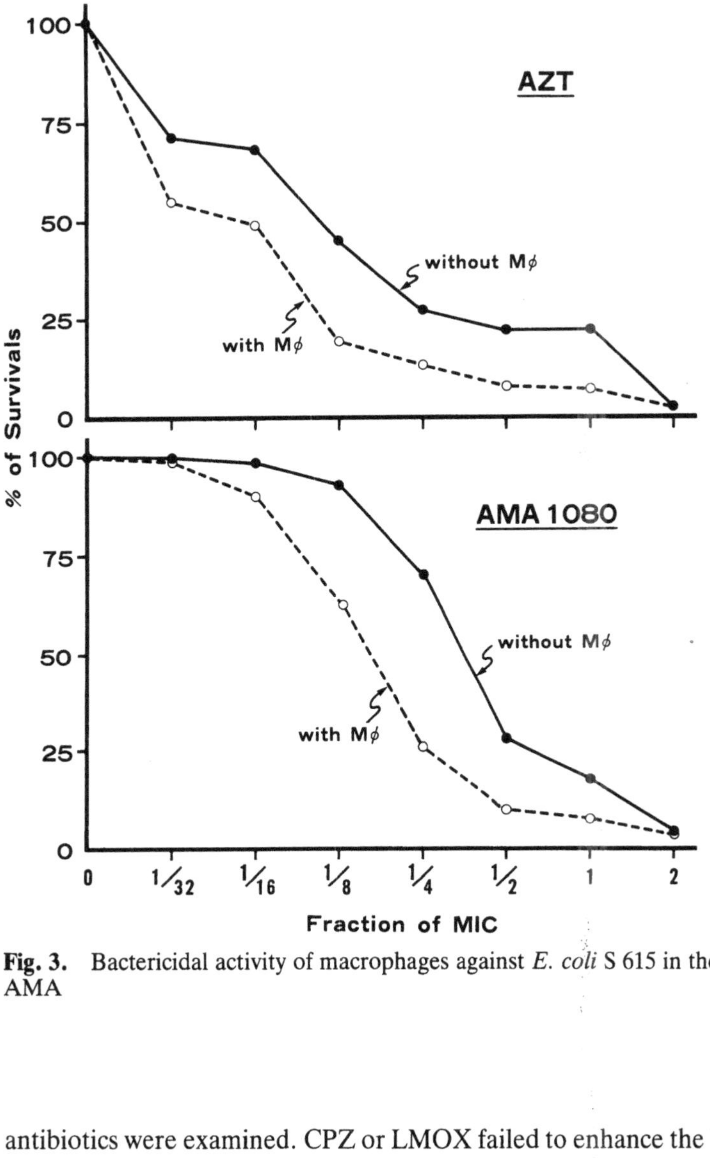

Fig. 3. Bactericidal activity of macrophages against *E. coli* S 615 in the presence of AZT or AMA

antibiotics were examined. CPZ or LMOX failed to enhance the bactericidal activity of macrophages whereas AC1370 enhanced the activity as reported by other investigators (9) (Table 1). Enhancement of bactericidal activity of macrophages was also evaluated with monobactams for *Pseudomonas aeruginosa, Klebsiella pneumoniae* and *Serratia marcescens*(data not shown).

Intracellular killing activity of macrophages in the presence of monobactams

As shown in Table 2, enhancement of intracellular killing was observed in the presence of sub-MICs of monobactams. Pretreatment of *E. coli* with monobactams for 2 hr before phagocytosis resulted in more efficient enhancement of intracellular

Table 2. Effect of Monobactams on Intracellular Killing of *E. coli* S615 by Macrophages

Antibiotics	Conc. (MIC)	% Survival after 1 hr	
AZT	½	48.4	
	¼	72.0	36.0[a]
	⅛	81.0	51.1[a]
	1/16	83.1	
AMA 1080	½	41.1	
	¼	57.1	34.3[a]
	⅛	89.4	78.9[a]
	1/16	90.8	

The results are expressed as percentages of the zero-time sample surviving after 1 hr.
[a] Before phagocytosis bacteria were pretreated with antibiotics for 2 hr.

killing by macrophages. This result indicates that the pretreated bacteria could become more susceptible to killing by macrophages. This might be due to the direct damage on bacteria caused by monobactams.

Effect of monobactams on phagocytosis

Uptake of *E. coli* by macrophages after 60 min incubation with sub-MICs of AZT increased by 20–30% in comparison to the control without AZT. In the case of AMA, uptake also increased slightly (data not shown).

Macrophage viability

After phagocytosis of *E. coli*, viability of macrophages was examined. Even after 8 hr, more than 90% of macrophages survived in the presence of ¼ MIC of AZT or AMA. Contrarily, without monobactams, the number of surviving macrophages decreased with increasing incubation period; about 80% of macrophages survived after 3 hr and 60% after 8 hr.

Effect of monobactams on macrophages

Pretreatment of macrophages with monobactams before phagocytosis resulted in no enhancement of killing by macrophages in any concentrations, even as high as 4 MIC (data not shown). From these results, it is likely that the enhancement of bactericidal activity is due to increased phagocytosis and intracellular killing and that monobactams directly damage the bacterial cells. Monobactams had no effect on macrophages themselves.

Synergistic effect of lysozyme or macrophage lysate with monobactams

Antibiotics could act synergistically with the bactericidal substances of macrophages such as various enzymes. Such synergistic effects of monobactams and lysozyme, one of the bactericidal substances released from macrophages, was examined. It is known that *E. coli* cells are resistant to the bactericidal lytic action of lysozyme. Table 3 shows clearly the same result. Presence of sub-MICs of AZT or AMA resulted in an enhanced killing effect of lysozyme on *E. coli*, but the presence of CPZ resulted in no enhancement (Table 3). This result coincides with the enhancement of macrophages killing caused by monobactams but not by CPZ. To investigate the mechanism of the enhancement of bactericidal activities of macrophages, macrophage lysates were used instead of lysozyme. The addition of macrophage lysate failed to kill *E. coli* in antibiotic-free media. Contrarily, the addition of sub-MICs of either AZT or AMA in conjunction with macrophage lysate augmented

Table 3. Effect of β-lactam Antibiotics on Killing by Lysozyme

Antibiotics	Conc. (MIC)	Lysozyme	
		Without	With
None		1.0	0.98
AZT	½	1.0	0.76
	¼	1.0	0.72
	⅛	1.0	0.53
AMA 1080	½	1.0	0.81
	¼	1.0	0.65
	⅛	1.0	0.51
CPZ	½	1.0	1.0
	¼	1.0	0.97
	⅛	1.0	1.0

The results are expressed as the survival ratio of bacteria after 2 hr incubation with lysozyme and without lysozyme.

Table 4. Synergistic Effect of Monobactams on Killing of Macrophage lysate

Antibiotics	Conc. (MIC)	Surviving ratio after 2 hr			
		None	with MΦs	MΦ lysate	Lysozyme
None		1.0	0.91	0.97	0.96
AZT	¼	1.0	0.41	0.74	0.71
	⅛	1.0	0.26	0.52	0.52
AMA 1080	¼	1.0	0.41	0.57	0.75
	⅛	1.0	0.43	0.62	0.71

The results are expressed as the ratio of survival, after 2 hr-incubation with macrophages (MΦs), MΦ lysate or lysozyme and without them.

killing of bacteria (Table 4). From these results, it seems likely that the presence of monobactams renders *E. coli* susceptible to bactericidal substances released from macrophages.

Electron microscopic observations

The enhanced killing activity and preservation of macrophage viability in the presence of monobactams was supported by morphological observations under the electron microscope. In the absence of antibiotics, many *E. coli* cells were engulfed by the macrophages, while others remained attached to the surfaces; the macrophages lost their normal morphological characteristics (Fig. 4a). On the other hand, in the presence of ¼ MIC of AZT, many filamentous bacterial cells were phagocytized by macrophages which maintained their normal morphological characteristics (Fig. 4b). These results are in good agreement with the studies of macrophage viability (see above). Morphological alterations of *E. coli* in the presence of AZT were examined with a transmission electron microscope. In normally growing cells, the cytoplasmic membrane lay in an organized and in a regular manner under the cell wall (Fig. 5a). In AZT (¼ MIC)-treated *E. coli*, the structure of the cytoplasmic membrane was disorganized in an aberrant manner (Fig. 5b, c). Similar morphological alterations were observed on the specimen of cells treated with AMA.

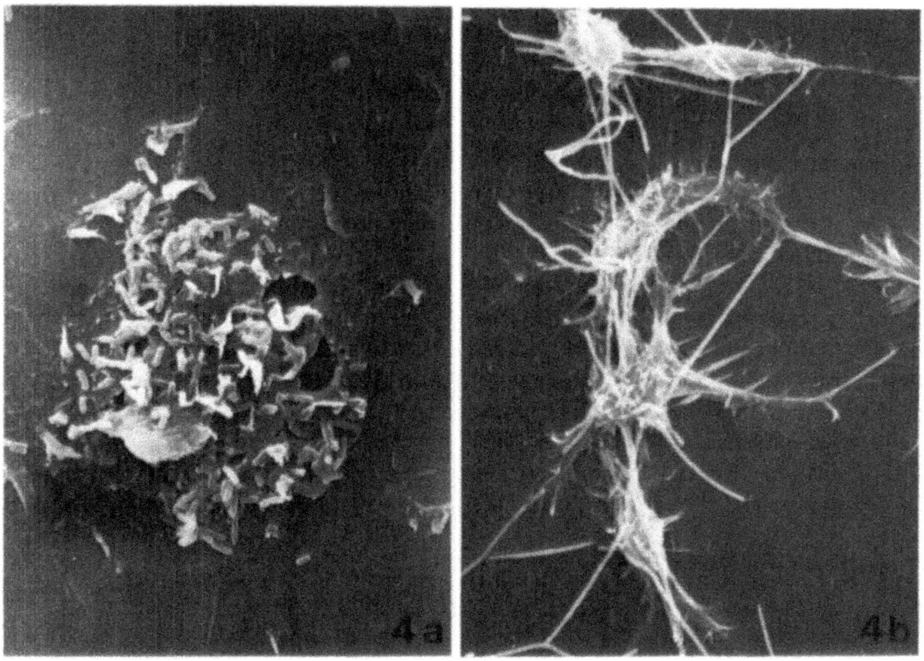

Fig. 4a, b. Scanning electron micrographs of macrophages incubated with *E. coli* S615 for 2 hr in the absence (**a**) and presence of ¼MIC of AZT (**b**). **a**; x 1,200, **b**; x 750

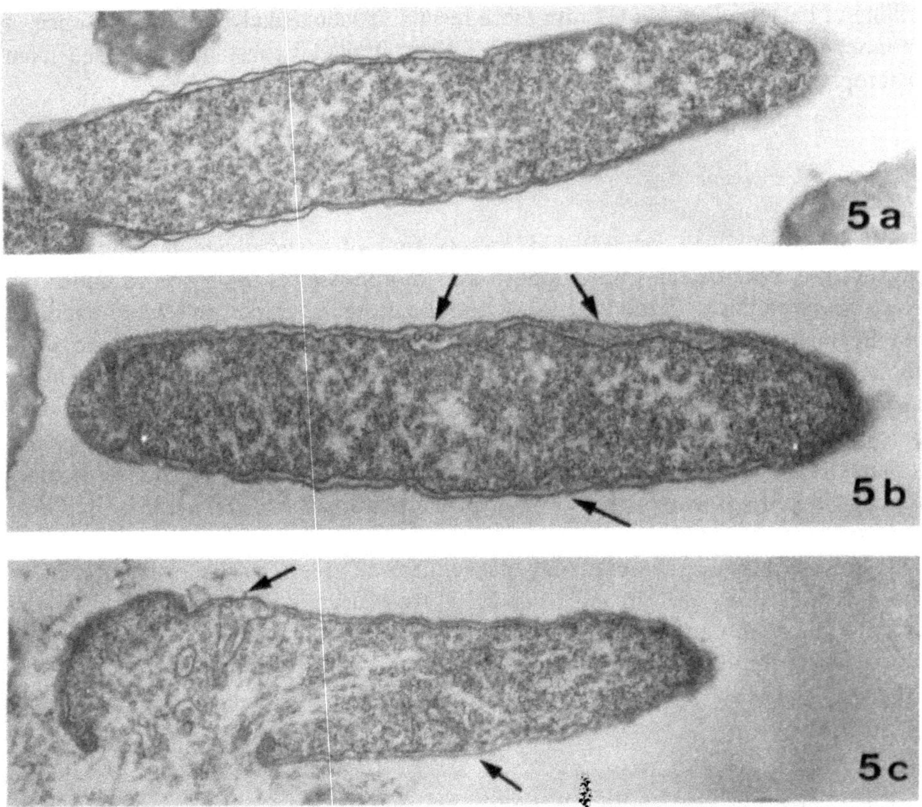

Fig. 5a–c. Transmission electron micrographs of *E. coli* S615 cells x 20,000. **a**) A normally growing cell; **b, c**) Cells treated with ¼MIC of AZT for 2 hr. Arrows indicate disorganization of cytoplasmic membrane

Discussion

The present investigation was designed to evaluate the effect of sub-MICs of monobactams on the bactericidal activity of macrophages. Our study was carried out with an established cell line of mouse macrophage. *E. coli* has been shown to be phagocytized and killed more efficiently by polymorphonuclear leukocytes than by macrophages [7]. However, we have employed an assay system using macrophages in order to clarify the effects of antibiotics on the host-parasite relationship, because phagocytosis and killing by polymorphonuclear leukocytes could be so efficient as to mask the effects of antibiotics. We have shown that sub-MICs of monobactams enhance macrophage killing of *E. coli, P. aeruginosa, K. pneumoniae* and *S. marcescens*. The enhancement was observed by pretreatment of the bacteria but not the macrophages with monobactams. It appears that these antibiotics act directly on bacteria to render them more susceptible to ingestion and more sensitive to killing by macrophages. It is known that the sub-MICs of β-lactam antibiotics induce gram-negative bacteria to form filaments by blocking septation [8], and that some of these filaments are phagocytized at a high rate [1]. However, this mechanism has not been fully clarified. After ingestion by phagocytes, many bacteria are killed. Metabolic products as well as lysosomal constituents contribute to this bactericidal activity in both polymorphonuclear leukocytes and macrophages. A number of components of the lysosomal granule function as bactericidal substances. It is also known that the bactericidal halogenating system, which utilizes myeloperoxidase, hydrogen peroxide and chloride, causes lethal halogenation of bacteria. *E. coli* S615 posseses catalase which catalizes the reduction of hydrogen peroxide, and in addition, it is resistant to lysozyme. Thus, *E. coli* cells are relatively resistant to intracellular killing in the absence of monobactams even they are phagocytized by macrophages. Consequently, the ingested cells may be able to proliferate. On the other hand, *E. coli* weakened by monobactams treatment may readily be damaged further and killed by substances released from macrophages. The results in Table 4 support this notion; the artificial macrophage lysate killed the bacteria damaged by monobactams more efficiently than untreated bacteria.

The penicillin-binding-protein (PBP) profile of AZT has indicated a very high affinity for PBP3 which controls cell division [13], a moderate affinity for PBP1a and poor affinities for PBP1b, 2,4 and $\frac{5}{6}$ [6]. It has been suggested that both PBP1b and 3 are responsible for the bactericidal effect of β-lactam antibiotics [3]. The fact that AZT is bactericidal [12] in spite of its binding only to PBP3 and not to PBP1b might be due to additional effects of AZT. Indeed, morphological alterations were caused by sub-MICs of monobactams. Disorganization and aberrant structure of cytoplasmic membranes were observed in AZT-treated *E. coli* (Fig. 5b, c), and other morphological changes have been reported in AZT-treated *P. aeruginosa* [5]. Furthermore, increasing susceptibility to lysozyme in the presence of monobactams may be induced by impairment of the outer membrane of *E. coli*, which is known to be the barrier for many substances including lysozyme.

E. coli cells damaged by sub-MICs of monobactams were significantly more susceptible to phagocytosis and to the bactericidal activity of macrophages. The exact mechanism of action of monobactams on enhanced macrophage phagocytosis and killing effects remains to be elucidated. The synergistic action of monobactams on macrophage bactericidal functions may play an important role in host defences against gram-negative bacteria.

Conclusion

The effects of sub-MICs of monobactams (Aztreonam and AMA1080) on the host-parasite relationship were studied with an *in vitro* system using an established mouse macrophage cell line. The presence of monobactams ($\frac{1}{32}$–1 MIC) enhanced the macrophage bactericidal activity against *E. coli* S615. Even 4 times the MIC of monobactams had no direct effect on macrophages. The synergistic bactericidal effects were also observed when *E. coli* cells were treated with sub-MICs of monobactams and lysozyme or macrophage lysate. From these results, it is suggested that the effects of monobactams are exerted not on macrophages themselves but on bacteria; in the presence of monobactams, *E.coli* cells become more susceptible to bactericidal substances released from macrophages. Electron microscopic observations support this idea.

(A description of these findings has been already submitted to "Antimicrobial Agents and Chemotherapy".)

Summary

The influences of sub MIC of several ß-lactam antibiotics on the killing of bacteria were studied *in-vitro* using of established mouse macrophage cell line. The presence of sub MIC of Monobactam group antibiotics, Azthreonam and AMA-1080, enhanced the bactericidal activity of macrophages against *E. coli*S615. Such effect could not be induced by some other cephalosporins. It is suggested from experiments of macrophages pretreated with these drugs and morphological observations that such effect could not be ascribed to the action of these drugs on macrophages but to that on bacteria.

References

1. Andreana A, Perna P, Utili R, Dilillo M, Ruggiero G (1984) Increased phagocytosis and killing of *Escherichia coli* treated with subinhibitory concentrations of cefamandole and gentamicin in isolated rat livers. Antimicrob Agents Chemother 25: 182–186
2. Burdett IDJ, Murray RGE (1974) Septum formation in *Escherichia coli:* characterization of septal structure and the effects of antibiotics on cell division. J Bacteriol 119: 303–324
3. Chase HA, Fuller C, Reynolds PE (1981) The role of penicillin-binding proteins in the action of cephalosporins against *Escherichia coli* and *Salmonella typhimurium.* Eur J Biochem 117: 301–310
4. Eisenstein BI, Ofek I, Beachey EH (1979) Interference with the mannose binding and epithelial cell adherence of *Escherichia coli* by sublethal concentrations of streptomycin. J Clin Invest 63: 1219–1228
5. Elliott TSJ, Greenwood D (1984) The morphological response of *Pseudomonas aeruginosa* to aztreonam, cefoperazone, ceftazidime and N-formimidoyl thienamycin. J Med Microbiol 17: 159–169
6. Georgopapadokou NH, Smith SA, Sykes RB (1982) Mode of action of azthreonam. Antimicrob Agents Chemother 21: 950–956
7. Leijh PCJ, Barselaar M, Zwet T, Dubbeldeman-Rempt I, Furth R (1978) Kinetics of phagocytosis of *Staphylococcus aureus* and *Escherichia coli* by human granulocytes. Immunol 37: 453–456
8. Lorian V (1975) Some effects of subinhibitory concentrations of antibiotics on bacteria. Bull N Y Acad Med 51: 1046–1055
9. Ohnishi H, Kosuzume H, Inaba H, Okura M, Mochizuki H, Suzuki Y, Fujii R (1983) Effects of AC-1370, a new semisynthetic cephalosporin, on phagocyte functions. Antimicrob Agents Chemother 23: 874–880
10. Petit JC, Dagnet GL (1981) Enhanced killing of *Pseudomonas aeruginosa* by human polymorphonuclear leukocytes in presence of subinhibitory concentrations of carbenicillin and ticarcillin. Biomedicine 34: 29–33
11. Root RK, Isturiz R, Molavi A, Metcalf JA (1981) Interactions between antibiotics and human neutrophils in the killing of Staphylococci. J Clin Invest 67: 247–259
12. Shah PM, Losert-Bruggner B, Still W (1981) Bacterial activity of SQ26,776. J Antimicrob Chemother 8(Suppl.E): 77–80

13. Spratt BG (1975) Distinct penicillin-binding proteins involved in the division, elongation and shape of *Escherichia coli* K12. Proc Natl Acad Sci U.S.A. 72: 2999–3003
14. Verhoef J, Peterson PK, Quie PG (1977) Kinetics of Staphylococcal opsonization, attachment, ingestion and killing by human polymorphonuclear leukocytes; quantitative assay using (^{3}H)-thymine labeled bacteria. J Immunol Methods 14: 303–311
15. Zerigs BJ (1981) Bactericidal and fungicidal assay. In: Horscowitz H, Holden HT, Bellanti JA, Ghaffen A (eds) Manual of macrophage methodology. Marcel Dekker, New York and Basel, p 251

Discussion

Daschner: Do you have any explanation for the longer survival of macrophages in the presence of aztreonam or other antibiotics?

Iida-Tanaka: E. coli cells have catalase and some other substances that may be toxic for macrophages, and also they are resistant to bactericidal action of macrophage lysozyme. Thus, E. coli cells are relatively resistant to intracellular killing systems of macrophages. In the absence of antibiotics, E. coli cells are intact and the bacteria ingested by macrophages may be able to proliferate in the macrophages. Consequently, macrophages degenerate and fail to survive. On the other hand, in the presence of antibiotics, macrophages can survive though phagocytizing a lot of filamentous E coli cells. Because E. coli cells are weakened by monobactam treatment, they may be readily damaged and killed by macrophages.

Ginsburg: One of the explanations for the lower toxicity to macrophages of the antibiotic-grown bacteria is the enhanced release of LPS by the antibiotic during growth.

The synergistic effect of lysozyme and antibiotics on the killing of the bacteria is an important phenomenon. I would like to suggest that proteases may further enhance the destruction of bacteria hit by antibiotics and lysozyme. This is based on our observation that proteases will markedly enhance the destruction of mammalian cells injured by cytotoxic antibodies. The last figure you showed suggests that the break in the cell walls is located adjacent to the cross walls where the murein hydrolases are localized, suggesting that these may be activated.

Iida-Tanaka: At this moment, we have no evidence. However, our speculation is the following: The penicillin-binding-protein profile of aztreonam has indicated a very high affinity to PBP3 which controls the septum formation of E. coli. Therefore, monobactams inhibit the septum formation of E. coli and induce filament formation. The septum region therefore may be much more susceptible to autolytic enzymes which contribute to septum formation of E. coli cells.

Martin (comment on question of Dr. Ginsburg): Why do aztreonam treated E. coli cells show breakdown of the cell wall restricted to the sites of cross wall formation?

As has been shown some time ago by Squibb workers, aztreonam binds preferentially to the enterobacterial penicillin-binding protein 3, a murein transpeptidase engaged in murein synthesis during septum formation of the bacterial cell wall. As a consequence of this specific inhibitory interaction *and* the continuing activity of the bacterial murein hydrolases ("autolytic" enzymes) localized breakdown of the murein and outer membrane will occur at those sites of the cell wall only where septum formation and cell division should have taken place. The openings thus created on the bacterial surface will also allow penetration of externally added lysozyme into the cell wall and further degradation of murein. This appears to explain the enhancement of bacteriolytic activity of lysozyme on E. coli in combination with some β-lactam antibiotics, as described by Dr. Iida-Tanaka and her colleagues.

IV. The Influence of Antibiotics on the Pathogenicity Factors and Virulence

Effects of Antibiotics on Streptococcal Adherence: Release of Fibronectin-Lipoteichoic Acid Complexes from Group A Streptococci by Sublethal Concentrations of Penicillin

E. H. BEACHEY[1,2], T. J. NEALON[1*], H. S. COURTNEY[1,2], and W. A. SIMPSON[1,2]

[1] From the Veterans Administration Medical Center and the Departments of Medicine
[2] Microbiology and Immunology, University of Tennessee Center for the Health Sciences, Memphis, TN 38104
* Present Address: Department of Research Immunology, San Antonio State Chest Hospital, San Antonio, TX 78223

Introduction

In a number of studies [1, 3, 10] reviewed in Beachey 1981 [4] and Beachey et al. [6], we have obtained evidence that surface lipoteichoic acid (LTA) mediates the attachment of group A streptococci to oral epithelial cells. We have also obtained evidence that fibronectin molecules residing on epithelial cells serve as receptors for the LTA mediated binding of group A streptococci [12]. Indeed, we demonstrated that the fibronectin molecule, a large glycoprotein found in plasma and various connective tissues, as well as on the surfaces of oral epithelial cells, contains specific fatty acid binding sites that recognize the glycolipid end of the LTA molecule [7]. The binding of radiolabeled fibronectin to cells of group A streptococci was inhibited specifically by LTA, but not deacylated LTA. Similarly, purified LTA, but not deacylated LTA, was capable of forming complexes with purified plasma fibronectin [7]. These studies provide compelling evidence for the notion that LTA mediates the attachment of group A streptococci to fibronectin molecules on the surface of oral epithelial cells and thereby initiates the events necessary for oropharyngeal colonization by these microorganisms.

In none of these studies, however, have we shown in a direct way that fibronectin forms complexes with LTA on the surface of the bacteria. Therefore, the present study was undertaken in an attempt to determine whether such complexes are formed *in situ*. We took advantage of our previous findings that group A streptococci incubated in the presence of sublethal concentrations of penicillin released their surface LTA; as they lost their surface LTA, the bacteria lost their ability to bind to oral epithelial cells [1]. We reasoned that if fibronectin binds to LTA, the bound fibronectin should be released along with the LTA in the presence of sublethal concentrations of penicillin.

Methods

The bacteria employed for these studies consisted of an M type 5 strain of *Streptococcus pyogenes* and the ATCC 25923 strain of *Staphylococcus aureus*. The latter

The Influence of Antibiotics on the
Host-Parasite Relationship II
Edited by D. Adam, H. Hahn, W. Opferkuch

sensitive fibronectin receptor on these organisms. The organisms were cultured in Todd-Hewitt broth (THB), washed and used in studies of the binding of radiolabeled fibronectin as described [12]. To study the effect of penicillin as compared to clindamycin, the cultured bacteria were sedimented by centrifugation and resuspended in the original volume of THB supplemented with serial 10-fold dilutions of 10 μg/ml of penicillin V or clindamycin, and incubated for the indicated time intervals at 37 °C. Fibronectin was purified from plasma and radiolabeled by reductive methylation with [^{3}H]formaldehyde as described [8]. LTA was isolated from *Streptococcus pyogenes* 1RP41 and purified as described [12]. Antisera against LTA, type 5 M protein (pep M5) and whole type 5 streptococci were raised in rabbits [9]. Goat anti-rabbit IgG was purchased from Cappel Laboratories, Cochranville, PA. Penicillin V (phenoxymethylpenicillinic acid) was obtained from Sigma Chemical Co., St. Louis, MO and clindamycin hydrochloride (Cleocin) from The Upjohn Co., Kalamazoo, MI.

Results

In order to determine the specificity of the formation of LTA fibronectin complexes on the surface of group A streptococci, we undertook comparative studies with a strain of *S. aureus*. The latter organisms are known to possess LTA, but appear to bind fibronectin mainly via a protein moiety that is sensitive to clindamycin [11]. In contrast, we have garnered considerable evidence that LTA is the major streptococcal receptor for fibronectin and have shown that LTA is released and adherence is lost from streptococci treated with penicillin [1] whereas streptococci treated with clindamycin gained adhering ability [5]. Thus streptococci and staphylococci were compared in the following studies.

In preliminary studies, we found that the viability of neither the streptococci nor the staphylococci was affected by the concentrations of penicillin or clindamycin employed in these studies [9]. As previously reported [1], increasing concentrations of penicillin caused a dose-response increase in the release of LTA from group A streptococci. We found a similar release of LTA from the test strain of staphylococci (Table 1). In contrast, clindamycin had no effect on LTA release from either organism.

Next, we investigated the effects of penicillin and clindamycin on the binding of fibronectin to streptococci and staphylococci. As expected, the binding of fibronectin to streptococci decreased in the presence of penicillin (Table 2) and the decrease corresponded to the loss of LTA (Table 1). Clindamycin had no effect on fibronectin binding to streptococci. In contrast, the binding of fibronectin to staphylococci was reduced by clindamycin, but not by penicillin. These results are consistent with the idea that LTA is the major receptor for fibronectin on streptococci, whereas a protein moiety is the major receptor for this molecule on staphylococci.

The specificity of the binding was demonstrated by the ability of unlabeled fibronectin to inhibit the binding of [^{3}H]fibronectin to streptococci or staphylococci [9]. Furthermore, LTA inhibited the binding of fibronectin to antibiotic treated streptococci equally as well, if not better, than to untreated control organisms. In

Table 1. Release of LTA from *S. pyogenes* and *S. aureus* induced by penicillin but not clindamycin

	Release of [^{3}H]LTA, cpm × 10^{-3} [a]			
	S. pyogenes		*S. aureus*	
Time, hrs.	Pen.	Clinda.	Pen.	Clinda.
0	0	0	0	0
1	19	1	30	0
2	23	2	67	7
3	29	1	93	8

[a] The results are expressed as CPM of [^{3}H]labeled material released from bacteria grown in [^{3}H]glycerol and washed as previously described [1]. The values obtained with control organisms grown in the absence of antibiotics have been substracted from each test value.

Table 2. Effect of penicillin and clindamycin on the binding of fibronectin to streptococci and staphylococci

	[^{3}H]fibronectin binding, % of control			
	S. pyogenes		*S. aureus*	
Times, hrs.	Pen.	Clinda.	Pen.	Clinda.
0	100	100	100	100
1	88	100	100	92
2	60	100	100	63
3	35	92	100	38

contrast, LTA failed to inhibit the binding of fibronectin to antibiotic treated or untreated control staphylococci [9].

The next question was whether or not the bound fibronectin could be released from the bacterial cells with penicillin or clindamycin. As can be seen in Table 3, when cells of *S. pyogenes* coated with [^{3}H]fibronectin were incubated with penicillin, most of the radiolabel was released. Clindamycin had no effect. Neither antibiotic released the bound fibronectin from cells of *S. aureus* (Table 3).

Finally, we analyzed the material released from the fibronectin coated streptococci for the presence of LTA-fibronectin complexes. Anti-LTA, but not

Table 3. Effect of penicillin and clindamycin on the release of [^{3}H]fibronectin bound to streptococci or staphylococci

	% of [^{3}H]fibronectin released from:	
Antibiotic	*S. pyogenes*	*S. aureus*
None	10	2
Penicillin	87	2
Clindamycin	5	2

Table 4. Immunoprecipitation of LTA-fibronectin complexes released from streptococci by penicillin

Serum[a]	% of released [^{3}H]fibronectin precipitated
Anti-LTA	
unabsorbed	92
absorbed with LTA-RBC[b]	2
Anti-pep M5 protein	1
Preimmune serum	3

[a] The supernatants of streptococci incubated with penicillin for 3 hrs were incubated with the various sera followed by goat anti-rabbit IgG. The amount of fibronectin in the precipitated complexes was estimated by scintillation counting.
[b] LTA-RBC, washed sheep erythrocytes coated with LTA as described [9].

preimmune nor anti-M protein serum followed by goat anti-rabbit IgG, precipitated almost all of the radiolabeled fibronectin released from the streptococci (Table 4). The immunoprecipitating activity of the anti-LTA was abolished by absorbing the antiserum with erythrocytes coated with LTA, but not by control uncoated erythrocytes. These results suggest that fibronectin is released in complex with LTA from streptococci treated with sublethal concentrations of penicillin.

Discussion

Our results support the notion that LTA is the major receptor for the binding of fibronectin to group A streptococci. Although staphylococci possess similar amounts of LTA on their cell surfaces, the molecule appears not to bind significant quantities of fibronectin. Rather, in agreement with Proctor et al. [11], our data support the idea that the receptor for fibronectin or staphylococci is composed of a protein moiety. The sensitivity of the latter receptor to clindamycin, but not penicillin, strengthens this concept.

The importance of LTA in the attachment of group A streptococci to mucosal surfaces of the upper respiratory tract has been well established (for review, see Beachey et al. [6]). However, these are the first studies demonstrating that the LTA is able to form complexes with its host cell receptor (fibronectin) *in situ*. The finding that sublethal concentrations of penicillin release not only the LTA, but also the receptor material already attached to it, may have important clinical implications, not only in regard to group A streptococcal infections, but also with regard to prevention and therapy at the earliest adherence and colonization steps of serious bacterial infections in general. Our studies provide new approaches to the use of antibiotics as antiadhesive agents that may prevent or disengage ligand-receptor interactions that mediate bacterial adherence.

Acknowledgement. The research studies of the authors are supported by Research funds from the US Veterans Administration and by Research Grants AI-10085 and AI-13550 from the US Public Health Service. Dr. Nealon was the recipient of a post-doctoral traineeship award (AI-07238) and Mr. Courtney was the recipient of a pre-doctoral traineeship award (AI-07238) from the US Public Health Service.

Summary

We investigated the effect of sublethal concentrations of clindamycin and penicillin on the binding of fibronectin to *Staphylococcus aureus* and *Streptococcus pyogenes*. We found that clindamycin increased the binding of fibronectin to *S. pyogenes* but decreased the binding to *S. aureus*. Subinhibitory concentrations of penicillin had the opposite effect; it decreased the binding of fibronectin to *S. pyogenes* but increased binding to *S. aureus*. The latter decrease in fibronectin binding to *S. pyogenes*was associated with a release of lipoteichoic acid from the streptococcal cells. These findings suggest that subinhibitory concentrations of antibiotics may influence the binding of these gram positive bacteria to fibronectin receptors which reside on mucosal surfaces.

References

1. Alkan ML, Beachey EH (1978) Excretion of lipoteichoic acid by group A streptococci: influence of penicillin on excretion and loss of ability to adhere to human oral mucosal cells. J Clin Invest 61: 671–677
2. Beachey EH, Dale JB, Simpson WA, Evans JP, Knox KW, Ofek I, Wicken AG (1979) Erythrocyte binding properties of streptococcal lipoteichoic acids. Infect Immun 23: 618–625
3. Beachey EH, Ofek I (1976) Epithelial cell-binding of group A streptococci by lipoteichoic acid on fimbriae denuded of M protein. J Exp Med 43: 759–771
4. Beachey EH (1981) Bacterial adherence: adhesion-receptor interactions mediating the attachment of bacteria to mucosal surfaces. J Infect Dis 143: 325–345
5. Beachey EH, Eisenstein BI, Ofek I (1982) Prevention of the adhesion of bacteria to mucosal surfaces: influence of antimicrobial agents. In: Eickenberg/Hahn/Opferkuch (eds) Influence of Antibiotics on the Host-Parasite Relationships, Springer Verlag, Munich, pp 171–182
6. Beachey EH, Simpson WH, Ofek I, Hasty DL, Dale JB (1983) Attachment of *streptococcus pyogenes* to mammalian cells. Rev Infect Dis 5: S670–S677
7. Courtney HS, Simpson WA, Beachey EH (1983) Binding of streptococcal lipoteichoic acid to fatty acid-binding sites on human plasma fibronectin. J Bact 153: 763–770
8. Grinnel F (1980) Fibroblast receptor for cell-substratum adhesion: studies on the interaction of baby hamster kidney cells with latex beads coated by cold insoluble globulin (plasma fibronectin). J Cell Biol 86: 104–112
9. Nealon TJ, Beachey EH, Courtney HS, Simpson WA (1985) Release of fibronectin-lipoteichoic acid complexes from group A streptococci with penicillin. Submitted for publication
10. Ofek I, Beachey EH, Jefferson W, Campbell GL (1975) Cell membrane-binding properties of group A streptococcal lipoteichoic acid. J Exp Med 141: 990–1003
11. Proctor RA, Hamill RJ, Mosher DF, Textor JA, Olbrantz PJ (1983) Effects of subinhibitory concentrations of antibiotics on *Staphylococcus aureus* interactions with fibronectin. J Antimicrob Chemother 12: Suppl. C. 85–95
12. Simpson WA, Beachey EH (1983) Adherence of group A streptococci to fibronectin on oral epithelial cells. Infect Immun 39: 275–279

Discussion

Ginsburg: As we have demonstrated LTA (cell-sensitizing agent) is prevented from binding to RBC by a variety of phospholipids. Presumably these act as receptors (binding sites) for LTA. It seems that outside the oral cavity phospholipids act as the major binding sites for LTA.

Beachey: There is no doubt that the lipid end of LTA must be able to intercalate into phospholipid bilayers that may be more available on erythrocytes or other naked cells. As you stated, the oral mucosal cells as well as other mucosal surfaces are covered by a number of secretory materials that constantly bath these surfaces. We believe that fibronectin is one of those secretory secretions that coat oral mucosal cells.

Gemmell: Can you tell us what happens when a mixture of *Strept. pyogenes* cells and *Staph. aureus* cells are allowed to interact with fibronectin? Do they compete for the same receptor site in the molecule and does penicillin at sub-MIC levels differentiate between the binding of staphylococci and streptococci?

Beachey: With respect to the coaggregation by fibronectin of streptococci and staphylococci, Dr. Itzhak Ofek and Andy Simpson have just begun these sorts of experiments, the idea being that if coaggregation occurs, the different species of organisms must bind to different sites of the fibronectin molecule. Indeed, they have been able to show coaggregation among several different species of bacteria. There has been some problem of reproducibility perhaps because the Fn molecule is really a dimer held together by disulfide bonds at one end. Early experiments suggest that reproducibility is better with reduced and alkylated Fn.

Döring: Is the M-protein very susceptible to proteolytic cleavage? Especially to PMN-derived proteases, e. g. PMN elastase or cathepsin G? Do you get reduced ahderence of Streptococci in high inflammatory states?

Beachey: The M protein is exquisitely sensitive to proteolytic enzymes. In fact, it produces its own proteinase that can completely denude the bacteria of M protein if the pH of the culture media is reduced below 6.8. As to whether or not the organisms adhere less well in inflammatory tissues where PMN derived proteases are high, I don't know. It would be difficult to test this question in vivo because of the large number of variables such as PMN content, pH, fibronectin content as well as the content of a number of other blood and serum factors.

Stephens: Do you find increased concentrations of M protein or other surface proteins in supernatants of group A Streptococci, exposed to subinhibitory concentrations of penicillin?

Beachey: With regard to release of M protein by antibiotics, we should really look at this more carefully now that the technology has been better developed. Using rather crude methods a number of years ago we were unable to detect release of M protein by immunoprecipitation methods. These studies should now be repeated with such methods as Western blots and ELISA inhibition.

Pulverer: You mentioned that the streptococcal surface is hydrophobic. Is this a general statement for all streptococci or are there differences from strain to strain as we found it when screening Coagulase negative staphylococci?

Beachey: I agree with you that there may be considerable variation in hydrophobicity among different strains. Perhaps even more importantly we have observed variation within the same strain depending on the phase of growth. For example, streptococci in the log phase of growth are highly encapsulated and are hydrophilic. However, such log phase organisms also fail to adhere either to epithelial cells or to macrophages. I should emphasize that we always used streptococci in the stationary phase of growth at which stage adherence is the best. With regard to the variability in hydrophobicity among strains of coagulase-negative staphylococci, Dr. Gordon Christensen has been examining a large number of coagulase-negative staphylococci for their relative hydrophobicity. Although he has not yet identified all of the contributing factors, it is well known that certain strains produce more slime than others. It appears that hydrophobicity may be related to slime production as well as other factors.

Influence of Subinhibitory Concentrations of Antibiotics on Aggregation and Adherence of *Streptococcus mutans*

M. Ghione[1], G. Meixelsperger[2], G. Pelizzoni[2], and T. Gliozzi[1]

[1] Istituto di Microbiologia Medica Universitá di Milano
[2] Clinica Odontoiatrica Universitá di Milano

Introduction

Adherence phenomena play a pivotal role in providing a selective connection between microorganisms and host structures. A good example is offered by the dental plaque, the pathologic structure involved in the inception and progression of dental caries and some periodontal diseases. The plaque formation starts owing to the aggregation of gram-positive, oral bacteria and their adherence to the smooth surface of tooth enamel.

The microorganisms endowed with this ability include some *Actinomycetes* and chiefly *Streptococcus mutans* [18,22,34].

Together with these strains a variety of gram-negative cocci and rods can enter into the composition of the plaque, forming a till now insufficiently explored ecosystem [36].

In spite of the widespread acceptance of the appellation the taxonomic position of *S. mutans* is not yet fully clarified.

This situation accounts for the difficulty frequently encountered in correctly classifying a certain number of isolates. For these reasons the operational term of *S. mutans* complex is adopted in the present context to assemble the groups of microaerobic or anaerobic oral *Streptococci* which in spite of their antigenic [3] genetic [5] and biochemical [33] heterogeneity exhibit a phenotypic homogeneity [4] and share the capability of synthesizing high-molecular-weight water-insoluble glucan polymers directly from sucrose through the activity of a cell bound enzyme (glucosyltransferase: GTF). Emphasis is put on this characteristic because of its relevance in the mechanism of aggregation, and smooth surface adherence. These phenomena together with the excretion of acidic fermentation products are indeed the major determinants of the pathogenic properties of this group of bacteria [23].

The key position held by *S. mutans* complex in oral pathology accounts for the abundant research centered on these microorganisms. Effort has been devoted to detect antibiotics or other antibacterial compounds capable of exerting a selective inhibiting activity without altering the delicate equilibrium of the oral microbial ecology.

The trials until now carried out to this effect either failed, or were explored only in limited cases or at the time of writing are still at preliminary stages [10,6,25,28,32]. A different approach aiming to obtain not the killing but the selective dislodgment of *S. mutans* complex bacteria from their niche has been also envisaged.

The Influence of Antibiotics on the
Host-Parasite Relationship II
Edited by D. Adam, H. Hahn, W. Opferkuch

Attempts have been in fact carried out to investigate the possibility of selectively inhibiting glucan synthesis in these bacteria using antibodies [11], specific enzyme inhibitors of microbial origin [9, 29, 26, 7] sucrose analogues [31] and other sugars [24]. Since evidence exists that low antibiotic levels can interfere with the adherence of group A streptococci and *E.coli* to mucosal cells [1, 2, 37], the present study was carried out with the aim of obtaining information on the influence of some representative antibiotics, employed in subinhibitory concentrations, on the aggregation and adherence to smooth surfaces of bacteria belonging to the *S. mutans* complex. Considering that saliva interacts in a complex way with the clearance and attachment of oral streptococci [35, 19, 8] parallel assays of antibiotic activity have been carried out both in absence and presence of saliva.

Material and Methods

Culture and cultural conditions

S. mutans ATCC 25175 and 23 strains of human oral origin (isolated on MS agar supplemented with sucrose 20% and bacitracin 0.2 μg/ml, see GOLD *et al.* [21], and cloned on blood agar) belonging to *S. mutans* complex according to their morpho-functional characteristics [12–17], as well as *S. salivarius* ATCC 25975 and other streptococcal strains isolated from human oral flora namely *S. mitis, S. milleri, S. faecalis, S. pneumoniae,* used in this study were maintained in Brain Heart Infusion Broth (BHIB) and subcultured anaerobically for at least ten passages in a chemically defined medium (CDM), sterilized by filtration, containing sorbitol, aminoacids, vitamins and fenol red as indicator.

This medium supports rapid abundant and transferable growth of the tested strains in anaerobic conditions whilst BHIB allows the bacterial multiplication also in aerobic conditions.

Preparation of saliva coated wells

Whole saliva from two young adult donors was collected in an ice chilled container.

The saliva was clarified by centrifugation, heated at 60 °C for 30 min.' to inactivate the enzymes and sterilized by filtration through Millipore membrane.

50 μl of a 12.5% dilution of saliva in sterile saline solution were put in each microtiter plate well.

After incubation for 18 hours at +4 °C the content of the wells was poured off, the wells were rinsed with cold sterile saline and the plates were immediateley used for testing as below specified.

Growth, aggregation and adherence assay

The color change of the indicator, contained in the medium, from red to yellow was taken as evidence of bacterial metabolism and growth.

This method has been found to be preferable to the turbidimetric or nephelometric ones that, at borderline growth level, gave unreliable results owing to the variable size of the streptococcal chains. The method gave satisfactory estimates of the value of the minimal inhibitory concentration (MIC).

Bacterial aggregation was determined after short time incubation in media containing sucrose in final concentration ranging from 1% to 10^{-8}% (w/v) and different amounts of antibiotics.

The degree of aggregation was determined turbidimetrically or visually and rated +++, indicating clumping of the largest majority of bacteria (i. e. turbidimetric values from 20 to 40); ++ indicating heavy aggregation (t.v. 40–60); + indicating weak but appreciable aggregation (t.v. 60–90); or – indicating absence of aggregation (t.v. 90–100).

Bacterial growth adherence tests were carried out in microtiter plates incubated overnight under strict anaerobic conditions. The untreated or saliva precoated wells contained 160 μl of CDM (either plain or enriched with sucrose 0.1% final concentration and/or diverse concentrations of antibiotics) and 40 μl of bacterial suspension.

Bacteria which did not adhere to the plastic wells were removed by pouring off the medium and repeatedly washing the plates under tap water jet. The plates blotted on adsorbent paper pads were then stained by filling the wells with hydroalcoholic solution of gentian violet, washed again, dried and the absorbance was automatically read with a Titertek Multiskan (Flow Laboratories). The antibiotic concentration capable of reducing the bacterial adherence to 50% of the control value (AIC_{50}) can be easily deduced from the Titertek Multiskan printouts. The ratio of the parameters ($R = MIC/AIC_{50}$) was calculated for each experimental condition.

Some preliminary or confirmatory tests were performed either with CDM or BHIB in glass test tubes or plastic plates with large size wells to carry out turbidimetric reading with EEL colorimeter and more detailed microscopic observation with inverted Leitz microscope.

Antibiotics

Samples of Penicillin G, Erythromycin, Chlortetracyclin, Chloramphenicol, Streptomycin, Rifampicin, Doxorubicin, Daunorubicin, Bleomycin were purchased from the commerce. Concentrated acqueous solution were diluted with CDM to a wide range of final concentrations.

Determination of water-insoluble glucan

After incubation in CDM supplemented or not with sucrose the cultures were sedimented at 2000 × g for 10 min, washed three times with distilled water stained with gentian violet 10% (w/v) ethanol solution for 1 min and washed again three times with distilled water.

The residual dye was extracted with absolute ethanol, and diluted with distilled water to the original volume of the culture. The optical density (OD) of the solution was determined at 305 nm. wavelength with a Beckman spectrophotometer. The amount of dye absorbed by bacterial structures other than glucan, determined on bacterial cultures grown in CDM without sucrose, was fairly low and its amount was subtracted from the whole culture value to obtain corrected OD (COD) values.

The glucan production was scored as absent (−): COD values lower than 0.05; weak (+): COD from 0.05 to 0.15; marked (++): COD from 0.15 to 0.45; heavy (+++): COD higher than 0.45.

Results

Unless otherwise specified the data hereafter given are generated by tests carried out in microtiter plates. No significant differences have been observed when glass tubes or plastic wells of different type and size have been used.

Aggregation

The experimental data indicate that the degree of aggregation was dependent on the interaction of two main factors namely the bacterial strain employed and the concentration of sucrose in the culture medium.

At constant sucrose concentration each strain displayed an individual capacity to produce glucan and a parallel ability to aggregate. In fact, as shown in Table 1 the value of the minimal concentration of sucrose (MSC) capable of supporting a microscopically detectable streptococcal aggregation in CDM culture varied with different strains.

Since this MSC value was found to be consistent in repeated experiments this parameter provided the basis for the subdivision of bacteria belonging to *S. mutans* complex in subgroups such as e. g. good and poor glucan producers.

The data are reproducible provided that repeated subcultures in CDM without sucrose are carried out before testing. If, on the other hand bacteria had undergone subcultures in sucrose containing media, such as the ones including extractive components, results were not consistent upon retesting. Streptococci isolated from human oral samples and not belonging ot the *S. mutans* complex were unable to produce glucan regardless of the sucrose concentration and did not aggregate, with the exception of a strain of *S. saldivarius* ATCC 25975 (Table 1).

The aggregation of *S. mutans* strains was inhibited by antibiotics at various concentrations depending on the antibiotic and the strain but irrespective of the sucrose concentration and glucan production.

The minimal aggregation inhibiting concentration was found to be superimposable to the minimal growth inhibiting concentration (Fig. 1).

In our experimental conditions subinhibitory concentrations of some antibiotics have been observed to be capable of enhancing the bacterial aggregation. Such unexpected behaviour, displayed by Chlortetracycline, Anthracyclines and Rifampicin, but not other antibiotics, did not appear to be related to the intrinsic growth inhibitory activity of the compounds (Fig. 1).

Table 1. Relationship among aggregation, adherence, glucan production and minimal sucrose concentration able to induce microscopic aggregation (MSC)

Strains	Aggregation	Glucan[c] production	MSC μ/ml	Adherence
S. mutans 29 c 1[a]	+++	+++	1–10	+++
S. mutans ATCC 25175	++	++	100	++
S. mutans 32 m 2[b]	+	+	1 000	+
S. salivarius ATCC 25975	±	±	10 000	±
S. mitis	–	–	>100 000	–
S. milleri	–	–	>100 000	–
S. faecalis	–	–	>100 000	–
S. pneumoniae	–	–	>100 000	–

[a,b] strains taken as representative of good (a) or poor (b) glucan producer strains
[c] in glass test tube, culture in CDM + sucrose 0.1%

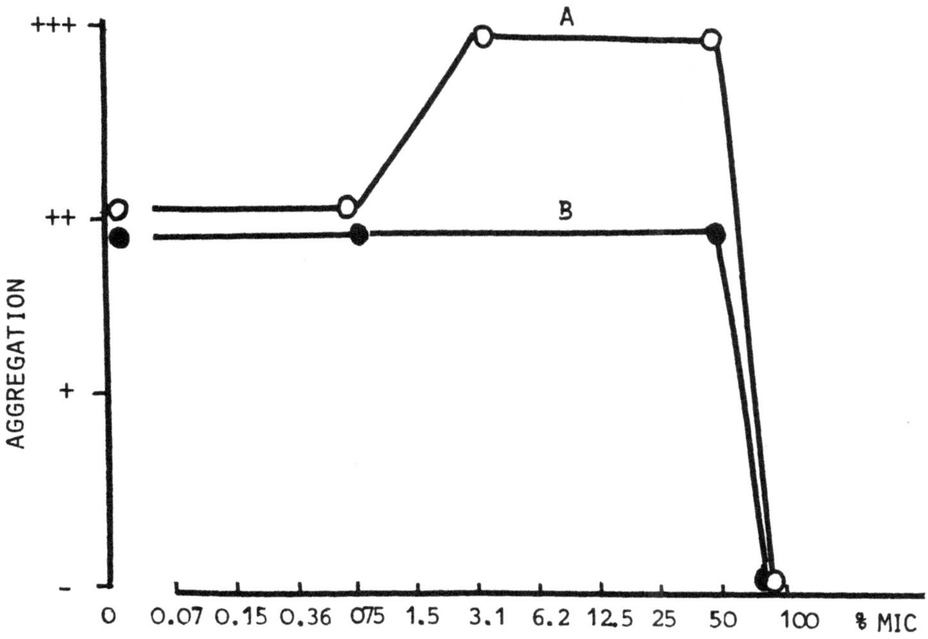

Fig. 1. Modification of *S. mutans* ATCC 25175 aggregation in CDM + sucrose 0.1% as a function of fractions of the MIC of various antibiotics. Curve A shows the activity of Chlortetracyclin (MIC = 0.05 μg/ml), Rifampicin (MIC = 0.006 μg/ml), Doxorubicin (MIC = 1.4 μg/ml), Daunorubicin (MIC = 6 μg/ml, whereas curve B represents Penicillin G (MIC = 0.01 μg/ml), Streptomycin (MIC = 3.5 μg/ml), Erythromycin (MIC = 0.02 μg/ml), Chloranphenicol (MIC = 0.9 μg/ml), Vancomycin (MIC = 0.05 μg/ml). Bleomycin did not inhibit *S.mutans* at the highest tested concentration (20 μg/ml). The data were generated by tests carried out contemporaneously under standardized and uniform conditions in glass tubes (16 × 160 mm). The actual MIC values can vary according to the test conditions

Adherence

Data similar to those described for aggregation have been observed for adherence.

At comparable levels of bacterial growth the amount of bacteria which adhered to glass or plastic smooth surfaces dependend on the sucrose concentration in the medium.

At identical sucrose concentrations the adherent cell fraction varied for different bacterial strains and depended on the individual strain capability to produce water-insoluble glucan polymers (Table 1).

A clear correlation has been observed between the MIC and AIC_{50} values. For different antibiotics the R (MIC/AIC_{50}) values have been found either to cluster around mean values (as it is the case of Penicillin G and Streptomycin) or to distribute along a correlation line. Representative data generated by a set of contemporaneous tests are given in Fig. 2.

The adherence phenomena appeared to be enhanced by subinhibiting concentrations of the same antibiotics which were found to be able to induce an increase in the aggregation tests. Illustrative data are given in Fig. 3.

The growth inhibitory activity of the antibiotics was only slightly modified in tests carried out in saliva pre-coated plastic wells.

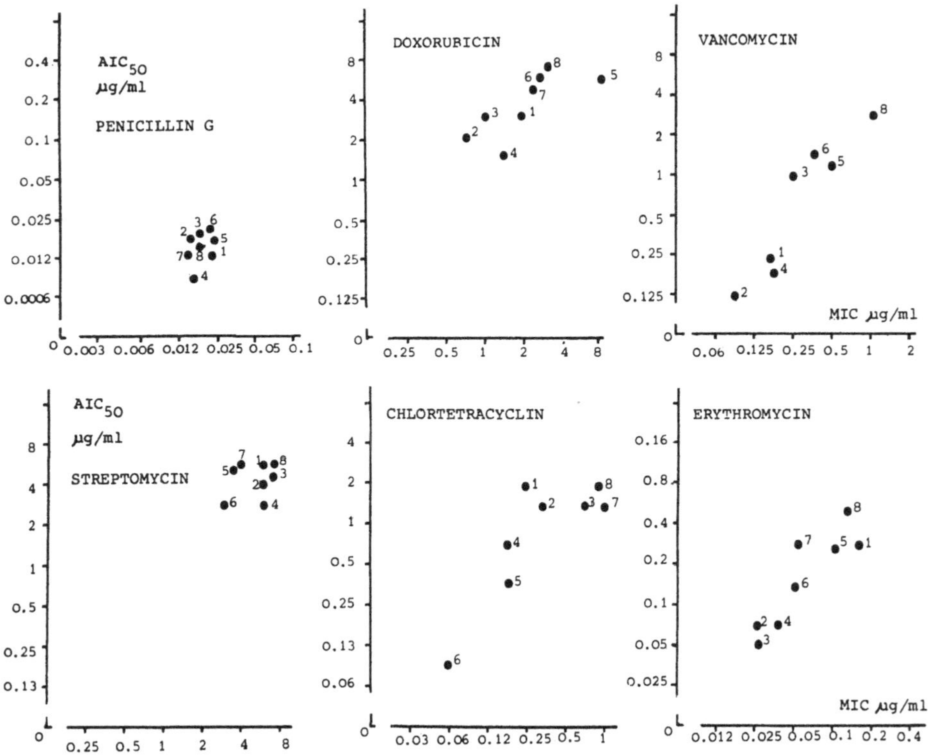

Fig. 2. R values observed for eight bacterial strains (1–8) belonging to *S. mutans* complex contemporaneously tested vs six antibiotics at different concentrations. MIC (µg/ml) values are given on the abscissa, AIC_{50} (µg/ml) values on the ordinate

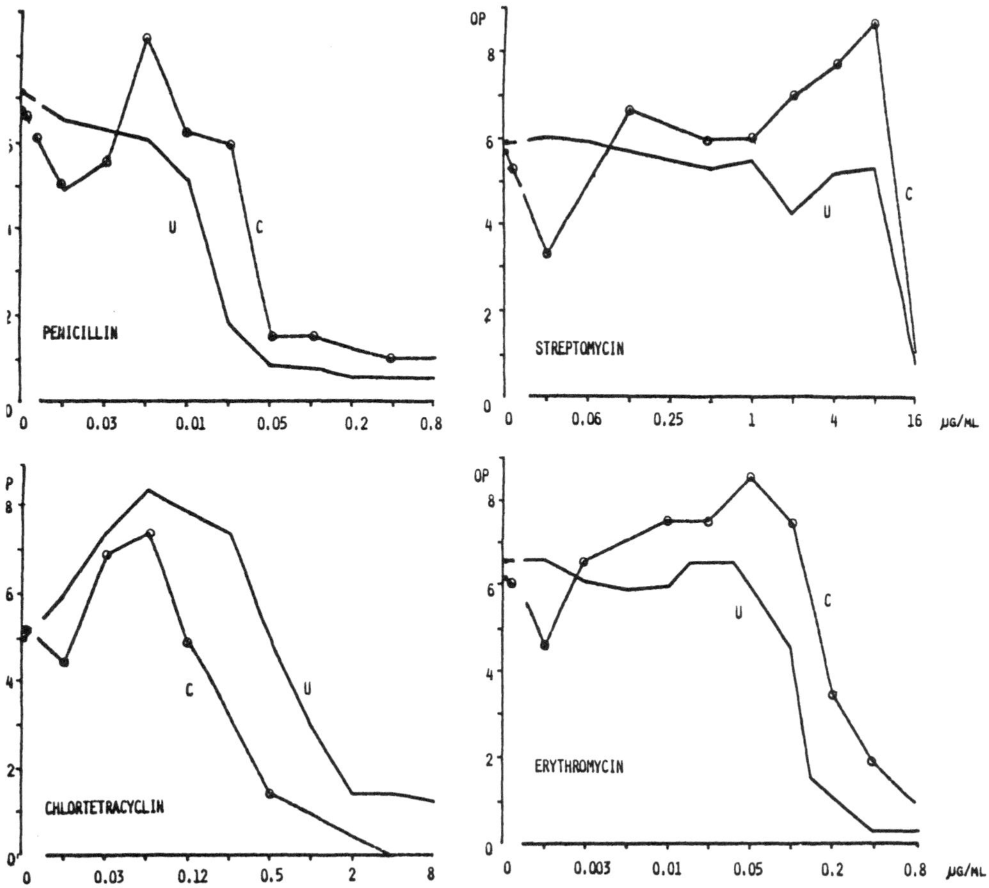

Fig. 3. Adherence curve of *S. mutans* ATCC 25175 cultures in CDM supplemented with 0.1% sucrose at different concentrations of diverse antibiotics (values on the abscissa). On the ordinate are given the opacimetric values (OP). The tests have been carried out in uncoated (U) or saliva precoated (C) plastic wells

The presence of saliva in the experimental conditions employed did not exert a significant influence on bacteria in terms either of growth or adherence in the absence of antibiotics. In saliva coated containers the capacity of subinhibitory concentrations of antibiotics to enhance adherence was increased even in the case of antibiotics inactive in terms of adherence enhancement in the absence of saliva. A limited, but consistent and reproducible abatement of bacterial adherence was induced by very low concentrations of antibiotics only in saliva coated vessels.

Discussion

Bacterial adhesion to host structures is an important factor in the colonization of cutaneous or mucosal surfaces as well as of deep tissues.

This phenomenon plays a peculiar role in oral microbiology, indeed some bacteria colonizing the tooth surface are endowed with particularly strong adhesive properties possibly ro resist removal from oral cleansing systems and to ensure their retention in a niche open to a sustained interchange with the environment such as the oral cavity. The formation of water insoluble glucan from sucrose is a major factor determining the adhesion of *S. mutans* complex bacteria to smooth surfaces.

This characteristic and other glucan depending properties such as the capability of forming clumps or agglomerating other oral bacteria are believed to be of relevance in the formation of dental plaque.

The production of glucan depends on the presence of sucrose. However it should be noted that detectable amount of glucan can be synthesized by bacteria belonging to the *S.mutans* complex even at sucrose concentrations as low as 0.0001%. The biological properties of glucan synthesizing bacteria and chiefly their surface characteristics relating to aggregation and adherence are greatly influenced by such environmental changes [38].

Since variable amount of sucrose, also exceeding the above quoted concentration by some order of magnitude, are contained in complex media, a sucrose free synthetic medium (either plain or enriched with known amounts of sucrose and/or inhibitors) has been employed to study some aspects of processes interfering with aggregation and attachment of *S. mutans* to smooth surfaces.

This research was carried out in the frame of studies aimed at searching means for inducing a selective hindrance to the adherence of cariogenic bacteria without altering the equilibrium of other bacterial species in the oral environment [20].

The present study has shown that the glucan-mediated aggregation and the adherence to smooth surface of bacteria of the *S.mutans* complex was not inhibited, but enhanced by subinhibitory concentration of frequently employed antibiotics in the presence of plastic-adherent saliva components, whilst very low concentrations of antibiotics exhibited a small but reproducible adherence inhibiting effect.

The occurrence of opposite sign interactions among oral streptococci and diverse salivary factors can explain the dual role played by saliva in promoting and inhibiting the bacterial adherence. This phenomenon is documented by experimental data [35, 19, 8] and we suppose that it can be selectively magnified by interaction with antibiotics in diverse range of concentration.

No attempt was carried out to isolate the saliva factor(s) responsible for these interactions. Literature data [30] and some explorative assays (data not shown) indicate the relevance of antibodies in this context.

Taken together our data may have a bearing on some puzzling aspects of the trials carried out with antibiotics in caries treatment.

The analysis of clinical and experimental data show indeed that there is a discrepancy between the potential inhibitory effect on *S. mutans* complex bacteria displayed by antibiotic concentrations easily obtainable in the oral cavity and the rather limited or temporary results obtained with antibiotics in caries treatment or prevention.

Abundant experience demonstrates that glucan producing oral streptococci seldom can be eradicated from the tooth surface even by topical administration of high concentrations of inhibitors [6, 25, 28].

On one hand our findings confirm and amplify the literature data [10, 27] on the overall fairly good intrinsic sensitivity of *S. mutans* to antibiotics.

On the other hand our findings suggest that one of the causes of the incomplete success or failure in reaching the control of caries disease with antibiotics could be attributed to the paradoxical enhancing effect of bacterial adherence to saliva coated surfaces displayed by antibiotics. This enhancement can be seen over a large range of concentrations.

Existing knowledge on the drug kinetic suggests that after either systemic or topic administration of even full doses of antibiotics levels of drugs inhibitory for growth and adherence will be maintained in the oral cavity for short period of time, whereas subinhibitory adherence-enhancing concentration will last longer.

In conclusion the data of the present study indicate that a number of antibiotics at subinhibiting levels and irrespective of their mechanisms of antibacterial action induce in glucan producing bacteria of the *S. mutans* complex enhancement of aggregation and bacterial adherence to smooth surfaces particularly evident when they are coated with saliva.

These phenomena are associated with morphologic modifications, in fact subinhibiting concentrations of antibiotics induce the appearance of long chains made of large bacterial bodies, that probably establish with saliva coated surfaces a different connection than the short chains of small bacterial bodies typical of normal cultures of streptococci of the *S. mutans* complex.

Summary

The phenomena of aggregation of oral Streptococci belonging to the *S. mutans* complex and their coaggregation with other microorganisms of the oral flora, chiefly *Actinomycetes,* play a pivotal role in conditioning their adherence to the enamel surface.

Experimental data show that some of the mechanisms supporting these phenomena are modified by subinhibitory concentrations of antibiotics, and some are resistant. The cause of these discrepances will be discussed.

References

1. Alkan ML, Beachey EH (1978) Excretion of lipoteichoic acid by group A streptococci: influence of penicillin on excretion and ability to adhere to human oral mucosal cells. J Clin Invest 61: 671–677
2. Beachey EH, Eisenstein BI, Ofek I (1982) Adherence of bacteria. Prevention of the adhesion of bacteria to mucosal surfaces: influence of antimicrobial agents. In: The influence of antibiotics on the host-parasite relationship. Eickenberg J *et al.* (eds) Springer Verlag p 171–182
3. Bratthall D (1970) Demonstration of five serological groups of streptococcal strains resembling *Streptococcus mutans*. Odontol Revy 21: 143–152
4. Carlsson J (1968) A numerical taxonomic study of human oral streptococci. Odontol Revy 19: 137–160
5. Coykendall AL (1971) Genetic heterogeneity in *Streptococcus mutans*. J Bacteriol 106: 192–196

6. De Paola PF, Jordan HV, Berg J (1974) Temporary suppression of *Streptococcus mutans* in humans through topical application of Vancomycin. J Dent Res 53: 108–114
7. Endo A, Hayashida O, Murakawa S (1983) Mutastein, a new inhibitor of adhesive-insoluble glucan synthesis by glucosyltransferase of *Streptococcus mutans*. J Antib 36: 203–207
8. Ericson D (1984) Agglutination of *Streptococcus mutans* by low-molecular-weight salivary components: effect of B_2-microglobulin. Infect Immun. 46: 526–530
9. Felgenhauer B, Trautner K (1981) Inhibition der Glucosytransferase von *Streptococcus mutans* durch a-Glucosidase-Inhibitoren. Dtsch Zahnärztl Z 36: 841–844
10. Fitzgerald RJ (1972) Inhibition of experimental dental caries by antibiotics. Antimicrob Agents Chemother 1: 296–302
11. Furuta T, Nisizawa T, Chiba J, Hamada S (1983) Production of monoclonal antibody against a glucosyltransferase of *S. mutans* 6715. Infect Immun 41: 872–875
12. Gianni' E, Salvato A, Calderini A, Lisa M, Rescaldani R (1983) Patogenicità cariogena e *S. mutans*: nuovo metodo di isolamento per uno studio epidemiologico. Minerva Stomatol 32: 481–483
13. Gianni' E, Salvato A, Calderini A, Lisa M, Gittu' G (1983) Patogenicità cariogena e *S. mutans*: indagine epidemiologica in soggetti anziani edentuli. Minerva Stomatol 32: 777–781
14. Gianni' E, Salvato A, Calderini A, Lisa M (1984) Patogenicità cariogena e *S. mutans*: indagine epidemiologica in soggetti adulti con dentatura permanente. Minerva Stomatol 33: 565–570
15. Gianni' E, Salvato A, Calderini A, Lisa M (1984) Patogenicità cariogena e *S. mutans*: indagine epidemiologica in bambini edentuli e con dentatura decidua. Minerva Stomatol 33: 413–416
16. Gianni' E, Salvato A, Lisa M, Calderini A (1984) Patogenicità cariogena e *S. mutans*: tipizzazioni di ceppi isolati dal cavo orale. Stomatol Mediter 4: 203–209
17. Gianni' E, Salvato A, Calderini A, Lisa M (1984) Patogenicità cariogena e *S. mutans*: epidemiologica e tipizzazione dei ceppi isolati dal cavo orale. Odontoiatria Oggi 1: 21–26
18. Gibbson RJ, van Houte J (1975) Bacterial adherence in oral microbial ecology. Am Rev Microbiol 29: 19–44
19. Gibbson RJ, Moreno EC, Etherden I (1983) Concentration-dependent multiple binding sites on saliva-treated hydroxyapatite for *Streptococcus sanguis*. Infect Immun 39: 280–289
20. Gliozzi T, Mattina R, Ghione M (1983) Fattori influenzanti i fenomeni di aderenza e aggregazione in *S. mutans*. Atti XX Congr Nazionale Soc Ital Microbiologia Gardone Riviera 12–15 maggio 1983
21. Gold OGH, Jordan NV, Van Houte J (1973) A selective medium for *Streptococcus mutans*. Arch Oral Biol 18: 1357–1364
22. Hamada S, Slade HD (1980) Biology immunology and cariogenicity of *Streptococcus mutans*. Microbiol Rev 44: 331–384
23. Hamada S, Koga T, Ooshima T (1984) Virulence factors of *Streptococcus mutans* and dental caries prevention. J Dent Res 63: 407–411
24. Imai S, Takeuchi K, Shibata K, Yoshikawa S, Kitahata S, Okada S, Araya S, Nisizawa T (1984) Screening of sugars inhibitory against sucrose-dependent synthesis and adherence of insoluble glucan and acid production by *Streptococcus mutans*. J Dent. Res 63: 1293–1297
25. Jordan HV, De Paola PF (1977) Effect of a prolonged topical application of Vancomycin on human oral *Streptococcus mutans* population. Arch Oral Biol 22: 193–199
26. Koga T, Hamada S, Murakawa S, Endo A (1982) Effect of a glucosyltrasferase inhibition on glucan synthesis and cellular adherence of *Streptococcus mutans*. Infect Immun 38: 882–886
27. Kral TA, Callaway MD (1984) Penicillin-induced lysis of *Streptococcus mutans*. Infect Immun 46: 442–447
28. Loesche W, Bradley DR, Woolfolk MP (1977) Reduction of dental decay in rampant caries individuals following short-term Kanamycin treatment. J Dent Res 56: 254–265

29. Okami Y, Takashio M, Umezawa H (1981) Ribocitin, a new inhibitor of dextran sucrose. J Antib 34: 344–345
30. Olson GA, Bleiweis AS, Small PA Jr (1972) Adherence inhibition of *Streptococcus mutans*: an assay reflecting a possible role of antibody in dental caries prophylaxis. Infect Immun 5: 419–427
31. Ooshima T, Izumitani A, Sobue S, Okahashi N, Hamada S (1983) Non-cariogenicity of the disaccharide palatinose in experimental dental caries of rats. Infect Immun 39: 43–49
32. Pallanza R, Scotti R, Beretta G, Cavalleri B, Arioli V (1984) In vitro activity of A-16686, a potential antiplaque agent. Antimicrob Agents Chemother 26: 462–465
33. Shklair IL, Keene HJ (1976) Biochemical characterization and distribution of *Streptococcus mutans* in three diverse populations. In: Microbial aspects of dental caries. Stiles J *et al.* (eds). Information Retrieval Washington p 201–209
34. Socransky SS, Hubersak C, Propas D (1970) Induction of periodontal destruction in gnotobiotic rats by a human oral strain of *Actinomyces naeslundi*. Arch Oral Biol 15: 993–995
35. Tabak LA, Levine MJ, Mandel ID, Ellison SA (1982) Role of salivary mureins in the protection of the oral cavity. J Oral Pathol 11: 1–17
36. Van Der Hoeven JS, De Jong MH, Rogers AH, Camp PJM (1984) A conceptual model for the co-existence of *Streptococcus spp.* and *Actinomyces spp.* in dental plaque. J Dental Res 63: 389–392
37. Vosbeck K (1982) Effects of low concentrations of antibiotics on *Escherichia coli* adhesion. In: The influence of antibiotics on the host-parasite relationship. Eickenberg J *et al.* (eds). Springer Verlag p 183–193
38. Wu-Yuan CD, Tai S, Slade HD (1979) Properties of *S. mutans* grown in a synthesis medium: binding of glucosyltransferase and in vitro adherence, and binding of dextran/glucan and glycoprotein and agglutination. Inf Immun 23: 600–608

Discussion

Ginsburg: Is the enhanced adherence of *Strept. mutans* to saliva coated plates due to the presence of saliva "agglutinins"? These agglutinins may function to eliminate bacteria from the oral cavity and may be part of the protection against colonization. Some of the "agglutinins" may be associated with a group of histidine-rich proteins in saliva. The prolonged use of antibiotics to eliminate caviogenic bacteria may be associated by acquisition of high resistance to antibiotics and thus may not be a practical approach for prolonged periods.

Ghione: Saliva contains different factors possibly active von *S. mutans*. They include secretory antibodies, enzymes, bioactive proteins, metabolites, etc. Some of them can increase and some can decrease the bacterial aggregations. Most probably the ratio of enhancing and inhibiting factors changes in diverse physiologic and pathologic conditions. It is almost impossible to give a comprehensive picture of the dynamic situations that can take place at the level of the dental plaque. As far as exogeneous factors are concerned, for sure the presence of sucrose is essential for plaque formation and the administration of antibiotics can play a role in enhancing the adherence processes.

Clostridium Difficile Colonization of Infant Hamsters

R. D. ROLFE* and J. P. IACONIS

* Department of Microbiology, Texas Tech University Health Sciences Center. Lubbock, Texas 79430. USA

Introduction

Clostridium difficile-Associated Intestinal Disease

Toxigenic *C. difficile* is the major cause of antimicrobial-associated pseudomembranous colitis (PMC) in humans and of ileocecitis in Syrian hamsters [1,25]. All major classes of antimicrobial agents have been reported to induce *C. difficile*-associated intestinal disease in humans. In addition to PMC, *C. difficile* is the cause of approximately 30% of cases of antimicrobial agent-associated non-specific colitis or diarrhea without colitis [7,8]. *C. difficile* also is an etiologic agent of PMC or diarrhea without colitis in patients with no history of antimicrobial therapy, of diarrhea in patients receiving antineoplastic agents, and of inflammatory bowel disease exacerbations [6,18,28].

The mechanisms by which *C. difficile* causes diarrhea or mucosal injury are not entirely understood. *C. difficile* produces two immunologically and biochemically distinct toxins: toxin A (enterotoxin) and toxin B (cytotoxin) [27]. It is not known which of these toxins is responsible for the pathological changes seen in *C. difficile*-associated intestinal disease.

Clostridium difficile Colonization of Newborns and Infants

An extremely interesting, and as of yet unexplained, phenomenon is the high percentage of healthy neonates which harbor both intestinal *C. difficile* and cytotoxin. *C. difficile* has been isolated from the feces of up to 60% of healthy infants less than 1 year of age [15,26]. Carrier rates for *C. difficile* fall sharply after 1 year of age, although in the second year they are still higher than the 4% carrier rate for adults [26]. Reasons for the fall in colonization rates for *C. difficile* are not fully understood but could be due in part to a broadening diet, acquisition of competing flora, and/or maturation of mucosal immunologic processes. Concentrations of *C. difficile* and cytotoxin in the feces of healthy infants are frequently similar to concentrations found in the intestinal tracts of adults with PMC [4,17]. In addition, investigators report no significant differences in the recovery of *C. difficile* between children with and without gastroenteritis [10]. Interestingly, *C. difficile* has been

The Influence of Antibiotics on the
Host-Parasite Relationship II
Edited by D. Adam, H. Hahn, W. Opferkuch

reported to occur numerically more often in non-antibiotic treated children than in those given such agents [10].

The phenomenon of *C. difficile* colonization of the newborn and infant intestinal tract has become even more complex in recent years because increasing numbers of infants with *C. difficile*-associated intestinal disease, including PMC, are being recognized [5, 29]. These pediatric cases have occurred in the presence and absence of prior antimicrobial therapy [12]. Thus, it appears that *C. difficile* colonization and toxin production can occur in both healthy and symptomatic infants. Nonetheless, most investigators agree than the incidence of antimicrobial-associated intestinal disease, and pseudomembranous colitis in particular, is much lower in infants and older children than adults.

Why *C. difficile* readily colonizes the intestinal tracts of infants and is relatively rare in healthy adults remains an enigma. A majority of the theories proposed to explain the mechanisms by which *C. difficile* overgrows in the intestinal tract consider the interactions which undoubtedly exist between *C. difficile* and the normal intestinal bacterial flora. Several investigators have presented experimental evidence to show that the normal bacterial flora of the gastrointestinal tract constitute an extremely important defense mechanism which effectively interferes with the establishment of many enteric pathogens [9]. The dramatic quantitative and qualitative fluctuations in the bacterial populations of the normal intestinal flora which occur immediately after birth up until the time the animal begins to sample solid food indicate that the normal flora are not well balanced. It has been suggested that this may contribute to some of the intestinal diseases seen in young children since the protective mechanisms of the normal flora are probably diminished or absent. This may contribute to the overgrowth of *C. difficile* in the intestinal tracts of many infants. There is indirect evidence to suggest that the intestinal bacterial flora may be important in protecting the host against *C. difficile*-associated intestinal disease [20, 22, 32].

Why there are usually no deleterious consequences resulting from *C. difficile* colonization of neonates is also unknown. Some investigators have suggested that the transient nature of *C. difficile* carriage in infants may account for the lack of clinical manifestations [2]. However, several investigators have been able to repeatedly isolate *C. difficile* over periods of several months from stool specimens of asymptomatic neonates [4, 26]. The asymptomatic colonization may be related to the site of *C. difficile* localization in the intestinal tract. In the infant, *C. difficile* may localize in regions of the intestinal tract where the effect of the toxin(s) on the epithelial cells is negligible. The lack of toxic symptoms in infants colonized with *C. difficile* may be due to low levels of toxin A in their intestinal tracts despite the proliferation of the cytotoxin (toxin B). Investigators have shown that the titer of fecal toxin B, as detected by tissue culture assay, present in human newborns varies considerably [24, 30]. Since toxin B is at least 1,000 times more cytopathic to tissue culture cells than toxin A, it is unknown at what levels toxin A is present in the intestinal tracts of most colonized infants. On the other hand, some investigators have demonstrated the presence of toxin A in the intestinal tracts of some infants asymptomatically colonized with *C. difficile* [16].

There are a number of other possible explanations for the absence of symptoms in infants colonized with *C. difficile*, despite the high levels of cytotoxin occasionally present. Intestinal epithelial receptors for toxin A and/or toxin B may be blocked or not present in infants. *C. difficile* toxins may also be more readily inactivated in infants as compared to adults. It is possible that secretory IgA (or other immunoglobulins) is produced in the intestinal tracts of infants or is present in maternal milk that either inactivates *C. difficile* toxin or prevents its absorption. Several investigators have reported that intestinal *C. difficile* is found significantly more often in formula-fed than in breast-fed infants [3, 26]. Antibodies against *C. difficile* toxin have been demonstrated in colostral specimens obtained from post-partum women [14, 31]. Healthy infants who harbor high counts of *C. difficile* (with or without high titer of cytotoxin) may lack an essential component of the intestinal flora necessary for *C. difficile* (or it's toxins) to display pathogenicity. In addition, it is still not certain that strains of *C. difficile* isolated from healthy individuals are as virulent as those strains isolated from patients with intestinal disease.

More studies are definitely needed to adequately understand the exact importance of *C. difficile* and its toxins in the bowels of neonates and young infants. However, because of the low incidence of PMC in infants and the difficulties associated with regulating endogenous and exogenous parameters on the human host, these date are difficult to obtain. On the other hand, comprehensive studies employing an animal model of *C. difficile* symptomatic and asymptomatic colonization of the newborn intestinal tract can be employed to aid in resolving many of these issues. The administration of an antimicrobial agent to adult hamsters consistently results in the development of a fatal ileocecitis due to intestinal overgrowth by toxigenic *C. difficile*. This adult hamster model of antibiotic-induced *C. difficile*-associated intestinal disease has been an invaluable aid to our understanding of antimicrobial-induced diarrhea in humans. This animal model has permitted investigation of a number of aspects of antimicrobial-associated diarrhea and colitis by methods that are neither feasible nor ethical in humans. Investigations in our laboratory have centered around examining the ability of *C. difficile* to colonize the intestinal tracts of infant hamsters and the feasibility of using this animal model to study parameters associated with the asymptomatic and symptomatic colonization of infants.

Intestinal Colonization of Infant Hamsters with Clostridium Difficile

Asymptomatic Colonization

Preliminary investigations were performed in our laboratory to determine whether infant hamsters are colonized with *C. difficile* and, if colonized, what effect colonization has on the health of the animals [23]. Infant hamsters of different ages were administered *C. difficile* orogastrically and 72 hours later were sacrificed and their intestinal tracts examined for the presence of *C. difficile* and cytotoxicity to HeLa tissue culture cells. The results, summarized in table 1, show that hamsters have an age dependent susceptibility to *C. difficile* enteric colonization. *C. difficile* only colonized the intestinal tracts of hamsters between 4 and 11 days of age. Average numbers of *C. difficile* in the intestinal tracts of colonized hamsters ranged from 7.9×10^4 to 2.0×10^7 bacteria per gram intestinal homogenate. Toxin titers in the intestinal filtrates prepared from infant hamsters colonized with *C. difficile*

Table 1. Colonization of infant Hamsters with *Clostridium Difficile*

Age (Days) of Hamsters at Time of Challenge[a]	Percent of Hamsters Colonized[b]	Number of *Clostridium difficile*[c]	Titer of Cytotoxin[d]
1 (n=18)	0	0	0
2 (n=26)	0	0	0
3 (n=22)	0	0	0
4 (n=24)	58	6.5±1.7	1.7
5 (n=17)	73	6.6±2.3	2.2
6 (n=19)	77	7.1±0.6	3.4
7 (n=20)	100	7.2±0.8	4.8
8 (n=22)	82	7.3±1.1	4.2
9 (n=22)	50	6.7±1.9	1.8
10 (n=16)	25	5.1±1.7	0.9
11 (n=24)	25	4.9±2.5	0.5
12 (n=23)	0	0	0
13 (n=19)	0	0	0
14 (n=20)	0	0	0
15 (n=25)	0	0	0
16 (n=17)	0	0	0
>70 (n=10)	0	0	0

[a] Hamsters were challenged orogastrically with 10^7 viable cells of *C. difficile*. n, Number of hamsters tested.
[b] Hamsters were sacrificed 72 hours after *C. difficile* challenge and their intestinal tracts were examined for the presence of *C. difficile*
[c] Mean ($\log_{10}$) ± standard deviation per gram (wet weight) of intestine. Only hamsters colonized with *C. difficile* are included
[d] Reciprocal of the highest dilution causing actinomorphic changes of at least 50% of the cells in the monolayer ($\log_{10}$). Only hamsters with detectable levels of cytotoxin are included

ranged from undetectable to $10^{4.8}$. Intestinal filtrates prepared from hamsters not colonized with *C. difficile* were consistently negative for cytotoxicity. None of the infant hamsters receiving *C. difficile* displayed evidence of toxicity and all gained weight at a rate comparable to that of untreated controls.

Colonization of infant hamsters intestinal tracts with *C. difficile* was persistent in that *C. difficile* could be recovered from hamsters up to 8 days after intragastric challenge of 4 day old hamsters with this microorganism. The number of *C. difficile* required to colonize 50% of 7 day old hamsters was 18 viable cells. On the other hand, 10^8 viable cells of *C. difficile* failed to colonize the intestinal tracts of non-antibiotic treated adult hamsters.

Clindamycin-Induced Ileocecitis in Infant Hamsters

The administration of clindamycin to adult hamsters consistently results in the development of a fatal ileocecitis due to intestinal overgrowth by *C. difficile* and subsequent toxin elaboration. This presumably occurs as a result of clindamycin altering the intestinal flora permitting the intestinal colonization and/or overgrowth by toxigenic *C. difficile*. The administration of clindamycin to infant hamsters will also undoubtedly result in an alteration of the intestinal flora which in turn may influence the ability of *C. difficile* to colonize the intestinal tract as well as influence disease production. The experiments described above suggested that infant hamsters may be resistant to clindamycin induced ileocecitis since non-antibiotic treated infant hamsters are readily colonized with *C. difficile* with no deleterious consequences. The experiments described below were designed to examine the effect of clindamycin administration of *C. difficile* colonization of infant hamsters [13].

Hamsters of different ages were administered clindamycin and were sacrificed 72 hours after treatment. The intestinal tracts of all sacrificed animals were removed and tested for the presence of *C. difficile* and cytotoxicity to HeLa tissue culture cells.

The intestinal tracts of all hamsters receiving clindamycin that were ≥4 days of age were colonized with *C. difficile* whereas only 4 of 11 one day old hamsters receiving clindamycin were colonized with this microorganism. The results in table 2 show that all hamsters ≥4 days of age developed an ileocecitis following clindamycin challenge. The incidence of lethality increased with the age of the animals such that 92% of 16 day old animals and all adult animals (>70 days old) died within 3 days of clindamycin challenge. Fatal ileocecitis occurred in those infant hamsters receiving clindamycin that had the highest intestinal numbers of *C. difficile* and highest titers of intestinal cytotoxin. All strains of *C. difficile* isolated from infant and adult hamsters produced the same in vitro cytotoxin titers. As compared to non-antibiotic treated infant hamsters colonized with *C. difficile* (Table 1), the average numbers of *C. difficile* and cytotoxin titers were significantly higher in the intestinal tracts of clindamycin treated infant hamsters. However, some non-antibiotic treated infant hamsters asymptomatically colonized with *C. difficile* possessed concentrations of *C. difficile* and cytotoxin titers comparable to and even higher than the numbers and

Table 2. *Clostridium Difficile* Colonization of Clindamycin Treated Hamsters

Age (days) of Hamsters at Time of Clindamycin Challenge[a]	Percent of Hamsters with Ileocecitis[b]	Percent of Hamsters Dead[c]	Number of *Clostridium difficile*[d]	Titer of Cytotoxin[e]
1 (n=11)	0	0	2.8±0.7	0.3
4 (n=10)	100	0	7.2±0.5	4.9
7 (n=11)	100	38	7.4±0.4	5.1
10 (n=10)	100	50	7.1±0.2	5.0
13 (n= 7)	100	57	7.3±0.6	5.8
16 (n=12)	100	92	7.4±0.4	5.7
>70 (n= 5)	100	100	7.3±0.2	6.1

[a] Hamsters were treated orogastrically with clindamycin (3 mg per 100 grams body weight). n, Number of hamsters treated

[b] The intestinal tracts of hamsters were removed immediately from dead animals or within 72 hours of clindamycin challenge and were judged as having ileocecitis if their ceca were enlarged and hemorrhagic

[c] Hamsters dead within 72 hours after clindamycin challenge

[d] See table 1 footnote

[e] See table 1 footnote

titers present in the intestinal tracts of clindamycin-treated infant hamsters with *C. difficile*-associated ileocecitis. Therefore, the differences in intestinal cytotoxin titers and numbers of *C. difficile* between clindamycin treated and non-antibiotic treated hamsters can not alone account for the differences in clinical manifestations.

Many mechanisms could account for the differences in the toxicity of *C. difficile* for clindamycin and non-antibiotic treated infant hamsters. Since the cytotoxicity of toxin A is usually masked by toxin B in the tissue culture assay used in these investigations, the levels of toxin A in the intestinal tracts of infant and adult hamsters is unknown. Infant hamsters colonized with *C. difficile* may lack symptoms despite high cytotoxin levels because of the absence of or low titers of toxin A. This could be caused by intestinal colonization with enterotoxin deficient strains of *C. difficile* or decreased production or inactivation of toxin A in the intestinal tract.

Investigations have demonstrated that clindamycin stimulates the in vitro production of cytotoxin by *C. difficile* [11, 48]. It has been postulated that toxin stimulation could, in part, explain the occurrence of PMC in association with antimicrobial use. The occurrence of a similar phenomenon in infant hamsters could explain the higher average cytotoxin titers in clindamycin treated infant hamsters as a compared to non-antibiotic treated infant hamsters. Clindamycin may also increase the host cells response to *C. difficile* toxins our cause the exposure of toxin receptors on intestinal epithelium of infant hamsters which may be blocked in infants not receiving clindamycin. In addition, attachment of *C. difficile* to the intestinal mucosa of infants receiving clindamycin may be greater than the attachment which occurs in non-antibiotic treated hamsters resulting in greater exposure of the intestinal epithelium to the toxins of *C. difficile*.

Why clindamycin did not induce ileocecitis in hamsters 1 day of age is speculative. Some of these newborn hamsters challenged with clindamycin become colonized with low numbers of *C. difficile* and possessed low titers of cytotoxin. Cecal microbial flora studies have shown that anaerobic bacteria are not present in high numbers in the ceca of hamsters 1 day of age (Table 3) [21]. This suggests that the environmental conditions (e. g. oxidation-reduction potential) in the intestinal tracts of these newborn hamsters are not conducive for anaerobic growth and may explain the failure of *C. difficile* to attain high numbers in clindamycin treated 1 day

Table 3. Cecal Flora of Hamsters

Age (days) of Hamsters[a]	Total Aerobic Cecal Flora[b]	Total Anaerobic Cecal Flora[b]
1	6.86	5.90
4	9.18	8.95
7	7.58	10.02
10	6.08	9.76
13	6.34	9.88
16	6.11	9.45
>70	6.18	9.40

[a] Three hamsters at each age were sacrificed and the aerobic and anaerobic cecal flora was determined with selective and non-selective agar media

[b] Mean (log_{10}) number of bacteria per gram (wet weight) of intestine

old hamsters. Why clindamycin permitted some infant hamsters 1 day of age to become colonized with *C. difficile*, whereas in the absence of antibiotic treatment infant hamsters 1 day of age are completely resistant to *C. difficile* intestinal colonization, is unknown.

Conclusions

During the last few years it has been documented by several investigators that *C. difficile* intestinal colonization and toxin production can occur in both healthy and symptomatic human infants. Concentrations of *C. difficile* and it's cytotoxin in the feces of healthy infants are frequently similar to those in the intestinal tracts of adult humans with PMC. The mechanism of this commensal intestinal colonization of newborns and infants with toxigenic *C. difficile* is unknown.

Non-antibiotic treated infant hamsters have an age dependent susceptibility to non-lethal enteric *C. difficile* colonization similar to the restricted age distribution of human newborn colonization. Intragastric administration of *C. difficile* to infant hamsters results in recovery of the organism in the intestinal tracts of 4 to 11 day old animals. These colonized hamsters remain asymptomatic despite cytotoxin titers in some hamsters approximating those found in the intestinal tracts of adult humans and hamsters with *C. difficile*-associated intestinal disease. On the other hand, clindamycin administration to hamsters ≥ 4 days of age results in the development of *C. difficile*-associated ileocecitis which is lethal to many of the infant hamsters.

This infant hamster model of asymptomatic and symptomatic *C. difficile* intestinal colonization may help delineate the mechanisms by which *C. difficile* colonizes the intestinal tracts of human infants in the presence and absence of disease. Results from this animal model could also serve as a useful comparative system for the identification of those factors in the large bowel of adults which contribute to the etiology of PMC and other intestinal diseases caused by *C. difficile*.

Acknowledgments. The author's research described in this manuscript was supported by United States Public Health Services grant RO1-A121489 from the National Institute of Allergy and Infectious Diseases.

We thank Woosun Song and Kerri Weiser for excellent technical assistance.

Summary

Toxigenic *C. difficile* is the major cause of antimicrobial-associated colitis in humans and of ileocecitis in Syrian hamsters, the latter serving as an animal model of this disease. Intestinal disease due to *C. difficile* presumably occurs as a result of antimicrobial agents altering the intestinal flora permitting the intestinal colonization and/or overgrowth by toxigenic *C. difficile*. During the last few years is has been demonstrated by several investigators that *C. difficile* intestinal colonization and toxin production can occur in both healthy and symptomatic infants. Factors which permit asymptomatic colonization and toxin production can occur in both healthy and symptomatic infants. Factors which permit asymptomatic colonization in a majority of infants yet allow a few infants to develop severe intestinal disease remain obscure. Experiments in our laboratory have shown that infant hamsters display an age-restricted susceptibility to *C. difficile* intestinal colonization similar to human infants. Intragastric administration of *C. difficile* to non-antibiotic treated infant hamsters results in multiplication of the organism in the intestinal tracts of 4-to toxicity despite concentrations of *C. difficile* and cytotoxin titers in some infant hamsters comparable to those found in the intestinal tracts of adult humans and hamsters with *C. difficile*-associated intestinal disease. On the other hand, infant hamsters 4 days of age develop a fatal ileocecitis following clindamycin challenge. Parameters associated with the asymptomatic and symptomatic colonization of infant hamsters with *C. difficile* will be discussed in this presentation.

References

1. Bartlett JG, Moon N, Chang TW, Taylor NS, Onderdonk AB (1978) The role of *Clostridium difficile* in antibiotic-associated pseudomembranous colitis. Gastroenterology 75: 778–782
2. Borriello SP, Larson HE (1981) Antibiotic and pseudomembranous colitis. J Antimicrob Chemother 7 (Suppl A): 53–62
3. Cooperstock M, Riegle L, Woodruff CW, Onderdonk A (1983) Influence of age, sex, and diet on asymptomatic colonization of infants with *Clostridium difficile*. J Clin Microbiol 17: 830–833
4. Donta ST, Myers MG (1982) *Clostridium difficile* toxin in asymptomatic neonates. J Pediatr 100: 431–434
5. Donta SM, Stuppy MS, Myers MG (1980) Neonatal antibiotic-associated colitis. Am J Dis Child 135: 181–182
6. Fainstein V, Bodey GP, Fekety R (1981) Relapsing pseudomembranous colitis associated with cancer chemotherapy. J Infect Dis 143: 865
7. George WL, Rolfe RD, Finegold SM (1982) *Clostridium difficile* and it's cytotoxin in feces of patients with antimicrobial agent-associated diarrhea and miscellaneous conditions. J Clin Microbiol 15: 1049–1053
8. George WL, Rolfe RD, Harding GKM, Klein R, Putnam CW, Finegold SM (1982) *Clostridium difficile* and cytotoxin in feces of patients with antimicrobial agent-associated pseudomembranous colitis. Infect 10: 1–3
9. Hentges DJ (1970) Enteric pathogen-normal flora interactions. Am J Clin Nutr 23: 1451–1456
10. Holst E, Helin I, Mardh P (1981) Recovery of *Clostridium difficile* from children. Scan J Infect Dis 13: 41–45
11. Honda T, Hernandez I, Kath T, Miwatani T (1983) Stimulation of enterotoxin production of *Clostridium difficile* by antibiotics. Lancet 1: 655
12. Hyams JS, Berman MM, Helgason H (1981) Nonantibiotic-associated enterocolitis caused by *Clostridium difficile* in an infant. J Pediatr 99: 750–752
13. Ianconis JP, Rolfe RD (1985) *Clostridium difficile*-associated ileocecitis in clindamycin treated infant hamsters. In preparation
14. Kim K, Pickering LK, Dupont HL, Sullivan N, Wilkins TD (1984) In vitro and in vivo neutralizing activity of human colostrum and milk against purified toxin A and B of *Clostridium difficile*. J Infect Dis 150: 57–62
15. Larson HE, Barclay FE, Honour P, Hill ID (1982) Epidemiology of *Clostridium difficile* in infants. J Infect Dis 146: 727–733
16. Libby JM, Donta ST, Wilkins TD (1983) *Clostridium difficile* toxin A in infants. J Infect Dis 148: 606
17. Mardh PA, Helin I, Colleen I, Oberg M, Holst E (1982) *Clostridium difficile* toxin in faecal specimens of healthy children and children with diarrhea. Acta Pediatr Scand 71: 275–278

18. Moskovitz M, Bartlett JG (1981) Recurrent pseudomembranous colitis unassociated with prior antibiotic therapy. Arch Intern Med 141: 663–664
19. Nakamura S, Mikawa M, Tanabe N, Yanakawa K, Nishida S (1982) Effect of clindamycin on cytotoxin production by *Clostridium difficile*. Microbiol Immunol 26: 985–992
20. Onderdonk AB, Cisneros RL, Bartlett JG (1980) *Clostridium difficile* in gnotobiotic mice. Infect Immun 28: 277–282
21. Rolfe RD (1984) Role of volatile fatty acids in colonization resistance to *Clostridium difficile*. Infect Immun 45: 185–191
22. Rolfe RD, Helebian S, Finegold SM (1981) Bacterial interference between *Clostridium difficile* and normal fecal flora. J Infect Dis 143: 470–475
23. Rolfe RD, Iaconis JP (1983) Intestinal colonization of infant hamsters with *Clostridium difficile*. Infect Immun 42: 480–486
24. Sherertz RJ, Sarubbi FA (1982) The prevalence of *Clostridium difficile* and toxin in a nursery population: a comparison between patients with necrotizing enterocolitis and an asymptomatic group. J Pediatr 100: 435–439
25. Silva J (1979) Animal models of antibiotic-induced colitis. In: Schlessinger D (ed) Microbiology-1979. American Society for Microbiology, Washington DC, p 258
26. Stark PL, Lee A, Parsonage BD (1982) Colonization of the large bowel by *Clostridium difficile* in healthy infants: quantitative study. Infect Immun 35: 895–899
27. Taylor NS, Thorne GM, Bartlett JG (1981) Comparison of two toxins produced by *Clostridium difficile*. Infect Immun 34: 1036–1043
28. Trnka YM, LaMont JT (1981) Association of *Clostridium difficile* toxin with symptomatic relapse of chronic inflammatory bowel disease. Gastroenterology 80: 693–696
29. Viscidi RP, Bartlett JG (1981) Antibiotic-associated pseudomembranous colitis in children. Pediatr 67: 381–386
30. Viscidi R, Willey S, Bartlett JG (1981) Isolation rates and toxigenic potential of *Clostridium difficile* isolated from various patient populations. Gastroenterology 81: 5–9
31. Wada N, Nishida N, Iwaki S, Ohi H, Miyawaki T, Taniguchi N, Migita S (1980) Neutralizing activity against *Clostridium difficile* toxin in the supernatants of cultured colostral cells. Infect Immun 29: 545–550
32. Wilson KH, Silva J, Fekety FR (1981) Suppression of *Clostridium difficile* by normal hamster flora and prevention of antibiotic associated cecitis. Infect Immun 34: 626–628

Discussion

Haag: You suggested that development of ileocecitis may be due to demasking of receptors by clindamycin. Can you rule out the possibility that clindamycin induces the production of another factor by *C. difficile,* e. g. a "helper" toxin or an enzyme which by themselves facilitates access of the toxin to its target?

Rolfe: Preliminary evidence in our laboratory suggests that the susceptibility of infant hamsters to clindamycin-induced ileocecitis may be due to "demasking" of toxin receptors on the intestinal epithelium by clindamycin. However, at the present time we can not rule out the possibility that clindamycin induces the production of other factors in the infant intestinal tract which facilitates the toxicity of *Clostridium difficile*.

Kroll: Is it possible that this disease is an immunopathological disease?

Rolfe: Although it is possible that *Clostridium difficile*-induced ileocecitis is an immunopathological disease, there is no evidence that I am aware of which shows this is the case. An immunological response to *Clostridium difficile* intestinal overgrowth can be demonstrated in humans: however, there appears to be no correlation between serum antibody to this microorganism and the development of disease. This does not rule out the possibility that a local immunological response in the intestinal tract plays a role in this disease.

Gillissen: Are there dose-related effects of the antibiotic?

Rolfe: We have not examined dose-related effects on clindamycin in infant hamsters. Investigations in adult hamsters have shown that only a relatively small amount of clindamycin (or most other antimicrobial agents) is required to induce *Clostridium difficile*-associated ileocecitis in hamsters. For the most part, if the concentration of antibiotic is sufficient to permit toxigenic *Clostridium difficile* intestinal overgrowth, then the hamsters will develop intestinal disease. The severity of the disease can not be correlated with the amount of antimicrobial agent administered.

Gentamicin and Tobramycin selectively inhibit the secretion of *Pseudomonas aeruginosa* elastase

R.L. Warren

Department of Microbiology and Immunology, College of Science and Engineering, School of Medicine, Wright State University

Introduction

The secretion of proteases by *Pseudomonas aeruginosa* grown in the presence of subinhibitory concentrations of gentamicin and tobramycin was studied. Both gentamicin and tobramycin selectively inhibited the secretion of these proteases. Conditions that inhibited the energy-dependent uptake of gentamicin, high magnesium ion concentration and high ionic strength, did not antagonize the ability of these aminoglycosides to block secretion of the proteases. The binding of gentamicin to the outer membrane of *P. aeruginosa* has been shown to cause a rapid disruption of this membrane. However, under the conditions of these experiments, neither gentamicin nor tobramycin increased the permeability of the outer membrane.

Another reported effect of the binding of the aminoglycoside, streptomycin, to the outer membrane of *Escherichia coli* is the reduction of cellular levels of cyclic 3', 5'-adenosine monophosphate (cAMP). This reduction in cAMP causes a selective inhibition in the synthesis of b-galactosidase, a catabolite repressible cytoplasmic enzyme. The addition of exogenous cAMP eliminates the selective inhibition of b-galactosidase. Unlike *E. coli,* the addition of 2.5 mM cyclic 3', 5'- guanosine monophosphate (cGMP) but not cAMP restored the ability of *P. aeruginosa* treated with tobramycin to secrete proteases.

These results suggested that the drugs inhibited the secretion of extracellular proteases by binding to an outer membrane target site which in turn reduced the cellular levels of cGMP. This reduction in cellular levels of cGMP blocked the secretion of extracellular proteases. Since the production of *P. aeruginosa* proteases does not appear to conform to the definition of a catabolite repressible enzymes, the relationship between the cellular levels of cGMP and secretion of these proteases is unknown.

Chronic lung infections with *Pseudomonas aeruginosa* is the primary cause of mortality in individuals with cystic fibrosis [12]. In addition the incidence of *P. aeruginosa* pneumonias is increasing in patient populationes receiving either immunosuppressive or respiratory therapy [6, 18, 22]. Aggressive antibiotic therapy with aminoglycosides given either singly or in combination with B-lactam antibiotics has been shown to improve lung function of the cystic fibrosis patient but fails to eliminate the bacteria [12].

The reason for the improved clinical condition of the patients following treatment is unknown, it is possible that either of these drugs could selectively inhibit the

The Influence of Antibiotics on the
Host-Parasite Relationship II
Edited by D. Adam, H. Hahn, W. Opferkuch

production of *P. aeruginosa* virulence factors without killing the bacteria. Although not all the virulence factors or their roles in the disease process are understood, it is clear that *P. aeruginosa* secrete a variety of extracellular products that interact to produce the lung infection and disease [2, 4, 7, 15, 17, 21, 24, 27]. These extracellular products aid in attachment to epithelial surfaces, prevent clearance (immune or mechanical), and cause tissue damage. If the aminoglycoside drugs selectively inhibit the secretion of these extracellular virulence factors, then these drugs could reduce damage to the lungs and improve lung function.

In this study, subinhibitory concentrations of gentamicin and tobramycin were shown to inhibit the secretion of *P. aeruginosa* elastase and alkaline protease. In addition, conditions that antagonize the uptake and bactericidal effects of these drugs do not affect the ability of these drugs to inhibit secretion of extracellular proteases. The addition of cGMP but not cAMP to the cells restores their ability to secrete these extracellular proteases.

Materials and Methods

Bacterial strain and media

The *Pseudomonas aeruginosa* strain used in this study was PAO 1161 (leu^-, $EP2^+$) obtained from B. Holloway, Department of Genetics, Monash University, Clayton, Victoria, Australia. Peptone broth dialysate medium (PBD) was prepared by dissolving 20 g of Bacto peptone (Difco, technical) in 100 ml of distilled water and dialyzing this solution against 900 ml of deionized distilled water overnight at 4 C. Then 1.4 g of $MgSO_4$, 10 g KSO_4, and 20 ml of glycerol were added to the peptone dialysate and the volume was brought to 1 liter and the pH adjusted to 8.0. The *P. aeruginosa* cells were grown overnight in PBD broth. The cells were harvested and washed twice in PBD broth. The cells were then resuspended to 200 Klett units (red filter) in PBD and the ^3H-leucine was added. The cells were then split into flasks containing PBD medium with the appropriate supplements. The cells were incubated with vigorous shaking at 37 C. Samples of cells were removed every 30 minutes and assayed for incorporation of ^3H-leucine into acid insoluble material. After 150 minutes of incubation the cells were harvested by centrifugation (10,000 $\times$ g for 10 minutes) and the culture supernatant fluid was passed through a 0.22 u filter to remove residual cells. The culture supernatant fluid was then assayed for protease activity, elastase and alkaline protease.

Competition Elisa

A two step competition enzyme linked immunoassay (Elisa) was used to quantitate the amount of *Pseudomonas* elastase and alkaline protease in culture supernatants. The method was used as described by Warren et al. [27]. Each sample was run in triplicate. Anti-elastase and anti-alkaline protease was prepared by injecting New Zealand white rabbits with either purified elastase or purified alkaline protease (Nagase Ltd., Osaka, Japan) as described previously [13].

Assay for total proteolytic activity

Proteolytic activity was determined using the azocasein assay as previously described [13]. The activity was expressed as the change in A_{390} units after 60 minutes of incubation at 37 C. The activity was reported as the mean of six determinations per sample with the standard deviation.

Measurement of protein synthesis

The washed cells were resuspended in 5 uCi of [^{3}H−]leucine (4,5−^{3}H; specific acticity, 2 C per mmole; New England Corp., Boston, Ma.). Three 50 ul samples of the cells were removed, spotted on Whatman 3MM filter paper and the acid insoluble counts were determined after washing these filters with 10% trichloroacetic acid. The results were expressed as the means of three samples with the standard deviation.

Statistical analysis

The rates of incorporation of ^{3}H-leucine into protein was determined by linear regression analysis. The rate of incorporation is expressed ± a 95% confidence limit. This value was obtained by multiplying the standard error of the regression coefficient by the t value at $P = 0.05$ with $n-2$ degrees of freedom. A significant difference between the rates of ^{3}H-leucine incorporation was determined by a Student's t test. The t value was obtained by dividing the difference between the regression coefficient by the standard error of the difference.

The values for azounits/ml, elastase and alkaline protease are expressed as the mean ± the standard deviation. Difference between the rates was determined by the Student's t test.

Antibiotics and reagents

Gentamicin sulfate was purchased from Sigma Chemical Co. St. Louis, Mo. Tobramycin was a generous gift from Eli Lilly Co., Indianapolis, Ind.

Results

Secretion of extracellular proteases by P. aeruginosa

The secretion of extracellular proteases by *P. aeruginosa* strain PAO 1161 resuspended in PBD broth is shown in figure one. Although there was a steady increase in the optical density of the cells the number of viable cells after 240 minutes did not increase more than two fold. The total protease activity found in the culture supernatant was primarily due to the activity of elastase. The incorporation of ^{3}H-leucine into protein was linear für 150 minutes and reached a maximum incorporation between 210 and 240 minutes (data not shown).

Effects of tobramycin on the production of extracellular proteases

In PBD broth the MIC of tobramycin was 4 to 8 µg/ml and the MBC was 125–250 µg/ml for strain PAO 1161. Resuspension of these cells in 0.5 ug tobramycin per ml of PBD had no significant effect on the rate of incorporation of ^{3}H-leucine into protein (Table 1). Both the total protease activity and the amount of elastase in culture supernatants was significantly reduced (P = 0.05). Similar results were obtained with gentamicin (data not shown).

Effect of tobramycin on the outer membrane

Hancock et al. [10, 11, 19, 20] demonstrated that a very early event associated with the binding of gentamicin to the outer membrane of *P. aeruginosa* was a rapid increase in the permeability of the outer membrane. It is possible that this disruption of the outer membrane is the mechanism by which these drugs block the accumulation of extracellular proteases. To test if tobramycin increased the permeability of the outer membrane of *P. aeruginosa* under the culture condition used, lysozyme was added to culture medium. Lysozyme, a protein of approximately 14000 molecular weight, is excluded from the peptidoglycan by an intact outer membrane. However, if the drugs increase the permeability of the outer membrane then lysozyme can digest the cell wall which causes the cells to lyse. The addition of lysozyme and tobramycin to cells resuspended in nutrient broth or PBD did not result in the lysis of the cells (Fig. 2a and 2b). However, the addition of lysozyme and gentamicin to cell resuspended in nutrient broth resulted in the lysis of the cells (Fig. 2c). Resuspending the cells in PBD blocked the ability of gentamicin and lysozyme to lyse the cells (Fig. 2d). The addition of either lysozyme or the drugs alone did not result in the lysis of the cells. Resuspension of *P. aeruginosa* in PBD

Table 1. Effects of tobramycin on the accumulation of extracellular protase by *P. aeruginosa* cells growing in PBD medium

Tobramycin (µg/ml)	^{3}H-leu incorporation[a] CPM	Azounits[b] per ml	Elastase[c] µg/ml
0	30.6±8.7 (9)	5.6±1.9 (9)	0.95±0.4 (7)
0.5	30.4±21.5 (5)	1.5±1.8 (5)	0.33±0.3 (4)

[a] The rate of incorporation of ^{3}H-leucine into acid insoluble material was determined as described in the Materials and Methods

[b] An azounit was the difference in the absorption at 390nM of azocasein substrate in one hour at 37 C

[c] The amount of elastase antigen in a 150 minute culture supernatant fluid sample was determined by competition ELISA as described in the Materials and Methods

[d] The number in the () represents the number of independent experiments

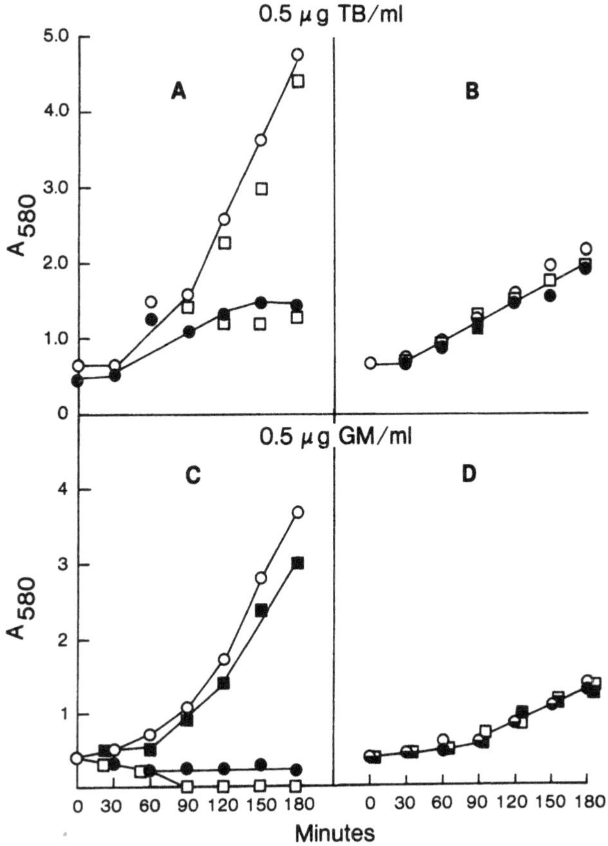

Fig. 1. Secretion of Extracellular Proteases by *Pseudomonas aeruginosa* Strain PAO 1161. An overnight culture of cells was washed twice and resuspended in resh PBD. The cells were incubated at 37 C and samples were taken at various times. The optical density (●) was measured for each sample, then cells were removed by centrifugation. The total protease activity (○) in the supernatants was determined by the azocasein assay. The amount of elastase (□) and alkaline protease (■) antigen in the supernatants was determined by competition Elisa

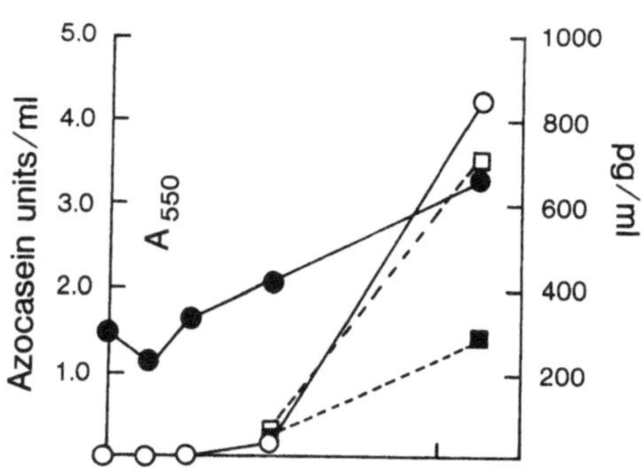

Fig. 2. Aminoglycoside-mediated lysozyme lysis of *P. aeruginosa*. Overnight cells were prepared as described in Fig 1. Half of the cells were resuspended in PBD broth (b and d) and half in NB (a and c). The medium was used without supplementation (○), with 20 μg lysozyme per ml (□), with 0.5 μg tobramycin or gentamycin per ml (●), or with drug and lysozyme (■). The optical density (OD) was determined every 30 minutes

medium protects them from aminoglycoside-promoted lysozyme lysis, suggesting that the drugs did not selective inhibit of extracellular proteases by altering the permeability of the outer membrane.

The addition of 15 mM magnesium ion to nutrient broth was shown to block the aminoglycoside-promoted lysozyme lysis of *P. aeruginosa* [26]. Both gentamicin and tobramycin inhibited the secretion of extracellular proteases by cells resuspended in nutrient broth with 15 mM magnesium [27].

Effect of cyclic nucleotides on the inhibition of protease secretion

The ability of tobramycin to block the secretion of extracellular proteases appears to a function of its association with the outer membrane rather than selective inhibition of the biosynthesis of these enzymes. Culture conditions that have been shown to antagonize both energy dependent uptake phase I and II of the drug, do not affect the drugs ability to inhibit the secretion of proteases. Artman et. al. demonstrated that the initial binding of the drug with the outer membrane causes the release of cyclic nucleotides [1]. This reduction in cyclic nucleotide levels resulted in the selective inhibition of the synthesis of B-galactosidase, a cytoplasmic enzyme that is dependent upon c-AMP for expression. If the selective inhibition of the *P. aeruginosa* extracellular proteases was due to a reduction in the level of cellular cyclic nucleotides, then the addition of cyclic nucleotides to the culture medium should reverse the effect of the drug on secretion. As shown in Figure 3, the addition of cGMP but not cAMP restores the ability of cells to secrete protease when treated with tobramycin. Neither GMP nor AMP affect the ability of tobramycin to inhibit secretion of proteases.

Discussion

P. aeruginosa is an opportunistic pathogen which can cause severe infection in the immunocompromised host [6] and in individuals with cystic fibrosis [12,18,22].

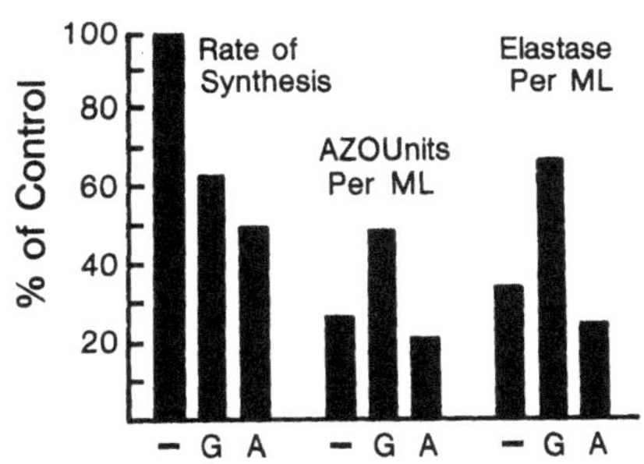

Fig. 3. Effect of cyclic nucleotides on the ability of tobramycin to inhibit secretion of elastase. Overnight cells were prepared as described in Fig. 1. The cells were resuspended in PBD broth with or without 0.5 μg tobramycin per ml. The PBD broth was supplemented with 2.5 mM cAMP (A), 2.5 mM cGMP (G), or was not suppllemented (–). The cells were incubated for 150 minutes at 37 C. Samples were removed every 30 minutes to determine the amount of incorporation of radioactive leucine into protein. The rate of protein synthesis was determined by linear regression analysis. After 150 minutes the celsl were harvested and the protease activity of the supernatant was determined by azocasein assay. The amount of elastase antigen in the 150 minute culture supernatant was determined by competition Elisa. The results were presented as the % of the cells not grown in the presence of tobramycin

P. aeruginosa secretes a number of extracellular enzymes that contribute to its virulence [17]. Elastase has been shown to contribute to the virulence of this microorganism by both in vivo [2, 4, 21, 24, 28] and in vitro [7, 15, 17] studies. Baker [2] used hamster tracheal organ cultures to study the contribution of exotoxin A and proteases in infections of the respiratory tract. He showed that tracheal explants infected with *P. aeruginosa* strains that produced elastase caused exfoliation of the epithelial layers. Treatment of the infected tracheal explant with gentamicin protected the tissues from damage. In subsequent studies (R. Warren, J. Johnson, T. Geers, and N. R. Baker, Abstr. Annu. Meet. Interscience Conference on Antimi crob. Agents and Chemother. 1985, 97, p. 107.) subinhibitory concentrations of gentamicin and tobramycin were shown to protect the tracheal explants from tissue destruction. It is possible that the protection afforded these tracheal explants was due to the selective inhibition of elastase and other extracellular virulence factors by the aminoglycosides and not due to the ability of these drugs to kill the bacteria.

The secretion of elastase was inhibited by sublethal concentrations of tobramycin and gentamicin. Culture conditions that have been shown to antagonize the uptake and bactericidal action of these drugs did not interfere with their ability to inhibit the secretion of the proteases. These results suggested that these drugs inhibit the secretion of proteases by some interaction with the outer membrane.

Hancock et al. [10, 11] demonstrated that the binding of gentamicin to the outer membrane *P. aeruginosa* cause a rapid disruption of the outer membrane. The ability of the drug to disrupt the outer membrane was antagonized by 1 mM magnesium ion (10). Disruption of the outer membrane could interfere with the export of elastase. However, under the conditions used in this study the concentration of magnesium ion in PBD medium prevented the disruption of the outer membrane by the drugs.

Another early event associated with the binding of gentamicin to the outer membrane of *E. coli* is a reduction in the cellular levels of cAMP [1]. Artman et. al. [1] showed that this reduction in cAMP was responsible for this selective inhibition of the synthesis of b-galactosidase. Althoug the selective inhibition of elastase by tobramycin was reversed by the addition of cGMP to the culture medium suggests a role for cGMP in the regulation of the synthesis of elastase, the mechanism is unclear. Whooley et al. [27] showed that the expression of *P. aeruginosa* exoproteases was produced in response to growth limitation by carbon, nitrogen, phosphate or oxygen and repressed by readily available energy sources. They concluded that the production of elastase was regulated by repression, but not by a classical catabolite repression as found for the expression of b-galactosidase in *E. coli*. It is possible that growth limiting conditions result in an increased level of cyclic nucleotides in the cell. This seems unlikely because Phillips et al. [23] demonstrated that the levels of cAMP in *P. aeruginosa* did not vary when the cells were grown on different carbon sources. They did demonstrate that *P. aeruginosa* histidase was repressed by the addition of succinate to the growth medium. This repression was not due to a reduction in cellular cAMP. It is also possible that cGMP and not cAMP regulates the expression of catabolite repressible enzymes in *P. aeruginosa*.

In summary the addition of sublethal concentrations of gentamicin or tobramycin inhibit the secretion of extracellular proteases by *P. aeruginosa*. These drugs appear to block either the secretion or synthesis of the proteases by interacting with the outer membrane which in turn reduces the cellular levels of cGMP. The addition of exogenous cGMP reverses the effect of these drugs.

Acknowledgements. We thank Wendy Campbell for her excellent technical assistance. This work was supported in part by a Biomedical Seed Grant from Wright State University, College of Science and Engineering and from a grant-in-aid from Dista Corp., Indianapolis, Ind.

Summary

The secretion of proteases by *Pseudomonas aeruginosa* grown in the presence of subinhibitory concentrations of gentamicin and tobramycin was studied.

Subinhibitory concentrations of gentamicin and tobramycin protects hampster tracheal ring explants infected with *P. aeruginosa* from distruction. The tissue distruction observed in these experiments is in large part due to the production of the extracellular protease, elastase. It is possible that the subinhibitory levels of the drugs protected the organ explants by selectively blocking the secretion of elastase.

References

1. Artman M Werthamer S Gleb P (1972) Catabolite repression inhibition of B-galactosidase synthesis by *Escherichia coli* in the presence of agents producing translation errors. Antimicrob Agents and Chemother 2:449–455
2. Baker NR (1982) Role of exotoxin A and proteases of *Pseudomonas aeruginosa* in respiratory tract infections. Can J Microbiol 28:248–255
3. Beggs WH, Andrews FA (1976) Role of ionic strength in salt antagonism of aminoglycoside action on *Escherchi coli* and *Pseudomonas aeruginosa*. J. Infect Dis 134:500–504
4. Blackwood LL, Stone RM, Iglewski BH, Pennington JE (1983) Evaluation of *Pseudomonas aeruginosa* Exotoxin A and elastase as virulence factors in acute lung infection. Incect. Immun, 39:198–201
5. Engvall E (1980) Enzyme immunoassay ELISA and EMIT. Methods in Enzymology 70:419–439
6. Feigin RD, Shearer WT (1975) Opportunistic infection in children. In the compromised host. J Pediatr 87:677–694
7. Gray L, Kreger A (1979) Microscopic characterization of rabbit lung damage produced by *Pseudomonas aeruginosa* proteases. Infect Immun 23:150–159
8. Hancock REW (1981) Aminoglycoside uptake and mode of action—with special reference to streptomycin and gentamicin I. Antagonists and Mutants J Antimicrob Chemother. 8:249–276
9. Hancock REW (1981) Aminoglycoside uptake and mode of action—with special reference to streptomycin and gentamicin II. Effect of aminoglycosides on cells. J Antimicrob Chemother 8:429–445
10. Hancock REW, Raffle VJ, Nicas TI (1981) Involvement of the outer membrane in gentamycin and streptomycin uptake and killing in *Pseudomonas aeruginosa*. Antimicrob. Agents Chemother 19:777–785
11. Hancock REW, Wong PGW (1984) Compounds which increase the permeability of the *Pseudomonas aeruginosa* outer membrane. Antimicrob Agents and Chemother 26:48--52
12. Hoiby N (1977) *Pseudomonas aeruginosa* infection in cystic fibrosis. Acta Pathol. Microbiol Scand Suppl 262:58–71
13. Jagger KS, Bahner DR, Warren RL (1983) Protease phenotypes of *Pseudomonas aeruginosa* isolated from patients with cystic fibrosis. J Clin Microbiol 15:1054–1058
14. Jensen SE, Fecyez IT, Stemke GWCJN (1980) Demonstration of a cell-associated, inactive precursor of exocellular protease production by *Pseudomonas aeruginosa*. Can J Microbiol 26:87–93

15. Johnson DA, Carter-Hamm B, Dralle WM (1982) Inactivation of human bronchial mucosal proteinase inhibitor by *Pseudomonas aeruginosa* elastase. Amer Rev Resp Dis. 126:1070–1073
16. Lory S, Tai PC, Davis BD (1983) Mechanism of protein excretion by gram-negative bacteria: *Pseudomonas aeruginosa* exotoxin A. J Bacteriol 156:695–702
17. Morihara, K, Tsuzuki H, Oda K (1979) Protease and elastase of *Pseudomonas aeruginosa:* Inactivation of human plasma a1-proteinase inhibitor. Infect Immun 24:188–193
18. Nathan P, Holder IA, MacMillan BG (1973) Burn wounds: microbiology, local host defenses and current therapy. Crit Rev Clin Lab Sci 4:61–100
19. Nicas TI, Hancock REW (1983) Alteration of susceptibility to ethylenediaminetetraacetate, polymyxin B, and gentamicin in *Pseudomonas aeruginosa* by divalent cation regulation of outer membrane protein H1. J Gen Microbiol 129:509–517
20. Nicas RI, Hancock REW (1980) Outer membrane protein H1 of *Pseudomonas aeruginosa:* Involvement in adaptive and mutational resistance to ethylenediaminetetraacetate, polymyxin B, and gentamicin. J Bacteriol 143:873–878
21. Pavlovskis OR, Wretlind B (1979) Assessment of protease (elastase) as a *Pseudomonas aeruginosa* virluence factor in experimental mouse burn infection. Infect Immun 24:181–187
22. Pennington JE, Reynolds HY, Carbone PP (1973) *Pseudomonas* pneumonia: a retrospective study of 36 cases. Am J Med 55:155–160
23. Phillips AT, Mulfinger LM (1981) Cyclic adenosine 3', 5'-monophosphate levels in *Pseudomonas putida* and *Pseudomonas aeruginsosa* during induction and carbon catabolite repression of histidine synthesis. J Bacteriol 145:1286–1292
24. Snell K, Holder IA, Leppla SA, Saelinger CB (1978) Role of exotoxin and protease as possible virulence factors in experimental infections with *Pseudomonas aeruginosa.* Infect Immun 19:839–845
25. Stapleton MJ, Jagger KS, Warren RL (1984) Transposon mutagenesis of *Pseudomonas aeruginosa* exoprotease genes. J Bacteriol 157:7–12
26. Warren RL, Baker NR, Johnson J, Stapleton MJ (1985) Selective inhibition of theaccumulation of extracellular proteases of *Pseudomonas aeruginosa* by gentamicin and tobramycin. Antimicrob Agents Chemother 27
27. Whooley MA, O'Callaghan JA, McLoughlin AJ (1983) Effect of substrate on the regulation of exoprotease production by *Pseudomonas aeruginosa* ATCC 10145. J Gen Microbiol 129:981–988
28. Woods DE, Cryz SJ, Friedman RL, Iglewski BH (1982) Contribution of toxin A and elastase to virulence of *Pseudomonas aeruginosa* in chronic lung infections of rats. Infect Immun 36:1223–1228

Discussion

Kadurugamuwa: Did you observe any difference in membrane bound ribosomes and cytoplasmic ribosomes?

Warren: I did not measure the amount of membrane or cytoplasmic ribosomes in these experiments.

Döring: If you get inhibition of protease secretion and not inhibition of protease synthesis, do you get high intracellular protease accumulation? Did you look after exotoxin A inhibition?

Warren: The subinhibitory concentrations inhibited the synthesis of elastase. The amount of cell associated elastase was the same in both drug treated and untreated cells.
No, I have not examined the effects of this drug on exotoxin A. We are currently developing a competition ELISA for exotoxin A.

Subinhibitory concentrations of antibiotics modulate amount of Shiga-like toxin produced by *Escherichia coli*

H. Karch[1,4], P. Goroncy-Bermes[1], W. Opferkuch[1], H.-P. Kroll[2], and A. O'Brien[3]

[1] Lehrstuhl für Med. Mikrobiologie und Immunologie, Ruhr-Universität Bochum, FRG
[2] Pharma-Forschungszentrum, Bayer AG, Institut für Chemotherapie, Wuppertal, FRG
[3] Uniformed Services University of the Health Sciences, Bethesda, USA
[4] Institut für Medizinische Mikrobiologie und Immunologie, Universitäts-Krankenhaus Eppendorf

Introduction

Many strains of *Escherichia coli* that cause enteric disease in a different way from the classical mechanism of enteroinvasiveness or elaboration of heat-labile or heat-stable enterotoxins produce a cytotoxin with biochemical and biological properties similar to the toxin of *Shigella dysenteriae* 1 (Shiga's bacillus) [11]. This toxin is enterotoxic for rabbit intestinal segments, lethal for mice and rabbits, and cytotoxic for tissue culture cells [10]. Its potential medical relevance in humans is suggested by the strong correlation between toxin production and the ability of *E. coli* to cause bloody diarrhea, hemorrhagic colitis or hemolytic-uremic syndrome [9].

The goal of this study was to assess the influence of subinhibitory concentrations of various antimicrobial agents with different mode of action on Shiga-like toxin production by *E. coli* 0157:H7 strain 933. The amount of toxin produced by this strain was shown to be equivalent to that of *Shigella dysenteriae* 1 [10, 12, 13], and phages isolated from the strain can convert *E. coli* K-12 to a highly toxinogenic organism [14]. Therefore we also examined the number of phages released from *E. coli* after growth in the presence and absence of antibiotics.

Material and Methods

Bacteria and culture conditions

E. coli 0157:H7 strain 933 was kindly provided by Dr. M. Cohen (CDC, Atlanta, Georgia) and cultivated at 37 °C with shaking in iron-depleted, glucose-syncase broth [11]. The growth kinetics was recorded spectrophotometrically at 583 nm and samples were taken at 4, 6, 12 and 24 hours of culture to determine the colony forming units (cfu) in the presence or absence of antimicrobial agents. Samples were centrifuged and the sedimented bacteria lysed by sonic disruption in order to assess the intracellular toxin content. For plaque assays a substrain of *E. coli* K-12 was used as the indicator strain and cultured in modified Luria broth as recently described [14].

The Influence of Antibiotics on the
Host-Parasite Relationship II
Edited by D. Adam, H. Hahn, W. Opferkuch

Cytotoxicity assay

A HeLa cell microcytotoxicity assay was used to quantitate the amount of toxin. This assay was adapted from [3].

Cultures were fed with growth medium that contained RPMI 1640 medium (Gibco, Grand Island, New York) supplemented with L-glutamine (3.2×10^{-3} M; Flow Laboratories, McLean, Virginia), sodium pyruvate (1×10^{-3} M; Boehringer Mannheim, Indianapolis, Indiana), penicillin (100 U/ml; Sigma), streptomycin (100 U/ml; Sigma) and 10% fetal calf serum (P. Dunn Labortechnik, Meckenbach, West Germany).

The toxin titer of a sample was taken as the reciprocal of the highest dilution that gave an absorption value less than 50% of the average value of untreated cells.

Plaque assay

The plaque assay was performed as recently described [14]. Plaques were counted after incubating the plates for 20 hours at 37 °C.

Antimicrobial agents

The quinolone ciprofloxacin was obtained from Bayer AG and all other antimicrobial agents were purchased from Sigma, St. Louis; Missouri. Cotrimoxazole was prepared by mixing 1 part trimethoprim with 19 parts sulfamethoxazole. The minimal inhibitory concentration (MICs) of the various agents for *E.coli* 933 grown in culture broth were as follows (in µg/ml): 0.016 for ciprofloxacin, 0.25 for trimethoprim, 0.50 for cotrimoxazole, 4 for sulfamethoxazole, 2 for ampicillin, 8.0 for colistin, 1 for gentamicin, and 512 for lincomycin. The final concentration of each antimicrobial agent in culture broth was ¹⁄₁₆ of its MIC.

Results

In Table 1, the yield of cell-associated and cell-free cytotoxin produced by *E. coli* 933 and the optical density of the culture are shown at various times over a 24 hour period. Between 4 and 6 hours (logarithmic growth phase) the amount of toxin remained constant. Yields of toxin production were maximal in the stationary phase of growth and toxin biosynthesis did not appear to increase between 12 and 24 hours of culture.

The results of a representative experiment (one of three) on the effect of ¹⁄₁₆ the MIC of different antibiotics on the amount of Shiga-like toxin produced by *E. coli* 933 at 12 hours of culture are shown in Table 2. The amount of both cell-associated and extracellular toxin was essentially the same during the first 6 hours for control and all antibiotic-treated cells (not shown). The yield of toxin was also unaffected by exposure of the bacteria to ampicillin or polymyxin E. However, bacteria grown for 12 hours in the presence of lincomycin or gentamicin produced less toxin. The most

Table 1. Relationship between growth of *E. coli* 0157:H7 strain 933 and synthesis of Shiga-like toxin

	time (hours)			
	4	6	12	24
OD 583	0.15	0.90	1.12	1.35
Yield of toxin[a]:				
intracellular	2500	2500	320000	320000
extracellular	160	160	10000	10000

[a] Toxin titers were determined by a microcytotoxicity assay and are given as the reciprocal of the highest dilution that gave an absorption value less than 50% of untreated cells

Table 2. Influence of antibiotics on Shiga-like toxin production by *E. coli* 0157:H7 strain 933

Antibiotic treatment[a]	Colony forming units/ml	Absorbance (584nm)	Yield of toxin[b] intra-cellular	extra-cellular
None	$5{,}8\times10^8$	1,12	320000	10000
Ampicillin	$5{,}6\times10^8$	1,05	320000	10000
Polymyxin E	$6{,}0\times10^8$	0,88	320000	10000
Lincomycin	$4{,}4\times10^8$	1,00	40000	5000
Gentamicin	$4{,}3\times10^8$	0,91	80000	5000
Ciprofloxacin	$6{,}0\times10^8$	1,00	20000	1250
Sulfamethoxazole	$5{,}0\times10^8$	0,95	640000	1280000
Trimethoprim	$4{,}5\times10^8$	0,95	2560000	2560000
Cotrimoxazole	$5{,}3\times10^8$	0,95	1280000	2560000

[a] The bacteria were grown in modified glucose-syncase broth with (1/16 MIC) or without antibiotics
[b] Toxin was quantitated after 12 hours of growth

dramatic reduction of toxin synthesis was seen when cells were grown in the presence of ciprofloxacin. By contrast, *E. coli* exposed to sulfamethoxazole, co-trimoxazole and trimethoprim increased toxin yield, especially in the presence of trimethoprim.

Since high-level toxin production by *E. coli* 933 is phage regulated [14] we speculated that the increase in toxin synthesis could be due to phage induction in the presence of the antifolate drugs. However, when we determined the kinetics of the spontaneous release of phages from control and trimethoprim treated cells we found that there was no spontaneous release of phage after 12 hours of culture. Additional studies were performed to determine the number of phages after induction with ultraviolet light. Lysates of bacteria grown in the presence of trimethoprim contained the same plaque-forming units per milliliter (1×10^5) as lysates of control cells.

Alterations in the outer membrane composition could affect secretion and regulation of toxin. Therefore an analysis of the outer membrane composition of bacteria grown with or without toxin-modulating antibiotics was performed. Quantitative studies of isolated outer membranes revealed that about twice as much KDO (a marker for lipopolysaccharide) was incorporated into the membranes of sulfamethoxazole–and cotrimoxazole-exposed bacteria than in membranes of controls. No changes in the quantity and arrangement of membrane constituents were seen between trimethoprim treated and untreated cells (Table 3). The influence of toxin enhancing antibiotics on the outer membrane protein profile was also examined (Fig. 1).

Table 3. Quantitative analysis of outer membranes after exposure to subinhibitory concentrations of various antimicrobials agents[a]

	Composition of outer membranes (OM)[b]			
Antimicrobial agent	% protein		mg KDO/mg (dry wt. OM)	nmol Pl/mg (dry wt. OM)
None	43	41	0.043, 0.048	575, 529
Ampicillin	48	45	00.42, 0.045	570, 568
Polymyxin E	42	46	0.046, 0.041	608, 564
Lincomycin	45	47	0.040, 0.047	584, 522
Gentamicin	43	46	0.046, 0.050	607, 589
Sulfamethoxazole	35	38	0.081, 0.087	820, 884
Trimethoprim	43	44	0.042, 0.038	605, 590
Cotrimoxazole	38	36	0.076, 0.080	880, 854

[a] *E. coli* 933 was grown 12 hours in 50 ml of iron-depleted glucose-syncase broth supplemented with 1/16 the MIC of each antimicrobial agent. The outer membranes were prepared from the bacterial cells by the procedure of [1]

[b] The values presented represent analyses from two separate experiments

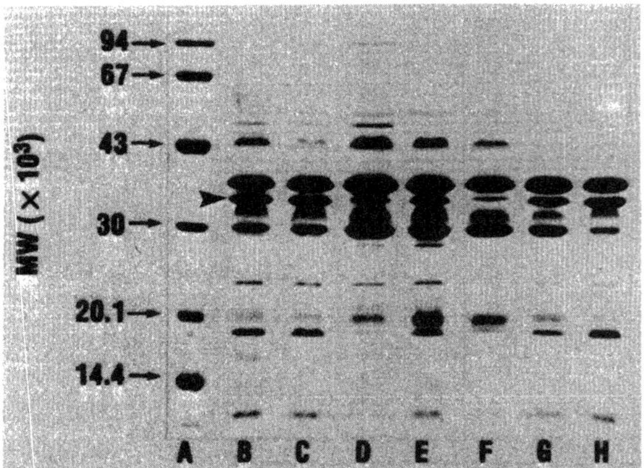

Fig. 1. SDS-PAGE of the outer membrane protein profile of *E. coli* 0157:H7 strain 933 grown for 12 hours in iron-depleted glucose syncase broth alone (lane B) or supplemented with 1/16 the MIC of lincomycin (lane C), trimethoprim (lane D), sulfamethoxazol (lane E), cotrimoxazole (lane F), polymixin E (lane G), or gentamicin (lane H). Lane A contains the molecular weight protein standards phosphorylase B, 94K; bovine serum albumin, 67K; ovalbumin, 43K; carbonic anhydrase, 30K; soy bean trypsin inhibitor, 20.1K, and α-lactalbumin, 14.4K. These standards were purchased from Pharmacia Fine Chemicals, Uppsala, Sweden as a low molecular weight protein kit. The band in the control lane to which the arrow points represents the 34K matrix protein

The intensity of the 34K protein which represents the matrix protein that anchors the outer membrane to the peptidoglycan [15] is reduced in the membranes exposed to trimethoprim (lane D) and cotrimoxazol (lane F) when compared to controls (lane B). Beside this alteration a number of differences in the protein patterns are seen. At the present it is unclear whether these variations are related to the effects on toxin biosynthesis and release.

Discussion

Diverse effects of antimicrobial agents on synthesis of Shiga-like toxin by a strain of *E. coli* 0157:H7 causing hemorrhagic colitis were observed. Two of the drugs used in this study that are known to inhibit protein synthesis, lincomycin and gentamicin, decreased the total amount of toxin at $\frac{1}{16}$ of their growth inhibiting doses. The reduction in toxin biosynthesis after growth in the presence of lincomycin is in contrast to the reported stimulatory effect of this agent on synthesis of classical heat labile (LT) enterotoxin [5]. The quinolone ciprofloxacin because of its mode of action can also affect protein biosynthesis. As shown by Dalhoff and Döring (this volume) extracellular proteases and toxins of *Pseudomonas aeruginosa* can be inhibited very effectively by ciprofloxacin. This effect was seen in vivo and in vitro. We also could show a significant reduction in the Shiga-like toxin production by subinhibitory concentrations of ciprofloxacin which surpasses the inhibitory effects of lincomycin and gentamicin. The decrease of virulence for highly toxinogenic *E. coli* (Kroll and Karch, manuscript in preparation) after treatment with subinhibitory concentrations of ciprofloxacin may be correlated to the effect of this drug on toxin biosynthesis.

The basis for the increase in toxin production after growth in antifolate drugs is not established. It is known that phages control production of Shiga-like toxin. Thus we speculated that the increase in toxin production is due to induction of toxin-converting phages. We tested this possibility but neither under conditions of repression nor under augmented toxin production we could detect spontaneous phage induction, or differences in the phage-titers after UV-induction. We could show that the presence of toxin enhancing drugs affected the quantitative and (or) qualitative composition of the outer membrane. Alterations in the outer membrane structure could influence the permeability and thus the secretion of toxin. A dramatic increase of extracellular toxin was seen after growth in the presence of toxin-enhancing drugs. Thus the level of toxin within a cell may be regulated by a feed-back control mechanism. If toxin is released from the cell at a faster rate than normal, the cell may compensate this loss by increasing its synthesis.

The ability of antimicrobial agents to modulate biosynthesis of Shiga-like toxin may also occur in vivo and potentiate or impair the virulence of the strain.

Acknowledgement: The authors thank Vera Wendt for help in preparation of the manuscript.

Summary

Shiga-like toxin has been defined as a cytotoxin for selected sublines of HeLa or Vero cells that can be neutralized by antibody prepared against purified *Shigella dysenteriae* (Shiga's bacillus) toxin. Shiga-like toxin is produced by different serogroups of *Escherichia coli* and has the same biological activities as purified Shiga toxin. The goal of this study was to assess the influence of subinhibitory concentrations of various antimicrobial agents on Shiga-like toxin production by *E. coli* 0157:H7 strain 933. Bacteria were cultured in 50 ml of iron-depleted glucose syncase broth with (1/16 MIC) or without antimicrobial agents. A modified HeLa cell microcytotoxicity assay was used to quantitate toxin in cell lysates and culture supernatants. The results indicated that none of the antibiotics altered the kinetics of toxin production. However, lincomycin, gentamicin and ciprofloxacin actually reduced the yield of toxin, whereas ampicillin and colistin had no effect on toxin production. By contrast trimethoprim markedly stimulated toxin production while sulfamethoxazole and cotrimoxazole moderately increased the amount of toxin. Toxin-enhancing antimicrobial agents promoted release of toxin from viable cells, a phenomenon that was associated with changes in the outer membrane structure and composition of *E. coli* 933.

References

1. Achtman M, Mercer A, Kusecek B, Pohl A, Heuzenroeder M, Aaronson A, Sutton and A, Silver RP (1983) Six widespread bacterial clones among *Escherichia coli* Kl isolates. Infect Immun 39:315–335
2. Folch J, Lees M and Sloane-Stanley GH (1957) A simple method for the isolation and purification of total lipids from animal tissues. J Biol Chem 226:497–505
3. Gentry MK and Dalrymple JM (1980) Quantitative microtiter cytotoxicity assay for Shigella toxin. J Clin Micro 12:361–366
4. Karkhanis YD, Zeltner JY, Jackson JJ and Carlo D (1978) A new and improved microassay to determine 2-keto-3-deoxyoctonate in lipopolysaccharides of gram-negative bacterial. Analytical Biochem. 85:595–601
5. Levner M, Wiener FP and Rubin BA (1977) Induction of *Escherichia coli* and *Vibrio cholerae* enterotoxins by an inhibitor of protein synthesis. Infect Immun 15:132–137
6. Lowry OW, Roberts NR, Leiner KY, Wu ML and Farr AL (1954) The quantitative histochemistry of brain I chemical methods. J Biol Chem 207:1–17.
7. Lugtenberg B, Meijers J, van der Hoek P, and van Alphen (1975) Electrophoretic resolution of the major outer membrane protein of *Escherichia coli* K12 into four bands. FEBS Lett 58:254–258
8. Markwell MA K, Haar SM, Bieker LL, and Tolbert WE (1978) A modification of the Lowry procedure to simplify protein determinations in membrane and lipoprotein samples. Anal Biochem 87:206–210
9. O'Brien AD, Chen ME, Wells JA, Farmer J, and Wachsmith TK (1985) Epidemiology of *Shigella dysenteriae* 1-like toxin production by *Escherichia coli*. In Advances Research on Cholera and Related Diarrheal Diseases. KTK Scientific Publisher, Tokyo, Japan (in press)
10. O'Brien AD, and LaVeck GD (1983) Purification and characterization of a *Shigella dysenteriae* 1 like toxin by *Escherichia coli*. Infect Immun 40:675–683
11. O'Brien AD, LaVeck GD, Thompson MR, and Formal SB (1982) Production of *Shigella dysenteriae* type 1 like cytotoxin by *Escherichia coli*. J Infect Dis 146:675–683

12. O'Brien AD, Lively TA, Chen ME, Rothman SW, and Formal SB (1983) *Escherichia coli* 0157:H7 strains associated with haemorrhagic colitis in the United States produce a *Shigella dysenteriae* 1 (Shiga) like cytotoxin. Lancet *i*:702
13. O'Brien AD, Lively TA, Chang TW, and Gorbach SL (1983) Purification of *Shigella dysenteriae* 1 (Shiga)-like cytotoxin from *Escherichia coli* 0157:H7 strain associated with hemorrhagic colitis. Lancet *ii*:573
14. O'Brien AD, Newland JW, Miller SF, Holmes RK, Smith HW, and Formal SB (1984) Shiga-like toxin converting phages from *Escherichia coli* strains that cause hemorrhagic colitis or infantile diarrhea. Science 226: 694–696
15. Rosenbusch JP, (1974) Characterization of the major envelope protein from *Escherichia coli*. J Biol Chem 249:8019–8029

Discussion

Pulverer: Is the responsible phage a temperate or virulent phage? Has trimethoprim some influence on lysogenic state of the E. coli strain? Is the phage an E. coli or a Shigella phage?

Karch: A temperate phage. We could not detect any spontaneously released phage, but we can induce phage with UV light. We have no evidence to support the theory that trimethoprim induces phage. We have not found inducible phage in Shigella. We believe it is an E. coli phage.

Kadurugamuwa: Chelex-100 will remove all divalent metal cations. Did you measure the amount of metal cations in the growth media before and after Chelex treatment? What metal cations did you add back into the culture media prior to cultivation of the organism? Nutrient depletion may also have influence on certain extracellular production!

Karch: No, we did not measure the contents of those ions. The culture medium was supplemented with $MgSo_4$ and $MnCl_2$.

Kadurugamuwa: Have you measured the amount of toxin which related to a dry weight of cells? Output of toxin into the supernatant is increased as the population size increases. Therefore, it would be a better idea to relate the toxin/dry weight of cells.

Karch: We have measured toxin content per wet weight. Thank you for this comment!

Kadurugamuwa: The loss of phage suggests an alteration on the membrane. Could this be due to be an iron regulated membrane protein? Certain iron regulated membrane proteins are receptors for certain phages.

Karch: We are investigating the nature of the phage receptor.

Interference of ciprofloxacin with the expression of pathogenicity factors of *Pseudomonas aeruginosa*

A. DALHOFF[1] and G. DÖRING[2]

[1] BAYER AG, Pharma-Research Centre, Institute for Chemotherapy, P. O. Box 101709, 5600 Wuppertal 1, FRG
[2] Hygiene-Institut der Universität Tübingen, Silcherstraße, 7400 Tübingen

Introduction

The target site for the 4-quinolones is the DNA gyrase which has been shown to be involved in a number of cellular processes other than supercoiling of the chromosome; DNA replication, transcription and repair as well as recombination have been shown to be DNA-gyrase dependent [3–7]. Thus, it seems to be conceivable that interference with the enzymatic activity of the DNA-gyrase may affect gene expression. Evidence has been accumulated indicating that, at higher doses, 4-quinolones and/or novobiocin interfere with transcription-translation of genes sensitive to catabolite repression but that they have no effect on constitutively expressed genes [8–16].

The expression of those genes in *Pseudomonas aeruginosa* coding for extracellular proteases and exotoxin A is considered to be of utmost importance for its pathogenicity [17]. The majority of clinical isolates already investigated produce proteases and/or exotoxin A in variable amounts in vitro [18–25].

Proteases secreted by *P. aeruginosa* inactivate immunoglobulins [26] and are active on complement component C3, fibrinogen, and elastin [27–29]; furthermore, Pseudomonas proteases destroy the two main inhibitors of eucaryotic proteases, the α1-antiprotease and the bronchial mucosal proteinase inhibitor [30–34] and inhibit neutrophil chemotaxis [35]. Exotoxin A causes a depletion of functional elongation factor 2 with subsequent inhibition of protein synthesis [36–37].

Inhibition of these multifunctional pathogenicity factors of *P. aeruginosa* protected animals against *P. aeruginosa* infections [38–41]. Thus, it may be speculated that ciprofloxacin, too, might affect the virulence of *P. aeruginosa*; because of its mode of action, ciprofloxacin may interfere with the synthesis of these pathogenicity factors. We therefore assessed the effect of ciprofloxacin on the synthesis of elastase and exotoxin A by *P. aeruginosa*. An aminoglycoside and a β-lactam antibiotic respectively served as positive and negative controls.

The Influence of Antibiotics on the
Host-Parasite Relationship II
Edited by D. Adam, H. Hahn, W. Opferkuch

Materials and Methods

Organisms

Four mucoid *P. aeruginosa* strains originally isolated from the sputum of patients with cystic fibrosis were used in this study. Strains were morphologically homogenous and were not cloned before commencement of the experiments so that the complexity and range of biological properties of these isolates was maintained.

Antibiotics

Ciprofloxacin (BAY 0 9867) and sisocimicin were obtained from Bayer AG. Leverkusen and cefsulodin from Grünenthal GmbH, Stollberg.

Expression of pathogenicity factors

Bacteria were pregrown overnight and then diluted tenfold into fresh prewarmed medium containing a constant drug concentration. The concentrations used for these tests corresponded to the 4-fold MIC value of each drug against each strain as determined *before* commencement of the experiments (day 0). Bacteria were exposed to either drug continuously for 5 days; each culture was diluted tenfold every 24 h. The cultures were incubated in Erlenmeyer flasks with 4 buffles containing 10 ml broth at 35 °C using a New Brunswick reciprocal water bath shaker at speed setting 5. The number of CFU/ml were determined daily immediately after dilution and at 2, 4, 6, 8 and 24 hours byplating serial dilutions onto Mueller Hinton agar plates either presoaked with B. cereus β-lactamases in order to inactivate cefsulodin or containing 20 mM Mg^{2+} in order to inactivate sisomicin and to reduce the antibacterial activity of ciprofloxacin.

Mueller-Hinton broth was used in order to assess synthesis of elastase and iron depleted trypticase soy broth was used in order to monitor synthesis of exotoxin A.

Determination of elastase, alkaline protease and exotoxin A

Elastase concentrations of the strains were determined as described previously [18, 19]. Briefly, every 24 hrs. a sample was withdrawn from the in vitro cultures; bacteria were harvested by centrifugation and protease concentrations in the supernatant fluids of the cultures were determined by radioimmunoassay. Exotoxin A was quantitated by radioimmunoassay as well.

In vivo studies

The granuloma-pouch technique in rats was used as a model for a chronic, local infection. This model was chosen as it offers the advantage that individual samples can be withdrawn repeatedly from the same animal so that a time dependent process can be monitored.

Results

The antibacterial effect of cefsulodin, sisomicin and ciprofloxacin against *P. aeruginosa* is summarized in Fig. 1. Cefsulodin and sisomicin exhibited a transient and

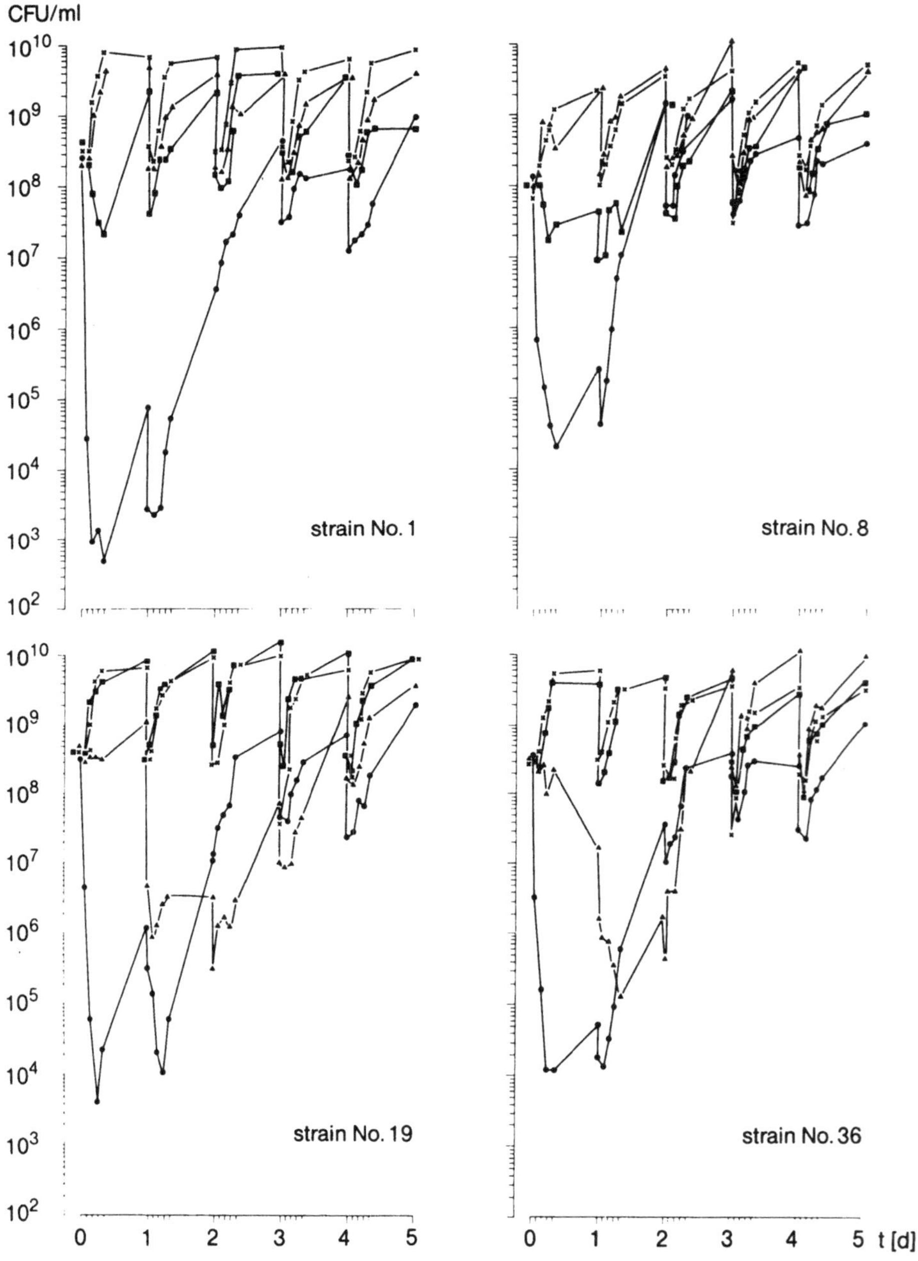

Fig. 1. Antibacterial activity of ciprofloxacin (●), cefsulodin (▲) and sisomicin (■) against 4 clinical isolates of P. aeruginosa. (x control) Strains were continuously exposed to a constant drug concentration equalling to 4-times the initial MIC-values

moderate effect against some of the isolates tested whereas ciprofloxacin affected all the 4 isolates within the first 48 hours of exposure; the numbers of CFU/ml were reduced by 4 to 5 $\log_{10}$ titers. Because of this marked reduction of viable counts by ciprofloxacin and the small amounts of elastase and exotoxin A produced being close to the limit of detectability cultures were not diluted on the second day in order to facilitate growth of those variants being less sensitive to the drug. Consequently bacterial counts increased and were approximately within the same order of magnitude as the drug-free controls and those cultures being exposed to cefsulodin and sisomicin respectively (Fig. 1). Periodical dilutions would have prevented the pronounced rise in viable counts (data not shown). However, it was not the aim of this study to compare the antibacterial efficacy of different drugs but to analyze their effect on the synthesis of pathogenicity factors produced by *P. aeruginosa*.

Determinations of the minimal inhibitory concentrations (MIC) on day 5 revealed that MIC-values were 8–32 times higher than before commencement of the study. Thus, bacteria were exposed to subinhibitory drug concentrations from the second or third day onwards because of the continuously decreasing susceptibility of the four isolates against the agents tested.

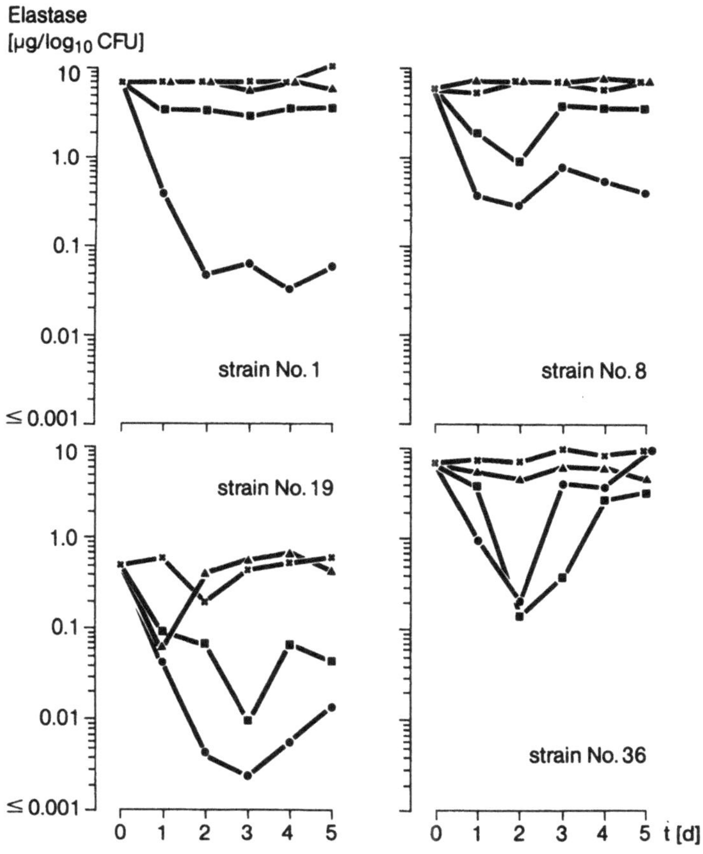

Fig. 2. In vitro effect of ciprofloxacin (●), cefsulodin (▲) and sisomicin (■) on elastase synthesis by 4 different strains of P. aeruginosa (x control)

In vitro synthesis of elastase and exotoxin A respectively was monitored daily. Cefsulodin did not affect elastase synthesis whereas sisomicin reduced elastase concentrations by approximately 50% (Fig. 2). In contrast ciprofloxacin reduced elastase synthesis in strains No. 1 and No. 19 more than hundredfold and in strain No. 8 approximately tenfold as compared to the drug free controls; the effect of ciprofloxacin on elastase production of strain No. 36 was not as marked (Fig. 2). Analogous effect were noted if strains were incubated in an iron free medium (Fig. 3). This iron-free medium was used in order to assess the effect of ciprofloxacin on exotoxin A production; ciprofloxacin reduced exotoxin A concentrations up to hundredfold in strains No. 1 and No. 19 whereas the effect on the other two strains was less marked. The effect of sisomicin and cefsulodin on exotoxin A production was not analyzed as these antibiotics exhibited no or only a moderate effect on protease synthesis.

Analysis of oxygen consumption and CO_2 production by the 4 *P. aeruginosa* test-strains (data not shown) revealed that ciprofloxacin inhibited oxygen consumption and CO_2 production respectively in strain No. 8 in direct proportion to the inhibition of growth; both parameters were not affected under these experimental conditions in the three other strains. Furthermore, the respiratory quotients–defined as the

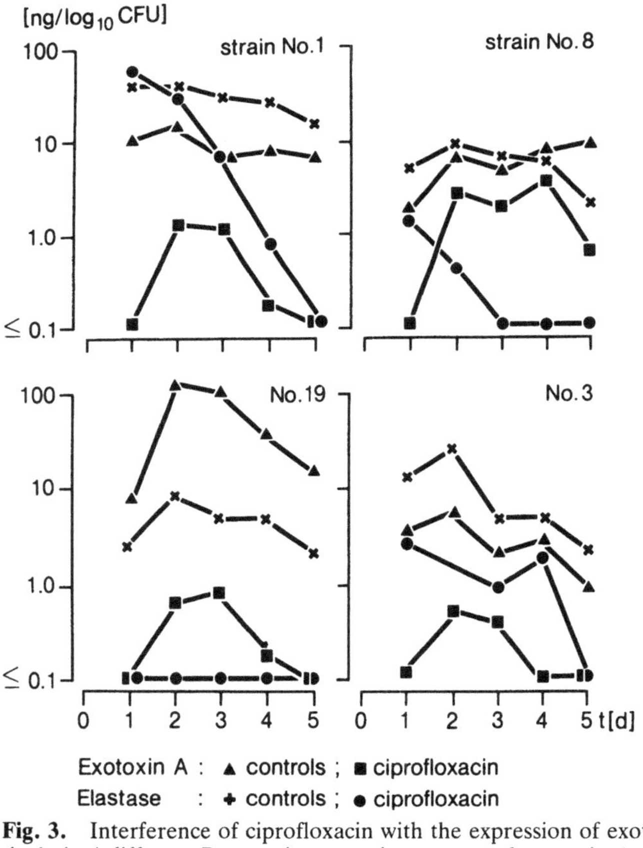

Fig. 3. Interference of ciprofloxacin with the expression of exotoxin A and elastase respectively in 4 different P. aeruginosa strains. ▲ control exotoxin A, ■ effect of ciprofloxacin on exotoxin A synthesis, x control elastase, ● effect of ciprofloxacin on elastase synthesis

Table 1. Effect of chemotherapy on the in vivo expression of proteases (AP = alkaline protease, Ela = elastase) and exotoxin A (ExoA) on the fourth day of treatment (μg/$\log_{10}$ CFU); (*) below the limit of detectability

Strain No.	Enzyme	Control	Ciprofloxacin	Sisomicin	Cefsulodin
	AP	7.23	– (*)	2.84	–
1	Ela	3.61	–	2.84	–
	Exo A	1.20	–	–	–
	AP	1.25	–	–	0.29
8	Ela	3.75	4.70	–	–
	Exo A	2.50	0.94	0.71	0.44
	AP	–	–	–	–
19	Ela	0.41	–	0.28	–
	Exo A	5.45	–	2.81	–
	AP	1.26	–	1.29	2.61
36	Ela	1.26	–	1.29	1.30
	Exo A	5.05	–	1.29	3.91

ratio CO_2 produced/oxygen consumed–equalled 1 in the 4 test strains irrespective of the absence or presence of ciprofloxacin. These in vitro findings were corroborated by animal experiments. Intravenous treatment of animals with cefsulodin (40 mg/kg b. i. d.) and ciprofloxacin (20 mg/kg b. i. d.) respectively eliminated strain No.1 and No. 19 from the infected foci whereas strains No. 8 and No. 36 were affected marginally. However, in vivo levels of proteases and exotoxin A were below the limit of detectability or markedly reduced due to ciprofloxacin treatment whereas the effect of cefsulodin was less pronounced (Table 1). Despite the negligible antibacterial effect of sisomicin treatment (10 mg/kg b. i. d.) (because of the anaerobic milieu within the infected pouches) synthesis of pathogenicity factors was reduced in any case to some extent. The general well-being of the animals improved in direct correlation to the interference with biosynthesis of proteases and exotoxin A; weight gain and body temperature normalized more rapidly (within 24 h) and was within the same range as the noninfected controls in those groups being treated with ciprofloxacin as compared to the sisomicin or cefsulodin treated animals (data not shown).

Discussion

P. aeruginosa causing chronic pulmonary infections in patients with cystic fibrosis exhibit a wide variety of biological properties and are genetically, physiologically and morphologically very heterogenous [42–47]. In agreement with data generated in the course of this study, differences in antibiotic susceptibility were on the one hand associated with heterogenous morphology, on the other hand, however, varying susceptibility to β-lactam antibiotics were noted even within the morphologically homogenous population of mucoid strains which were isolated from the same sputum [47–50]. Because of the heterogeneity of the initial inoculum those variants present before exposure to the drugs and exhibiting a reduced susceptibility against the corresponding agent were selected from the population during the incubation period of 5 days and became to predominate finally.

Thus, bacteria were exposed to subinhibitory drug concentrations from the second to third day onwards, exhibiting a marginal antibacterial effect only. In contrast, synthesis of proteases and exotoxin A respectively was significantly reduced by subinhibitory ciprofloxacin concentrations under in vitro as well as in vivo conditions. The degree of inhibition of protease and toxin production was strain-specific but independent from the inhibition of bacterial metabolism as indicated by the respiratory quotients and enzyme activity respectively; furthermore, the effect of ciprofloxacin on the synthesis of these pathogenicity factors was not correlated to growth inhibition as firstly in contrast to production of elastase and exotoxin A viable counts were most effectively reduced on day 1 and 2 whereas the effect on protease and toxin production was most pronounced on day 4 and 5 (Fig. 3); secondly, all results were related to one $\log_{10}$ titer of CFU.

In agreement with their mode of action, sisomicin affected elastase production only marginally and cefsulodin had no effect on elastase synthesis; thus we did not assess their effect on exotoxin A production.

A large body of evidence indicates that quinolones interfere with transcription-translation systems [8–16] and affect especially genes sensitive to catabolite repression [8]; it has been suggested that catabolite repressible promotors are particularly sensitive to DNA-supercoiling. Synthesis of extracellular proteases in Pseudomonas, too, was found to be controlled by a mechanism similar to catabolite repression; furthermore, it was reported that protein synthesis inhibitors suppress exoenzyme secretion more effectively than total protein synthesis. In agreement with these data it may be speculated that subinhibitory ciprofloxacin concentrations affect the expression of elastase and exotoxin A without reducing viable counts significantly. Despite the emergence of variants with reduced susceptibility against ciprofloxacin the reduced synthesis of the virulence associated factors may be of in vivo relevance as inhibition of pathogenicity factors of *P. aeruginosa* protected animals against experimental infections (38–41). Similarly, several authors reported that an increase in the pool of ciprofloxacin, β-lactam and/or aminoglycoside-resistant bacteria during chemotherapy of cystic fibrosis patients was not correlated with improvement in clinical signs and symptoms of infection [51–54, 56].

Thus, it may be speculated that in the chemotherapy of cystic fibrosis bacteriological response, i. e.–persistance or eradication of bacteria from the sputum is less important than interference with synthesis of pathogenicity factors.

References

1. Gellert M, Mizuuchi K, O'Dea MH, Nash HA(1976) DNA-gyrase: an enzyme that introduces superhelical turns into DNA. Proc Natl Acad Sci USA; 73: 4474–4478
2. Crumplin GC, Smith JT (1976) Nalidixic acid and bacterial chromosome replication. Nature 260: 643–645
3. Hays J, Boehmer S (1978) Antagonists of gyrase inhibit repair and recombination of UV-irradiated phage. Proc. Natl. Acad. Sci. USA; 75: 4125–4129

4. Kreuzer KN, Cozzardi NR (1979) Escherichia coli mutants thermosensitive for deoxyribonucleic acid gyrase subunit A: effects on deoxyribonucleic acid replication, transcription and bacteriophage growth, J Bacteriol 140: 424–435
5. Kubo M, Kano Y, Nakamura H, Nagata A, Imamoto F (1979) In vivo enhancement of general and specific transcription in Escherichia coli by DNA gyrase activity. Gene 7: 153–171
6. Mizuuchi K, Gellert M, Nash HA (1978) Involvement of supertwisted DNA in integrative recombination of bacteriophage lambda. J Mol Biol 121: 375–392
7. Orr E, Fairweather NF, Holland JB, Pritchard RH (1979) Isolation and characterization of a strain carrying a conditional lethal mutation in the con gene of Escherichia coli K12. Mol Gen Genet 177: 103–112
8. Sanzey B (1979) Modulation of gene expression by drug affecting deoxyribonucleic acid gyrase. J Bacteriol 138: 40–47
9. Smith CL, Kubo M, Imamoto F (1978) Promotor specific inhibition of transcription by antibiotics which act on DNA-gyrase. Nature (London) 275: 420–423
10. Sugino A, Higgins NP, Brown PO, Peebles CL, Cozzarelli NR (1978) Energy coupling in DNA-gyrase and the mechanism of action of novobiocin. Proc Natl Acad Sci USA 75: 4838–4842
11. Puga A, Tessman J (1973) Mechanism of transcription of bacteriophage S 13. II. Inhibition of phage specific transcription by nalidixic acid. J Mol Biol 75: 99–108
12. Falco SC, Zivin R, Rothman-Denes LB (1978) Novel template requirements of N4 iron RNA polymerase. Proc Natl Acad Sci USA 75: 3220–3224
13. Gomez-Eichelmann MC (1981) Effect of nalidixic acid and novobiocin on pBR 322 genetic expression in Escherichia coli mutants. J. Bacteriol 148: 745–752
14. Ostrowski J, Hulanicka D (1981) Effect of DNA gyrase inhibitors on gene expression of the cysteine regulon. Mol Gen Genet 181: 363–366
15. Brandsch R, Decker K (1982) The effect of gyrase inhibitors and cyclic AMP on induction and glucose repression of the 6-hydroxynicotine oxidases in Arthrobacter oxidans. Arch Microbiol 133: 274–277
16. Nordström K, Lindberg M (1978) Effects of streptomycin and novobiocin on S. aureus gene expression. J Bacteriol 133: 614–620
17. Sabath LD (1980) (Ed). Pseudomonas aeruginosa: the organism, disease it causes and their treatment. Bern: H. Huber
18. Döring G, Obernesser HJ, Botzenhart K (1982) Extracellular toxins of Pseudomonas aeruginosa. III. Radioimmunoassay for detection of alkaline protease. Zentral Bakteriol Mikrobiol Hyg (A) 252: 239–247
19. Obernesser HJ, Döring G (1982) Extracellular toxins of Pseudomonas aeruginosa. IV. Radioimmunoassay for detection of elastase. Zentral Bakteriol. Mikrobiol Hyg (A) 252: 248–256
20. Döring G, Obernesser HJ, Botzenhart K, Flehmig B, Høiby N, Hofmann A (1983) Proteases of Pseudomonas aeruginosa in patients with cystic fibrosis. J Infect Dis 147: 744–750
21. Hastie AT, Hingley ST, Kueppers F, Higgins ML, Tannenbaum CS, Weinbaum G (1983) Protease Production by Pseudomonas aeruginosa isolates from patients with cystic fibrosis. Infect. Immun. 40: 506–513
22. Gryz SJ, Iglewski BH (1980) Production of alkaline protease by Pseudomonas aeruginosa. J. Clin. Microbiol. 12: 131–133
23. Janda JM, Bottone EJ (1981) Pseudomonas aeruginosa enzyme profiling: predictor of potential invasiveness and use as an epidemiological tool. J. Clin. Microbiol. 14: 55–60
24. Hedberg M, Müller JK, Tompkins VN (1969) Elastase activity of Pseudomonas aeruginosa isolates from hospital patients. Am J Clin Pathol 52: 631–633
25. Müll JD, Callahan WS (1963) Estimation of elastolytic activity of strains of Pseudomonas aeruginosa. J Bacteriol 85: 1178–1179
26. Döring G, Obernesser HJ, Botzenhart K (1981) Extracellular toxins of Pseudomonas aeruginosa. II. Effect of two proteases on human immunoglobulins IgG, IgA and secretory IgA. Zentralbl Bakteriol Mikrobiol Hyg (A) 249: 89–98

27. Schultz DR, Miller KD (1974) Elastase of Pseudomonas aeruginosa: inactivation of complement components and complement-derived chemotactic and phagocytic factors. Infect Immun 10: 128–135
28a. Morihara K (1964) Production of elastase and proteinase by Pseudomonas aeruginosa. J Bacteriol 88: 745–757
28b. Scharmann W, Kraft W (1974) Hemmung der Blutgerinnung in vitro durch Protease von Pseudomonas aeruginosa. Blut 28: 90–99
29. Woods DE, Cryz SJ, Freidman RL, Iglewski BH (1982) Contribution of toxin A and elastase to virulence of Pseudomonas aeruginosa in chronic lung infections of rats. Infect Immun 36: 1223–1228
30. Moskowitz RW, Heinrich G (1971) Bacterial inactivation of human serum alpha-1 antitrypsin. J Lab Clin Med 77: 777–785
31. Morihara K, Tsuzuki H, Oda K (1979) Protease and elastase of Pseudomonas aeruginosa: inactivation of human plasma-proteinase inhibitor. Infect Immun 24: 188–193
32. Gadek JE, Fells GA, Zimmermann RL, Rennard SI, Crystal RG (1981) Anti-elastase of the human alveolar structures. Implications for the protease-antiprotease theory of emphysema. J Clin Invest 68: 889–898
33. Johnson DA, Carter-Hamm B, Dralle WM (1982) Inactivation of human bronchial mucosal proteinase inhibitor by Pseudomonas aeruginosa elastase. Am Rev Respir Dis 126: 1070–1073
34. Suter S, Schaad UB, Roux L, Nydegger UE, Waldvogel FA (1984) Granulocyte neutral proteases and Pseudomonas elastase as possible causes of airway damage in patients with cystic fibrosis. J Infect Dis 149: 523–531
35. Kharazmi A, Döring G, Hoiby N, Valerius NH (1984) Interaction of Pseudomonas aeruginosa alkaline protease and elastase with human polymorphonuclear leukocytes in vitro. Infect Immun 43: 161–165
36. Iglewski BH, Kabat D (1975) NAD-dependent inhibition of protein synthesis by Pseudomonas aeruginosa toxin. Proc Natl Acad Sci USA 72: 2284–2288
37. Iglewski B, Liu PV, Kabat D (1977) Mechanism of action of Pseudomonas aeruginosa exotoxin A: adenosine diphosphate-ribosylation of mammalian elongartion factor 2 in vitro and in vivo. Infect Immun 15: 138–144
38. Pavlovskis DR, Wretlind B (1979) Assessment of protease (elastase) as a Pseudomonas aeruginosa virulence factor in experimental mouse burn infection. Infect Immun 24: 181–187
39. Holder IA, Haidaris CG (1979) Experimental studies of the pathogenesis of infections due to Pseudomonas aeruginosa: extracellular protease and elastase as in vitro virulence factors. Can J Microbiol. 25: 593–599
40. Holder IA (1983) Experimental studies of the pathogenesis of infections due to Pseudomonas aeruginosa: Effect of treatment with protease inhibitors. Rev Infect Dis 5 (Suppl. 5): 914–921
41. Cryz SJ, Fürer E, Gerenanier R (1983) Protection against Pseudomonas infection in a murine burn wound sepsis model by passive transfer of antitoxin A, antielastase, and anti lipopolysaccharide. Infect Immun 39: 1072–1079
42. Williams RJ, Govan JRW (1973) Pyocine typing of mucoid strains of Pseudomonas aeruginosa isolated from children with cystic fibrosis. J Med Microbiol 6: 409–412
43. Diaz F, Mosovich LL, Neter E (1970) Serogroups of Pseudomonas aeruginosa and the immune response of patients with cystic fibrosis. J Infect Dis 121: 269–274
44. Fyfe JAM, Govan JRW (1980) Alginate synthesis in mucoid Pseudomonas aeruginosa: a chromosomal locus involved in control. J Gen Microbiol 119: 443–450
45. Fyfe JAM, Govan JRW (1981) A revised chromosomal locus for muc: a locus involed in the control of alginate production by mucoid Pseudomonas aeruginosa. Soc Gen Microbiol 8: 250–251
46. Piggott NH (1979) The biosynthesis of bacterial alginate. PhD diss, Edinburgh: University of Edinburgh
47. Govan JRW, Fyfe JAM, Baker NR (1983) Heterogeneity and reduction in pulmonary clearance of mucoid Pseudomonas aeruginosa. Rev. Infect Dis 5 (Suppl. 5): 874–880

48. Thomassen MJ, Demko CA, Boxerbaum B, Stern RC, Kuchenbrod PJ (1979) Multiple isolates of Pseudomonas aeruginosa with differing antimicrobial susceptibility patterns from patientd with cystic fibrosis. J Infect Dis 140: 873–880
49. Seale TW, Thirkill H, Tarpay M, Flux M, Renner OM (1979) 9: 72-8
50. Irvin RT, Govan JRW, Fyfe JAM, Costerton JW (1981) Heterogeneity of antibiotic resistance in mucoid isolates of Pseudomonas aeruginosa obtained from cystic fibrosis patients: role of outer membrane proteins. Antimicrob Agents Chemother 19: 1056–1063
51. Michalsen H, Bergan T (1981) Azlocillin with and without an aminoglycodide against respiratory tract infections in children with cystic fibrosis. Scand J Infect Dis (Suppl.) 29: 92–97
52. Godfrey AJ, Bryan LE, Rabin NR (1981) β-lactam resistant Pseudomonas aeruginosa with modified penicillin-binding proteins emerging during cystic fibrosis treatment. Antimicrob Agents Chemother 19: 705–711
53. Agostini M, Barlocco G, Bononi U, Cappelletti LM, Castellani L, Conforti M, Marazzani S, Martini N, Montemezzi P, Paulon G, Pederzini F, Saucassani L, Scroccaro G, Mastella G (1983) Alternative antibiotics against Pseudomonas infections in cystic fibrosis. In vitro activity pharmacokinetics, and double-blind randomized clinical trial with azlocillin, piperacillin, cefoperazone, ceftazidime, cefsulodin, cefotaxime, and moxotactam. Preliminary results. Drugs Exptl Clin Res 9: 671–686
54. McLaughlin FJ, Matthews WJ Jr, Strieder DJ, Sullivan B, Taneja A, Murphy P, Goldman DA (1983) Clinical and bacteriological responses to three antibiotic regimes for acute exacerbations of cystic fibrosis: ticarcillin–tobramycin, azlocillin–tobramycin, and azlocillin-placebo. J Infect Dis 147: 559–567
55. Boethling RS (1975) Regulation of extracellular protease secretion in Pseudomonas maltophilia. J Bacteriol 123: 954–961
56. Bender SW, Posselt HG, Wonne R, Stöver B, Strehl R, Shah PM (1984) Oral ciprofloxacin for pseudomonas bronchopneumoniae in CF. Proceedings of the 9th International Cystic Fibrosis Congress, Brighton, June 1984, Abstract No 4. 25.

Discussion

Vosbeck: What sub-MIC did you use in your in vitro experiments, and how do you explain the fact that the decrease of elastase production was seen only after 2 days of treatment?

Döring: In the initial two days the ciprofloxacin concentration was not subinhibitory, but when selection for resistant organisms started, MIC was changing all over the 5 days; after two days concentration was subinhibitory. Therefore, we could detect between days 2 to 5 a greater reduction of elastase than reduction of growth rate.

Pulverer: In your last slide you showed 2 different behaviours of strain 8 and 36 under ciprofoxacin influence. Have you done in vivo infection experiments with these two strains?

Döring: The results presented in the last slide were results from the infection of rats in the granuloma pouch model; we did not examine other animal infection models using these two strains.

V. Miscellaneous

In vivo Consequences of the Interaction between Antibiotic-damaged Pathogens and Host Animal Defences

C. G. Gemmell

Department of Bacteriology, Medical School, University of Glasgow, Royal Infirmary, Glasgow, Scotland

Streptococcus pyogenes is recognised as an important cause of infection in man and much of its virulence can be attributed to the possession of both structural and soluble antigens. Although the properties of pathogenic bacteria are often studied after *in vitro* growth, there is evidence that organisms grown *in vivo* may differ from those grown *in vitro* [1,2]. Under these circumstances it is conceivable that the host response may be determined by the expression of virulence factors of the pathogen. It is known that *S. pyogenes* cells can lose some of their normal complement of M antigen during *in vitro* culture. Only by exposing such cells to human or animal leukocytes [3] can one enrich the numbers of M-positive organisms within a culture. These bacteria are much less susceptible to serum opsonization and subsequent phagocytosis by leukocytes [4,5]. The presence of M protein as a surface "fuzz" prevents the activation of complement by the alternative pathway and renders the streptococcus resistant to phagocytic adherence and ingestion [6].

The biosynthesis of M protein has been shown recently to be influenced by the presence or absence of certain antibiotics in the culture medium at sub-growth inhibitory concentrations [7,8]. Much less M protein was present on bacterial cells grown in the presence of clindamycin or lincomycin and greater amounts of complement as C3b were deposited on the drug-treated streptococci during serum opsonization. This resulted in enhanced serum opsonization phagocytic ingestion and killing by polymorphonuclear leukocytes (PMN).

Normally *Streptococcus pyogenes* elaborates M protein, T protein and C substance (all forming part of the bacterial cell wall) and maximal amounts of each antigen appear within 12–16 hours' culture (see Table 1). Yields of antigen closely followed the growth cycle of the streptococcus. However, if the same cultural

Table 1. Antigen Synthesis by Streptococcus Pyogenes NCTC 100088 (M-25)

Time of Sampling (Hours)	M-Antigen Titre	T-Antigen Titre	Group-Specific Carbohydrate Antigen Titre
4	< 4	< 2	< 2
8	32	8	8
12	128	32	32
16	256	64	32
24	256	64	32

The Influence of Antibiotics on the Host-Parasite Relationship II
Edited by D. Adam, H. Hahn, W. Opferkuch

Table 2. Effect of Growth in the Presence of Various Antibiotics on Antigen Synthesis by Streptococcus Pyogenes NCTC 100088

Cultural conditions	Titre of Antigen		
	M-Protein	T-Protein	C-Substance
No Drug Present	256	64	32
Chloramphenicol-grown	256	64	32
Clindamycin-grown	< 2	< 2	< 2
Erythromycin-grown	128	64	16
Lincomycin-grown	< 2	< 2	< 2
Penicillin-grown	128	64	16

experiment is carried out in the presence of sub-MIC levels of certain antibiotics, primarily clindamycin and lincomycin, little or no M or T protein was present on the cells. It also appears likely that the antigenic nature of the C substance was altered by drug exposure (see Table 2).

These antigenic changes in *S. pyogenes* are reflected in changes in its relationship with the host's reticulo-endothelial system (R.E.S.). As well as altering the bacterium's susceptibility to phagocytosis, differences were observed in the ability of human PMN to develop directed movement (chemotaxis) towards the drug-damaged bacteria or their isolated cell-wall components. By measuring PMN chemotaxis under agarose [9] it was soon recognised [10] that the rate of directed movement towards streptococcal cell wall preparations grown in the presence of the lincosamines (clindamycin and lincomycin) was greater than that obtained with bacteria grown in the presence of chloramphenicol, erythromycin or penicillin (Table 3). The presence of M protein on the surface of *S. pyogenes* cells has been recognised to impair PMN chemotaxis [11] and repeat experiments with hydrochloric acid (HCl) or trypsin extracts of the bacterial cells grown in the presence or absence of these antibiotics showed that diminished M protein content of the damaged cells was probably responsible for the enhanced PMN chemotaxis (see table 4); no differences were detectable with the T antigen (trypsin extract) or C substance (hot HCl extracts).

Table 3. Effect of Pre-Treatment of *Streptococcus Pyogenes* with Various Antibiotics on the Activation of PMN Chemotaxis by Cell-Wall Preparations

Cultural Conditions	Chemotactic Index
no antibiotic present	1.5
Chloramphenicol-grown	1.5
Clindamycin-grown	5.5
Erythromycin-grown	1.6
Lincomycin-grown	4.2
Penicillin-grown	1.7

Table 4. Activation of PMN Chemotaxis by Extracts Containing M and T Proteins from *Streptococcus Pyogenes* Grown in the Prence or Absence of Various Antibiotics

Cultural Conditions	Chemotactic Index* of PMN towards	
	M-Antigen Extract	T-Antigen Extract
No Drug Present	1.5	1.1
Plus Chloramphenicol	1.5	1.1
Plus Clindamycin	4.8	1.3
Plus Erythromycin	1.6	1.2
Plus Lincomycin	3.6	1.2
Plus Penicillin	1.8	1.2

* mean of at least three experiments with different preparations of PMN

Although each of these experiments were performed with a strain of *S. pyogenes* whose antigenic mosaic had been previously selected by rotation through blood [3] any changes attributable to an action of the antibiotics at sub-MIC levels may not take place in an actual clinical infection. There exist a number of experimental models of infection in laboratory animals which demonstrate that *S. pyogenes* can release various exotoxins *in vivo* [12, 13] with subsequent antibody stimulation.

Significant levels of streptolysin S (SLS) and streptolysin O (SLO) were detectable within 24 hours and cell wall antibodies were measurable in the animals' serum 2 – 3 weeks following the multiplication of the streptococci in chambers implanted in the mouse peritoneum. Using a model skin lesion which could be developed in guinea pigs, necrosis occurred within 24–48 hours with doses of *S. pyogenes* ranging from 1.5×10^6 – 3.0×10^7 c.f.u.; the rate of growth attained *in vivo* is illustrated in Table 5. In addition the *in vivo*-grown bacteria also were able to elaborate a haemolysin antigenically similar to streptolysin O as well as a DNAase enzyme within the skin lesion.

Several features of the artificially induced *S. pyogenes* skin infections were investigated using antibiotic-damaged bacteria as the inoculum [14]. These parame-

Table 5. Growth of *Streptococcus Pyogenes* within Skin of Guinea Pigs with without Clindamycin Therapy 1 hour post Inoculation

Hours Post Inoculation	cfu/gram tissue of	
	control animal (no drug)	drug-treated animal (clindamycin)
4	4×10^7	3.6×10^7
8	6.6×10^7	5.6×10^7
24	9.9×10^7	8.25×10^7

3.3×10^7 cfu *Streptococcus pyogenes* (NCTC 9994) administered subcutaneously to each animal. Drug given I.V. in dose 0.5 µg/kg.

Table 6. Toxin Biogenesis within Skin of Guinea Pigs Following Inoculation of *Streptococcus Pyogenes* and Subsequent Therapy with Clindamycin

Toxin Measured	Biological Activity of Toxin in Tissues (as/Titre) Control (no Drug)	Drug-Treated Clindamycin
Streptolysin O	16	< 4
DNA-ASE	16	< 2
Hyaluronidase	4	< 2

3.3×10^7 cfu *Streptococcus pyogenes* (NCTC 9994) administered subcutaneously to each animal. Drug given I.V. in dose 0.5 μg/kg

ters included tissue histology, bacterial growth rate and toxin biosynthesis. Typical results are seen in Table 6 and in Plate 1. The development of lesions was confined to those bacterial inocula grown in the absence of any antibiotic or in the presence of chloramphenicol. Clindamycin-grown bacteria failed to produce necrotic lesions even when 3.0×10^7 c.f.u. were inoculated. Skin histology revealed that a less extensive subcutaneous abscess was present at the site inoculated with clindamycin-grown streptococci. There appeared to be greater infiltration of leukocytes and less cellular damage and necrosis in the latter. Measurement of certain toxins within the infected tissue also revealed differences between the different types of inocula. Streptolysin O, identifiable serologically, was only produced in significant amounts when normal bacteria were injected into the skin. These findings were also seen when clindamycin (0–05μg/kg animal) was administered to guinea pigs either one or six hours after the subcutaneous inoculation of 1.5×10^6 c.f.u. *Streptococcus pyogenes* (see Table 7). It is conceivable that drug therapy modified the ability of *Streptococcus pyogenes* to elaborate the various exotoxins and altered its surface topography to such an extent that PMN chemotaxis was accelerated due to the exposure of greater amounts of cell wall chemotaxin(s).

Indeed *in vitro* cultivation of certain strains of *S. pyogenes* in the presence of the lincosamines (clindamycin and lincomycin) altered the antigenic nature of the cell surface. Lower levels of M and T protein were present on the drug-treated bacteria.

Table 7. Streptococcus Pyogenes Infection in Guinea Pigs: Lesion Formation in Presence or Absence of Subsequent Antibiotic Therapy

Drug Treatment	Effect on Skin after 8 hours		24 hours	
	Erythema	Necrosis	Erythema	Necrosis
A	+	+	+	+
B	–	–	–	–
C	–	–	+	+

A = Guinea pig given no Antibiotic. B = Guinea pig given 0.05 μg/kg Clindamycin one hour after Inoculation of *S. Pyogenes* (1.5×10^6 cfu) s.c. C = Guinea pig given 0.05 μg/kg Clindamycin six hours after Inoculation of *S. Pyogenes* (1.5×10^6 cfu) s.c.

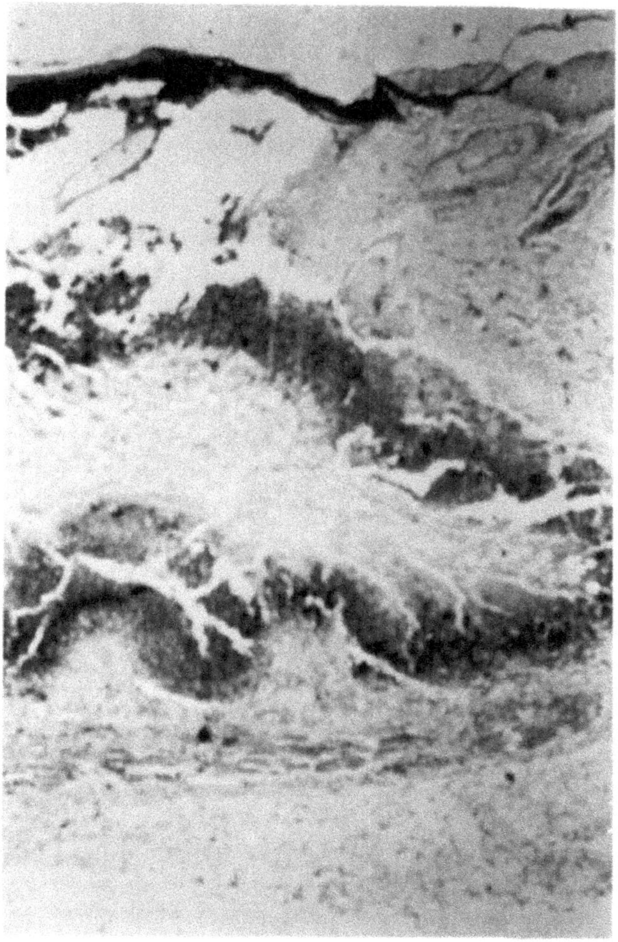

Fig. 1. Histological features of subcutaneous abscess induced in guinea pigs following the injection of 3×10^7 c.f.u. *Streptococcus pyogenes*. Clindamycin was administered i.v. (0.5 μg/kg) at the time of skin inoculation. Section stained by haematoxylin and eosin. Magnification × 100

The presence or absence of M protein on the surface of the cell was monitored following growth in the presence of sub-MIC levels of the antibiotic using serological detection methods with specific antibody and complement fixation and electron microscopy. It remains to be seen whether or not similar antigenic changes took place *in vivo* during the experimental infection model.

The growth environment of bacteria *in vivo* is complex and undoubtedly changes during the course of infection: such an environment cannot be duplicated *in vitro*. The chamber implant system with *S. pyogenes* [12] showed a rise in antibody response to streptolysin O beginning fourteen days following implantation and reaching a maximum after five weeks; the titre at this time was 128 units. These immunological changes followed the early elaboration of both streptolysin O and S *in vivo*. In addition, the production of antibody to cell surface antigens could be demonstrated indirectly with fluorescent antibody. Erythrogenic toxins A and C have been produced in tissue-cage fluids inoculated with *S. pyogenes* [13].

There exist few studies in which the direct measurement of phagocytosis of *in vivo* grown bacteria has been published. However, it is likely that the ability of bacteria to produce additional surface layers such as capsule (group B streptococci), and K antigen (*Escherichia coli*) or Vi antigen (*Salmonella typhi*) may alter the susceptibility of the producer bacteria to phagocytic uptake and killing by polymorphonuclear leukocytes [15].

Preliminary experiments in our laboratory in which *Staphylococcus aureus* has been passaged in mice several times before being used as target for phagocytosis *in vitro* demonstrated clear differences between the susceptibility of *S. aureus* cells grown *in vivo* from those grown *in vitro* to complement-mediated opsonization. The *in vivo* grown bacteria required more extensive opsonization before they were adequately phagocytosed by human PMN. Nevertheless, the *in vivo* grown bacteria were as susceptible to low doses of certain anti-staphylococcal drugs (clindamycin and fusidic acid) as described earlier [16]. Drug-damaged staphylococci became more susceptible to phagocytic ingestion and killing through reduced biosynthesis of protein A on the cell surface. Any changes in the virulence or antigenicity of the staphylococcus following *in vivo* passage failed to influence the ability of the host response (reticulo-endothelial system) to phagocytose the antibiotic-damaged target organism. It is possible that the changes observed earlier [17] in the host response to antibiotic damaged bacteria might be a consequence of at least some of these *in vitro* findings.

More definitive evidence for significant effects of low concentrations of certain antibiotics on the *in vivo* response to bacterial pathogens will have to await the development of more accurate methods of monitoring drug concentrations within the animal tissues. It is uncertain whether the drug concentration at the site of bacterium-PMN interaction is higher or lower than that needed to damage or kill the bacterial pathogen. Possibly the use of artificial liposomes as antibiotic-carriers might be beneficial in this respect. Otherwise research will need to be directed towards the development of a model which allows the free interaction of bacteria, antibiotic and the phagocytic cells. The anaerobic abscess involving *Bacteriodes fragilis* might be appropriate in this respect [18].

References

1. Beining PR, Kennedy ER (1983) Characteristics of a strain of *Staphylococcus aureus* grown *in vivo* and *in vitro*. J Bacteriol 85: 732–743
2. Finn TM, Arbuthnott JP, Dougan G (1982) Properties of *Escherichia coli* grown *in vivo* using a chamber implant system. J gen Microbiol 128: 3083–3 091
3. Becker CG (1967) Enhancing effect of type-specific anti-streptococcal antibodies on the emergence of streptococci rich in M protein. Proc Soc Exp Biol Med 124: 331–335
4. Bisno AL (1979) Alternate complement pathway activation by Group A streptococci: role of M protein. Infect Immun 25: 1172–1176
5. Fox EN (1974) M proteins of Group A streptococci. Bacteriol Rev 38: 57–86
6. Swanson J, Hsu KC, Gotschlich EC (1968) Electron microscopy studies on streptococci. I.M antigen. J Exp Med 130: 1063–1091
7. Gemmell CG, Peterson PK, Schmeling D, Kim Y, Mathews J, Wannamaker L, Quie PG (1981) Potentiation of opsonization and phagocytosis of *Streptococcus pyogenes* following growth in the presence of clindamycin. J Clin Invest 67: 1249–1256

8. Gemmell CG, Peterson PK, Schmeling D, Quie PG (1982) Studies on the potentiation of phagocytosis of *Streptococcus pyogenes* by treatment with various antibiotics. Drugs Exptl Clin Res 8: 235–240
9. Nelson RD, Quie PG, Simmons RL (1975) Chemotaxis under agarose: a new and simple method for measuring chemotaxis and spontaneous migration of human polymorphonuclear leukocytes and monocytes. J Immunol 115: 1650–1656
10. Gemmell CG, Amir MKA (1978) Effects of certain antibiotics on the formation of cellular antigens and extracellular products by group A streptococci. IN Pathogenic Streptococci (ed MT Parker) Reedbooks Ltd. England p 67–68
11. Gemmell CG, Amir MKA (1980) Antibiotic induced changes in streptococci with respect to their interaction with human polymorphonuclear leukocytes. IN Current Chemotherapy and Infectious Disease (eds. JD Nelson and C Grassi) Amer Soc Microbiol Washington DC. 2: 810–812
12. Duncan JL (1983) Streptococcal growth and toxin production *in vivo*. Infect Immun 40: 501–505
13. Knoll H, Holm SE, Gerlach D, Kohler W (1982) Tissue cages for study of experimental streptococcal infection in rabbits. I Production of erythrogenic toxins *in vivo*. Immunobiology 162: 128–140
14. Amir MKA (1980) Changes in the pathogenicity of streptococci induced by certain antibiotics relevant to the host parasite relationship. PhD Thesis University of Glasgow
15. Quie PG, Giebink S, Peterson PK (1981) Bacterial mechanisms for inhibition of ingestion by phagocytic cells. IN Microbial Perturbation of Host Defenses (ed. F O'Grady and H Smith) Acad Press London p 121–141
16. Gemmell CG, O'Dowd A (1984) Regulation of protein A biosynthesis in *Staphylococcus aureus* by certain antibiotics: its effect on phagocytosis by leukocytes. J Antimicrob Chemother 12: 587–597
17. Gemmell CG (1978) Effect of subinhibitory concentrations of antibiotics on experimental pyogenic infections in mice. IN Current Chemotherapy (ed.) Amer Soc Microbiol Washington DC p512–514
18. Joiner KA, Gelfand JA, Onderdonk AB, Bartlett JG, Gorbach SL (1980) Host factors in the formation of abscesses. J Infect Dis 142: 40–49

Discussion

Vosbeck: Clindamycin has been demonstrated to be accumulated in phagocytes. Its concentration at the site of infection might therefore be increased due to accumulation of phagocytes, so that in your model you might be dealing with local concentrations of antibiotic that are not sub-MIC at all. Have you measured the local antibiotic concentration?

Gemmell: In tissue extracts from the skin of the guinea pigs injected with *S. pyogenes* we looked biologically for any antibacterial activity. Although this does specifically detect clindamycin, we found no antibacterial activity in the extracts from the antibiotic-treated animals. The slower growth rates of the streptococci *in vivo* when antibiotic was given 1 and 6 hours later would suggest some local antibiotic being present however.

Ginsburg: Do you propose that we switch from penicillin to clindamycin in the treatment of upper respiratory infections with group A streptococci in view of your findings on the effect of this antibiotic on streptococci?

Your findings on the presence of streptolysin O (SLO), in the skin extracts is very interesting. Since the skin is very rich in lipids which can bind SLO I wonder whether you have established with certainty that the hemolysin you found was indeed SLO. Inhibition by cholesterol might resolve this question.

Gemmell:

1. In no way would I suggest that any changes be made to the use of penicillin as drug of choice in upper respiratory tract infections. The fact that clindamycin rather than penicillin impaired M antigen and toxin production merely reflects a difference in their mode of action, i.e. penicillin acts on cell wall biosynthesis whereas clindamycin acts on protein biosynthesis. However other antibiotics which impair protein synthesis do not resemble clindamycin action on the expression of virulence factors in streptococci.
2. I think that the finding of SLO is a reality in guinea pig skin injected with *S. pyogenes* on two counts— we were able to detect the haemolysin as an antigen by gel diffusion versus specific antiserum and we were also able to neutralize the haemolytic activity by specific antiserum. Inhibition by cholesterol would be a useful extra test to be sure of the toxin's presence.

The Influence of Streptomycin on the Ecology of the Intestine and Resistance to Infection

D. J. HENTGES, J. U. QUE, and S. W. CASEY

Department of Microbiology, Texas Tech University Health Sciences Center, Lubbock, Texas 79430, USA

Streptomycin administration increased the susceptibility of Swiss white mice to intestinal colonization with *Pseudomonas aeruginosa* and *Salmonella typhimurium* and facilitated translocation of these organisms extraintestinally. The antibiotic altered the composition of the mouse cecal flora, caused an increase in the pH and Eh of cecal contents and a decrease in the concentration of volatile fatty acids. In addition, the passage of contents was slower through the cecum of antibiotic treated than untreated animals. *P. aeruginosa* failed to multiply, under anaerobic conditions, in cecal contents from treated (T) or untreated (U) animals. A population decline was apparent in contents from untreated animals. *S. typhimurium* multiplied in contents from both groups of animals but its multiplication rate and total population were greater in contents from treated animals. *P. aeruginosa* multiplied in T-VFA broth, containing concentrations of volatile fatty acids detected in cecal contents of streptomycin treated mice, but failed to multiply in U-VFA broth, which contained acids in concentrations found in untreated mice. *S. typhimurium* multiplied in both types of broth but its multiplication rate and total population were greater in T-VFA broth than in U-VFA broth.

The results indicate that both *P. aeruginosa* and *S. typhimurium* are inhibited by volatile fatty acids in concentrations present in the ceca of untreated mice but that *P. aeruginosa* is more sensitive to the effects of the acids than *S. typhimurium*.

The influence of antibiotics on the resistance of animals to overgrowth of the intestine with non-indigenous bacteria has been the subject of a number of studies. Freter [3, 4] demonstrated that oral administration of antibiotics to mice and guinea pigs, which eliminated intestinal flora components, rendered the animals susceptible to infection with antibiotic resistant strains of *Shigella flexneri* and *Vibrio cholerae*. Bohnhoff et al. [1] reported that oral administration of streptomycin to mice greatly increased their susceptibility to *Salmonella enteritidis* infection. The antibiotic lowered the ID_{50} of *S. enteritidis* from 1 million viable bacteria to fewer than 10 bacteria [2]. Evidence from *in vitro* studies suggested that loss of flora components resulted in a decrease in the concentration of volatile fatty acids and an increase in pH of intestinal contents, providing conditions which favored multiplication of *S. enteritidis* [7].

Protection against colonization of the intestinal tract by non-indigenous bacteria was termed colonization resistance by van der Waaij et al. [9]. These investigators

The Influence of Antibiotics on the
Host-Parasite Relationship II
Edited by D. Adam, H. Hahn, W. Opferkuch

studied colonization resistance in mice before, during and after oral administration of streptomycin and neomycin. Mice were infected intragastrically with either *E. coli, Klebsiella pneumoniae*, or *P. aeruginosa* during each of three experimental periods. Before antibiotic treatment, colonization resistance was pronounced against all three organisms. During antibiotic administration, which virtually eliminated all flora components, there was a precipitous decrease in colonization resistance and the mice were easily colonized with the three organisms. After cessation of antibiotic treatment, the intestinal tract was repopulated with anaerobic organisms that had survived treatment and colonization resistance was gradually restored.

In our laboratory, ampicillin, streptomycin, kanamycin, clindamycin and metronidazole were given individually to Swiss white mice to determine their influence on colonization resistance against intragastrically administered *Pseudomonas aeruginosa* [6]. Despite differences in spectra of activity, all five antibiotics decreased the resistance of mice to intestinal colonization with *P. aeruginosa*. Streptomycin was the most effective of the antibiotics in both reducing colonization resistance and facilitating translocation of the challenge organism extraintestinally. Its effects were, therefore, investigated further. Streptomycin treated mice and untreated control mice were challenged with either *P. aeruginosa*, representing an opportunistic organism, or *Salmonella typhimurium*, representing an enteric pathogen [8]. The effect of streptomycin on the resistance of mice to infection and on the ecology of the intestine is the subject of this report.

Streptomycin resistant mutants of *P. aeruginosa* and *S. typhimurium* were used in these studies. The mice were given 5 mg/ml streptomycin sulfate (Sigma Chemical Co., St. Louis) in their drinking water for a period of three weeks. One week after initiation of antibiotic treatment, the treated mice and untreated controls were inoculated intragastrically with 0.1 ml of a suspension containing either 10^8 viable *P. aeruginosa* or 10^3 viable *S. typhimurium*. Concentrations of streptomycin in cecal contents and counts of the challenge organisms in homogenates of various organs were determined. The results of these experiments are summarized in Tables 1 and 2. Counts are expressed in the tables as mean $\log_{10}$ viable organisms recovered from colonized animals.

Table 1 shows that the cecum and small intestine of all streptomycin treated mice were colonized with *P. aeruginosa* at mean counts in the neighbourhood of 10^3 to 10^4

Table 1. Effect of Streptomycin Administration on the Establishment of *P. aeruginosa* in Mice-7 Days After Challenge[a]

Organ	*Untreated*		*Treated*	
Cecum	0	0/17	3.64±0.98[b]	17/17[c]
Small Intestine	0	0/17	4.14±0.99	17/17
MLN	0	0/17	2.33±0.64	9/17
Liver	0	0/17	2.16±0.95	9/17
Spleen	0	0/17	1.92±0.52	8/17

[a] Inoculum 10^8 Viable Organisms
[b] Mean Log_{10} Viable Counts ± SD per Gram Tissue
[c] Proportion of Mice harboring *P. aeruginosa*

Table 2. Effect of Streptomycin Administration on the Establishment of *S. typhimurium* in Mice-7 Days After Challenge[a]

Organ	*Untreated*		*Treated*	
Cecum	0	0/10	8.50 ± 0.78[b]	10/10[c]
Small Intestine	0	0/10	5.99 ± 1.66	10/10
MLN	0	0/10	5.52 ± 0.77	10/10
Liver	0	0/10	4.67 ± 1.16	10/10
Spleen	0	0/10	5.36 ± 1.12	10/10

[a] Inoculum 10^3 Viable Organisms
[b] Mean Log_{10} Viable Counts ± SD per Gram Tissue
[c] Proportion of Mice Harboring *S. typhimurium*

per gram. None of the untreated animals were colonized in these experiments. *P. aeruginosa* translocated to either the spleen, liver or mesenteric lymph nodes in 13 of 17 animals. Streptomycin activity was detected in filtrates of cecal contents of all treated animals.

Although the *S. typhimurium* inoculum was much smaller than the *P. aeruginosa* inoculum, the incidence of translocation of *S. typhimurium* and the population levels of this organism in the tissues of streptomycin treated mice were greater than when the mice were challenged with *P. aeruginosa* (Table 2). *S. typhimurium* failed to colonize untreated mice. In streptomycin treated animals, the mean *S. typhimurium* counts exceeded 10^8 per gram cecum and were almost 10^6 per gram small intestine. *S. typhimurium* translocated to the mesenteric lymph nodes, spleen and liver of all streptomycin treated mice. In every case, streptomycin activity was detected in filtrates of cecal contents of these animals.

Since the animals were far more susceptible to infection with *S. typhimurium* than with *P. aeruginosa*, we wanted to determine if mouse cecal contents more effectively supported multiplication of *S. typhimurium* than *P. aeruginosa*. Cecal contents from both treated and untreated mice were therefore removed in an anaerobic isolator, pooled separately and were heated at 121°C to kill viable microorganisms. The contents were then inoculated with either 10^6 *P. aeruginosa* or 10^5 *S. typhimurium* and were incubated anaerobically at 35°C. At intervals after inoculation, aliquots were removed, diluted and plated in order to count viable *P. aeruginosa* and *S. typhimurium*. Fig. 1 shows that *P. aeruginosa* failed to multiply in contents obtained from either treated or untreated mice but that, after 48 hours incubation, its population was approximately ten times greater in contents from streptomycin treated than untreated animals. *S. typhimurium*, on the other hand, multiplied in contents from both groups of animals but its population was approximately 50 times greater in contents from treated than untreated animals (Fig. 2). The results show that mouse cecal contents support the multiplication of *S. typhimurium* but not *P. aeruginosa*. They also show that factors are present in contents from untreated animals that inhibit both organisms. Inhibitory activity is diminished when the animals are treated with streptomycin.

How does streptomycin influence the ecology of the mouse cecum to cause a diminution of colonization resistance? Studies in this area of research are still in

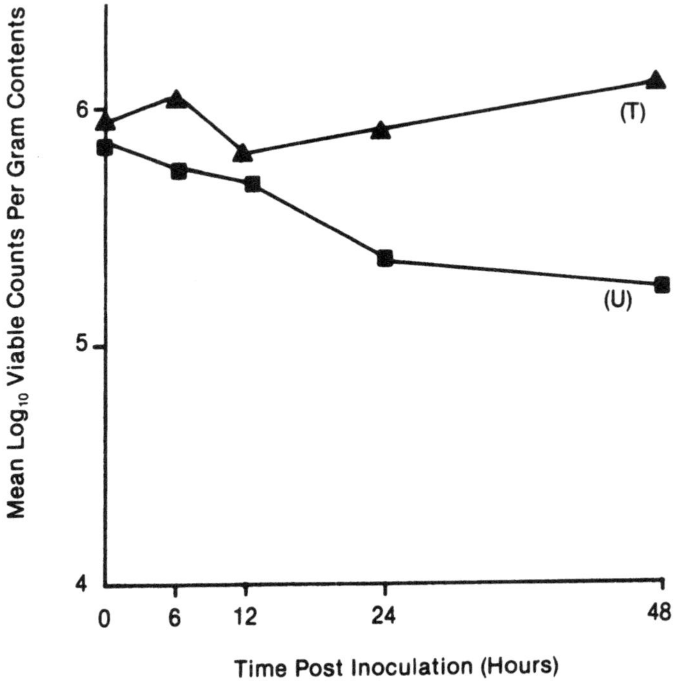

Fig. 1. Multiplication of *P. aeruginosa* in Cecal Contents Obtained from Streptomycin Treated (T) and Untreated (U) Mice

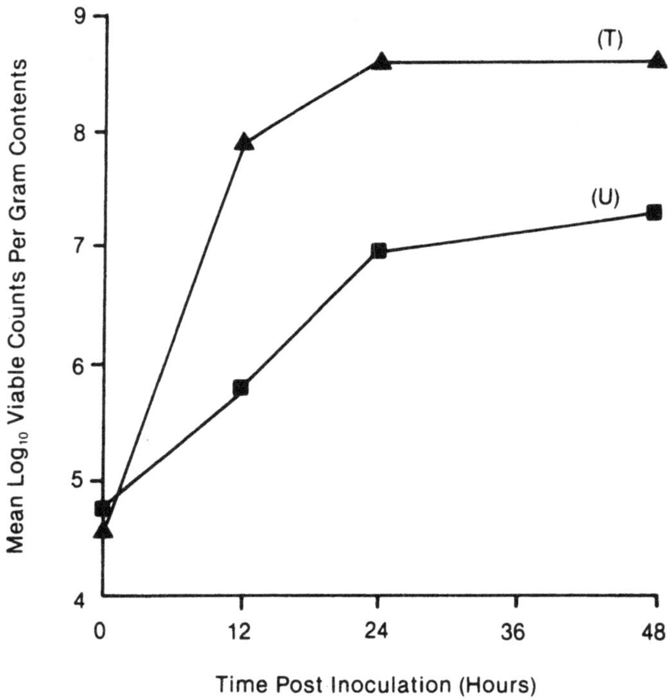

Fig. 2. Multiplication of *S. typhimurium* in Cecal Contents Obtained from Streptomycin Treated (T) and Untreated (U) Mice

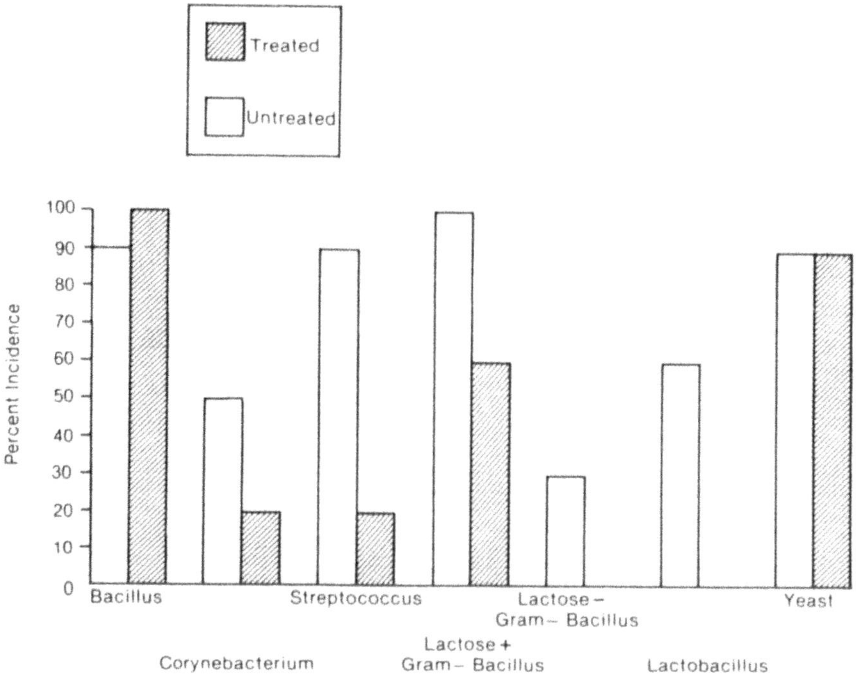

Fig. 3. Incidence of the Isolation of Aerobic Organisms from Cecal Contents of Treated and Untreated Mice

progress. To date, we have determined that the antibiotic alters cecal flora composition, the Eh, pH and volatile fatty acid concentration of cecal contents and the rate of passage of contents through the cecum.

Both aerobic and anaerobic components of the cecal flora were altered by the administration of streptomycin to the mice [5]. The results of experiments on the effects of streptomycin on cecal flora composition are illustrated in Fig. 3 and 4.

The incidence of isolation of various aerobic organisms from the cecal contents of 10 treated and 10 untreated mice is shown in Fig. 3. Streptomycin administration caused a decline in the frequency of the isolation of *Corynebacterium*, *Streptococcus* and *Lactobacillus*. Lactose-negative, gram-negative bacilli disappeared from the contents and lactose-positive gram-negative bacilli were partially replaced by streptomycin-resistant variants. The mean populations of the various organisms and the mean total populations recovered from the contents of treated and untreated animals were nearly the same, however.

Figure 4 illustrates the effects of streptomycin administration on the incidence of isolation of various genera of anaerobic bacteria from cecal contents of the mice. There was a striking decrease in the frequency of recovery of *Lactobacillus* and *Bifidobacterium* and a moderate decrease in the recovery of *Clostridium* as a result of antibiotic treatment. The total populations and population levels of the individual organisms that were recovered were approximately the same in the treated and

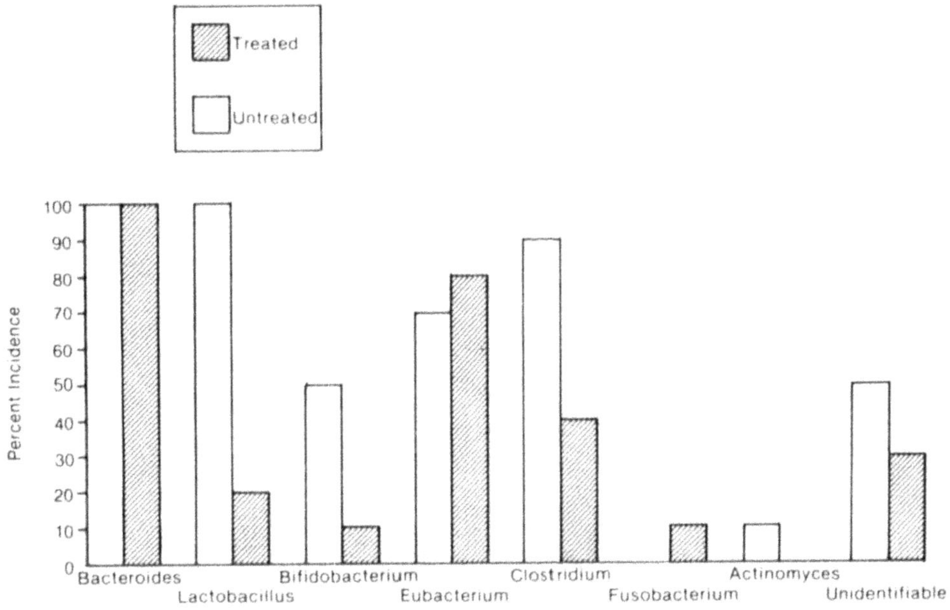

Fig. 4. Incidence of the Isolation of Anaerobic Organisms from Cecal Contents of Treated and Untreated Mice

untreated mice with the exception of *Lactobacillus* which had a population approximately 1.000-fold greater in the untreated animals.

Other changes in the ecology of the cecum occurred as a consequence of streptomycin treatment. The antibiotic caused an increase in the pH of cecal contents from a mean value of 6.42 in untreated animals to a mean value of 6.73 in treated animals (see Table 3). As a result of treatment, the concentrations of acetic acid, propionic acid and butyric acid in cecal contents dropped. Valeric acid was detected in low concentration in contents of untreated but not treated animals. Streptomycin treatment also caused a small increase in the Eh of mouse cecal contents, from a value of −129 mv in untreated animals to a value of −119 mv in treated animals.

Table 3. Effect of Streptomycin Administration on Volatile Fatty Acid Concentrations and pH of Cecal Contents of Mice

Acid	*Untreated*	*Treated*
Acetic	76.6±13.2[a]	47.9±7.7
Propionic	14.9±3.4	9.5±3.8
Butyric	34.7±8.4	9.5±1.9
Isovaleric	0	0
Valeric	4.8±1.1	0
pH	6.42±0.13	6.73±0.28

[a] μ Eq per Gram Cecal Content

The effect of streptomycin administration on the rate of passage of contents through the gastrointestinal tract of the mice was determined. Approximately 10^7 thermophilic *Bacillus* spores were given intragastrically to treated and untreated animals. The animals were sacrificed at intervals after spore administration and the number of spores isolated from various parts of the gastrointestinal tract were determined. Approximately 65% of the spores that entered the cecum of untreated animals passed through this organ within 1½ hours but only 49% passed through the cecum of treated animals during the same time period. These values are statistically significantly different indicating that passage of contents through the cecum is somewhat slower in streptomycin treated than untreated animals.

Volatile fatty acids were added to Brain Heart Infusion (BHI) broth in concentrations detected in cecal contents of treated (U-VFA) and untreated (U, VFA) animals and the pH values were adjusted to 6.73 and 6.42, respectively. The two types of broth were inoculated with between 10^5 and 10^6 per ml of either *P. aeruginosa* or *S. typhimurium* and were incubated at 35 °C under anaerobic conditions. Controls consisted of BHI broth without volatile fatty acids, at pH 6.73 or pH 6.42, which were similarily inoculated. Aliquots of the cultures were removed at intervals after incubation to determine colony counts.

Growth curves of *P. aeruginosa* are illustrated in Fig. 5. The organism attained populations of approximately 6×10^7 per ml by 24 hours in BHI broth without

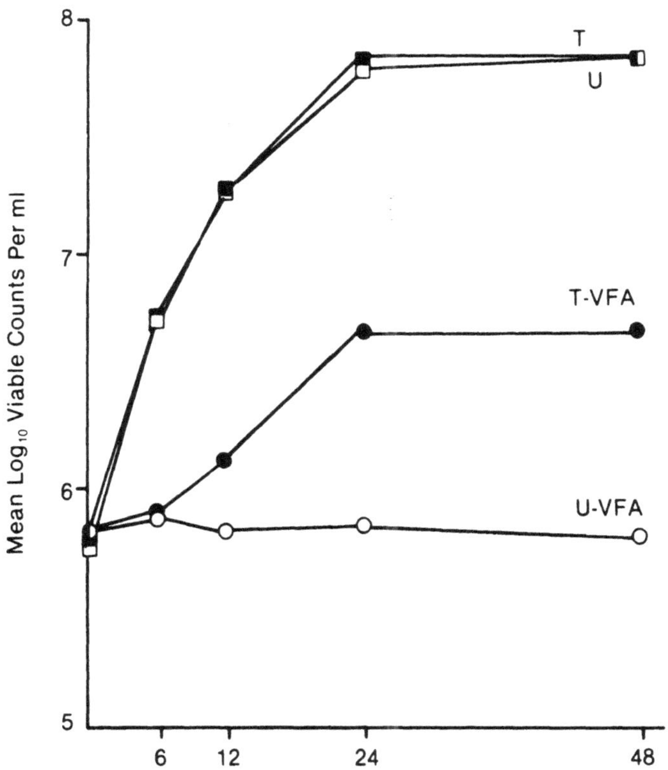

Fig. 5. Multiplication of *P. aeruginosa* in BHI Broth Containing Volatile Fatty Acids

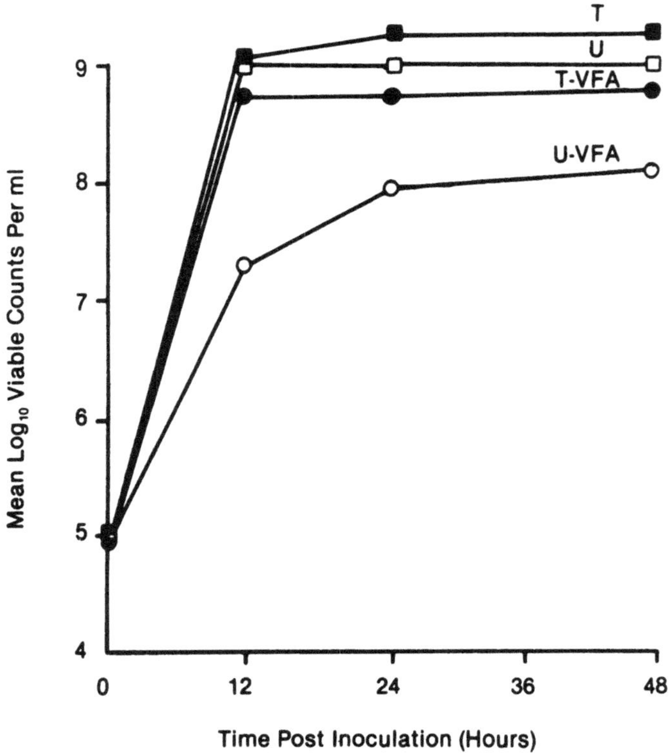

Fig. 6. Multiplication of *S. typhimurium* in BHI Broth Containing Volatile Fatty Acids

volatile fatty acids (T and U). Under conditions similar to those present in cecal contents of treated mice (T-VFA), total populations were approximately 10-fold smaller than controls. There was no evidence of *P. aeruginosa* multiplication, on the other hand, under conditions similar to those present in cecal contents of untreated mice (U-VFA).

S. typhimurium, which was less sensitive to the effects of volatile fatty acids than *P. aeruginosa*, multiplied in BHI broth containing the acids (Fig. 6). Its multiplication rate and total population after 24 hours incubation, however, were greater in broth adjusted to simulate conditions in cecal contents of treated mice (T, VFA) than untreated mice (U, VFA).

The study shows, therefore, that streptomycin administration decreases the resistance of mice to intestinal colonization with both *P. aeruginosa* and *S. typhimurium* and facilitates translocation of the organisms extraintestinally. The animals are much more susceptible to infection with *S. typhimurium* than *P. aeruginosa*. The ID_{50} for *S. typhimurium* in the untreated mouse is almost 10,000-fold lower than the ID_{50} for *P. aeruginosa* (Table 4). The reason for the greater susceptibility of mice to *S. typhimurium* is not understood but may be related to the greater resistance of this organism than *P. aeruginosa* to the effects of volatile fatty acids, which represent an important colonization resistance factor. The cecum of the animals is therefore, more receptive to colonization by *S. typhimurium* than *P. aeruginosa*.

Table 4. ID_{50} Values for Mice

Pseudomonas aeruginosa		*Salmonella typhimurium*	
Untreated	9.1×10^8	Untreated	1.8×10^5
Treated	9.1×10^4	Treated	1.3×10^0

Streptomycin administration slows the passage of contents through the cecum and alters mouse cecal flora composition. We have demonstrated the disappearance, or population decline of several key aerobic and anaerobic flora components in treated animals resulting in an environment which is less hostile to colonization by non-indigenous organisms. Presumably, the missing flora components contribute to the protective volatile fatty acid pool normally present in mouse cecal contents. Retarded passage of contents through the cecum, as a result of antibiotic treatment, also favors colonization by *P. aeruginosa* and *S. typhimurium*.

References

1. Bohnhoff M, Drake B L, and Miller C P, (1954) Effect of streptomycin on susceptibility of intestinal tract to experimental *Salmonella* infection. Proc Soc Exper Biol Med 86: 132–137
2. Bohnhoff M, and Miller C P (1962) Enhanced susceptibility to *Salmonella* infection in streptomycin-treated mice. J Infect Dis 111: 117–127
3. Freter R (1955) The fatal enteric cholera infection in guinea pig, achieved by inhibition of normal enteric flora. J Infect Dis 97: 57–65
4. Freter R (1956) Experimental enteric *Shigella* and *Vibrio* infections in mice and guinea pigs. J Exper Med 104:411– 418
5. Hentges D J, Que J U, Casey S W and Stein A J (1984) The influence of streptomycin on colonization resistance in mice. Microecol Ther 14: 53–62
6. Hentges D J, Stein A J, Casey S W and Que J U (1985) Protective role of intestinal flora against infection with *Pseudomonas aeruginosa* in mice: influence of antibiotics on colonization resistance. Infect Immun 47: 118–122
7. Miller C P, and Bohnhoff M (1963) Changes in the mouse's enteric microflora associated with enhanced susceptibility to *Salmonella* infection following streptomycin treatment. J Infect Dis 113: 59–66
8. Que J U and Hentges D J (1985) Effect of streptomycin administration on Colonization resistance to *Salmonella* typhimurium in mice. Infect Immun 48: 169–174.
9. van der Waaij D, Berghuis de Vries J M, and Lekkerkerk van der Wees J E C (1971) Colonization resistance of the digestive tract in conventional and antibiotic-treated mice. J Hyg, Camb 69: 405–411

Discussion

Kroll: What virulence characteristics have your Salmonella strains used to infect your animals? Depending on flagella, serum-sensitivity-patterns and so on there are tremendous differences in virulence.

Hentges: The strain used is very virulent for mice. It is motile but we have no additional information about virulence factors.

Ghione: The administration per os of oligosaccharide antibiotics is frequently associated in humans and animals with overgrowth of yeast. Did you observe this phenomenon in your experiments?

Hentges: Although we isolated yeasts from the animals we did not observe overgrowth as a consequence of streptomycin treatment.

Pulverer: Did you look at the serum protein patterns in treated and non-treated animals, are there missing or new bands correlated with the antibiotic treatment?

Hentges: We did not examine the effect of streptomycin administration on the serum of the animals. We are concentrating on the effects of the antibiotic on the ecology of the intestine.

Pichler: Have you got an explanation why translocation occurs so frequently in mice infected with streptomycin resistant *S. typhimurium*? Does this change of intestinal flora caused by streptomycin alter the permeability of the gut wall for Salmonellae?

Hentges: Is it possible that streptomycin affects the gut wall influencing permeability and translocation. Previous data show, however, that translocation is influenced by intestinal microbial numbers. The large numbers of Salmonella in the intestines of streptomycin treated mice might very well be responsible for the translocation observed.

Lorian: Could we assume that the reduction in normal flora allows overgrowth of Salmonella and, therefore, produces disease?

Hentges: Yes. Alteration of the normal flora by streptomycin results in conditions that favor multiplication of Salmonella in the intestine with subsequent disease production.

Lorian: What occurs in germ free mice?

Hentges: We did not challenge germ free mice with Pseudomonas or Salmonella. Based on other studies, I would assume these organisms would develop very large populations in the intestines of germ free animals and would produce extraintestinal infections.

Protection of tracheal explants infected with *Pseudomonas aeruginosa* by subinhibitory concentrations of aminoglycosides

N.R. BAKER and TERESA GEERS

Department of Microbiology, 484 West 12th Avenue, The Ohio State University, Columbus, Ohio, USA 43210

Introduction

Mortality rates for acute pneumonias caused by *Pseudomonas aeruginosa* may exceed 50% [11] and antibiotic therapy remains a difficult problem. Therapy frequently includes administration of an aminoglycoside alone or in combination with a beta-lactam antibiotic. Animal studies have shown that gentamicin or tobramycin were superior to carbenicillin or ticarcillin in controlling experimental *Pseudomonas* pneumonia [12]. Observations have also been made in cystic fibrosis (CF) patients with chronic *P. aeruginosa* lung infections which indicated that the lung functions of these patients improved following therapy with aminoglycoside antibiotics although the organisms were not cleared [9]. It seemed possible, therefore, that aminoglycoside antibiotics could prevent lung damage by a mechanism that did not involve inhibition of bacterial growth.

Several extracellular products of *P. aeruginosa* which are responsible for pathogenicity have been identified. The proteases, exotoxin A, and exoenzyme S have been shown to be important virulence factors in several models [1, 16]. The proteases were extremely destructive to lung tissue when they were injected directly into rabbit lungs [8] or when tracheal explants were exposed to purified elastase [1]. Furthermore, strains of *P. aeruginosa* that produced elastase caused extensive damage to the tracheal epithelium in an *in vitro* infection model [1]. Warren et al have shown that concentrations of aminoglycoside antibiotics that do not significantly inhibit growth rates or protein synthesis, inhibit the secretion of extracellular proteases [15]. Inhibition of the production of extracellular virulence factors of *P. aeruginosa* by subinhibitory concentrations of ciprofloxacin has been observed *in vitro* and *in vivo* [4, 5].

In the current study, hamster tracheal organ cultures were infected with protease producing strains of *P. aeruginosa* and treated with inhibitory and subinhibitory concentrations of beta-lactam or aminoglycoside antibiotics to determine the efficacy of these antibiotics in preventing tissue damage. The beta-lactams and aminoglycosides were also used in combination to test for synergistic effects.

The Influence of Antibiotics on the
Host-Parasite Relationship II
Edited by D. Adam, H. Hahn, W. Opferkuch

Materials and Methods

Bacteria

The strains of *P. aeruginosa* used in this study were obtained from John Govan, University of Edinburgh, Edinburgh, Scotland. Strains PAO 579 and PAO 581 are mucoid variants which were derived from the nonmucoid strain PAO 381 by carbenicillin selection [7]. Strains PAO 552 and PAO 553 are spontaneous nonmucoid revertants of PAO 579. These strains all produce elastase and other proteases. Mucoid strains were maintained by passage on Pseudomonas Isolation Agar, whereas nonmucoid isolates were passed on Trypticase Soy Agar (TSA). Stock cutures were stored at − 70 °C on TSA slants covered with Trypticase soy broth (TSB) containing 20% glycerol.

Tracheal organ culture

Tracheal organ cultures were prepared according to the method of Collier [3] as described previously [1]. Adult LSH/LAK male Syrian hamsters were sacrificed by intraperitoneal injection of 0.3 ml of 10% sodium brevitol. The trachea was excised and sectioned into rings between the cartilage with a sterile razor blade. The rings were placed in 35-mm tissue culture dishes, rinsed three times with minimal essential medium containing 1.2% HEPES buffer (MEM-HEPES) and 1 ml of MEM-HEPES was added to each culture. The rings were incubated for 18 h at 37 °C in 5% CO_2 air atmosphere and checked for ciliary activity by observation through an inverted microscope. Only rings displaying ciliary activity over 90% of the epithelial border were used.

Infection of tracheal cultures

Five ml of TSB was inoculated with the growth from an isolated colony on a TSA plate and incubated without shaking for 18 h at 37 °C. The concentration of the culture was estimated by reading the optical density at 540 nm of a 1 to 10 dilution of the culture and the inoculum was prepared by diluting the culture to 10^7 CFU/ml in MEM-HEPES. The rings were transfered to sterile 35-mm tissue culture dishes and 1 ml of the inoculum was added to each dish. The infected explants were incubated for 4 h at 37 °C in a 5% CO_2-air atmosphere and rinsed three times to remove free bacteria. Individual explants were transferred to separate wells of sterile 96 well tissue culture dishes for further treatment.

Antibiotic treatment and MIC determinations

Ceftazidime acid pentahydrate and tobramycin sulfate were obtained from Eli Lilly and Co., Indianapolis, Indiana and disodium carbenicillin and gentamicin sulfate were purchased from Sigma Chemical Co., St. Louis, Missouri. Stock solutions of antibiotics were prepared in sterile MEM-HEPES and diluted to the desired concentrations in MEM-HEPES.

The infected explants were treated with 200 μl of the appropriate antibiotic solution and incubated for 18 h at 37 °C in a 5% CO_2-air atmosphere. Controls consisted of infected explants in MEM-HEPES without antibiotics and uninfected explants incubated with the highest concentration of antibiotics. After the incubation the explant cultures were examined for the presence of bacterial growth in the cultures by looking for turbity. The explants were washed three times with sterile MEM-HEPES and prepared for examination by scanning electron microscopy.

The MIC of each antibiotic for *P. aeruginosa* strains infecting the tracheal explants was defined as the minimum concentration in which no turbity was seen following the 18 h incubation. Standard MIC's were determined by adding 50 ul of bacterial suspension containing 1×10^6 CFU/ml to 50 μl of serial twofold dilutions of the antibiotics in microtiter trays. The cultures were observed after 18 h incubation for visual turbity and the minimum concentration which prevented growth was recorded as the MIC. The standard MIC's were determined in MEM-HEPES and were the same as values obtained in Mueller-Hinton broth.

Synergy between the antibiotics was determined by testing strains PAO 381 and PAO 579 in checkerboard titration with twofold dilutions of gentamicin versus carbenicillin or ceftazidime and tobramycin versus carbenicillin or ceftazidime. The fractional inhibitory concentration (FIC) was calculated using the formula: FIC = [MIC (drug A in combination) / MIC (drug A alone)] + [MIC (drug B in combination) / MIC (drug B alone)]. Synergy was defined as an FIC less than 0.5. Antibiotics were tested on bacteria alone and on bacteria infecting tracheal explants.

Scanning electron microscopy

The washed explants were fixed in 5% glutaraldehyde in 0.1 M phosphate buffer, pH 7.2, for 1 h at 22 °C. They were washed three times in 0.1 M phosphate buffer and postfixed in 1% osmium tetroxide in 0.1 M phosphate buffer for 1 h at 4 °C. The explants were washed three times with 0.1 M phosphate buffer and dehydrated by successive 15 min washes in 30, 50, 75, 80, 90, 95 and 100% ethanol. The explants were critical point dried in a Bomar critical-point drier, mounted on carbon stubs, sputter coated with gold, and observed in a Hitachi S-500 scanning electron microscope.

Results

MIC determinations

The concentration of antibiotics required to inhibit the growth of the test strains of *P. aeruginosa* in infected tracheal explants was greater than the concentrations needed to inhibit the growth of those strains in suspension (Table 1). The MIC's for tobramycin and gentamicin were 2 to 8 fold higher in infected explants than in media alone whereas the MIC's for carbenicillin and ceftazidime were 10 to 40 times higher

Table 1. Comparison of MIC values for *P. aeruginosa* of tobramycin, gentamicin, carbenicillin, and ceftazidime determined by standard methods to MIC values of the antibiotics for *P. aeruginosa* infecting tracheal explants[a]

		MIC (μg/ml)	
Antibiotic	Strain	Standard	Organ culture
Tobramycin	381	2	8
	581	4	16
	552	2	8
	553	4	8
	579	2	16
Gentamicin	381	2	8
	581	2	16
	552	2	16
	553	8	16
	579	2	16
Carbenicillin	381	32	320
	581	32	320
	552	16	640
	553	32	320
	579	16	320
Ceftazidime	381	2	32
	581	1	16
	552	0.5	16
	553	0.5	8
	579	0.5	8

[a] In the standard MIC determination 10^5 CFU of *P. aeruginosa* were mixed with serial twofold dilutions of the antibiotics. In the infected organ culture assay, tracheal organ cultures were infected with *P. aeruginosa* for 4 h, washed to remove nonadherent bacteria, and exposed to serial twofold dilutions of the antibiotics. The cultures were observed for turbity after incubation for 18 h at 37 °C

in the infected explants than in suspension. Thus, the MIC values for the antibiotics in the infected tracheal explants were used in interpretation of the results of the protection studies.

Protection of infected tracheal explants

The uninfected tracheal explants and explants exposed to antibiotics maintained a normal epithelium during the 20 h culture period (Fig. 1a) but the epithelium was almost completely destroyed in explants infected for the same time with any of the test strains of *P. aeruginosa* (Fig. 1b). The infection could be stopped, however, by treatment of the infected explants with antibiotics after the 4 h infection period. Moreover, the aminoglycoside antibiotics protected the explants from damage at concentrations equal to or lower than the MIC for that antibiotc (Table 2 and Fig. 2). Following treatment with subinhibitory concentrations of tobramycin or

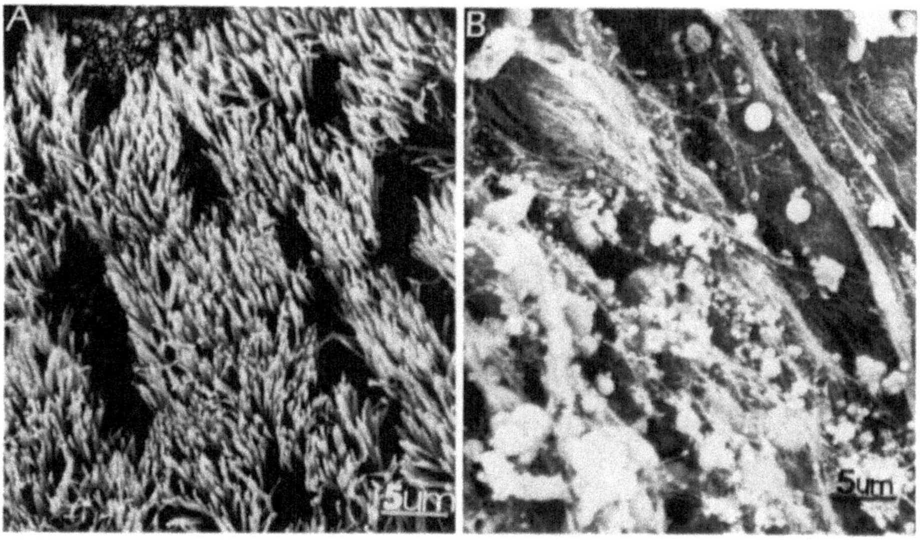

Fig. 1. Scanning electron micrographs of tracheal epithelium. **a**) Untreated control tracheal epithelium after 24 h in culture. **b**) Tracheal epithelium of an explant infected with *P. aeruginosa* for 4 h, washed, and incubated an additional 18 h. The epithelium was completely destroyed

Table 2. Comparison of the ability of aminoglycoside and β-lactam antibiotics to protect tracheal explants from damage following infection by several strains of *P. aeruginosa*

		Extent of damage[a] caused by *P. aeruginosa* strains				
Antibiotic	Conc. (μg/ml)	381	579	581	552	553
Tob	16	0	0*	0*	0*	0
	8	1*	3	1	2	0*
	4	3	3	3	3	3
GM	16	0	0*	0*	1*	0*
	8	1*	1	0	3	0
	4	3	3	NT	3	3
Cef	128	1	NT	0	0	0
	64	1	0	0	1	1
	32	2*	0	2	2	2
	16	3	1	3*	2*	3
	8	NT	3*	3	3	3*
Cb	640	NT	1	0	0*	0
	320	0	3*	3*	1	0*
	160	3*	3	3	2	3
	80	3	NT	NT	3	3

[a] Degree of damage: 0 = no damage; 1 = slight damage; 2 = moderate damage; 3 = extensive damage. NT = not tested. The asterix next to a number indicates the MIC for that strain of *P. aeruginosa* in the infected organ culture

gentamicin, it was possible to find clusters of organisms on the ciliated epithelium with no evidence of tissue destruction (Fig. 2a).

The beta-lactam antibiotics were only protective at high concentrations and tissue damage was observed frequently in infected explants treated with concentrations of antibiotic equal to or greater than the MIC for that antibiotic (Table 2 and Fig. 3). Filamentous forms of the bacteria were also observed over the epithelial surface, and were usually associated with areas of damaged epithelium (Fig. 3).

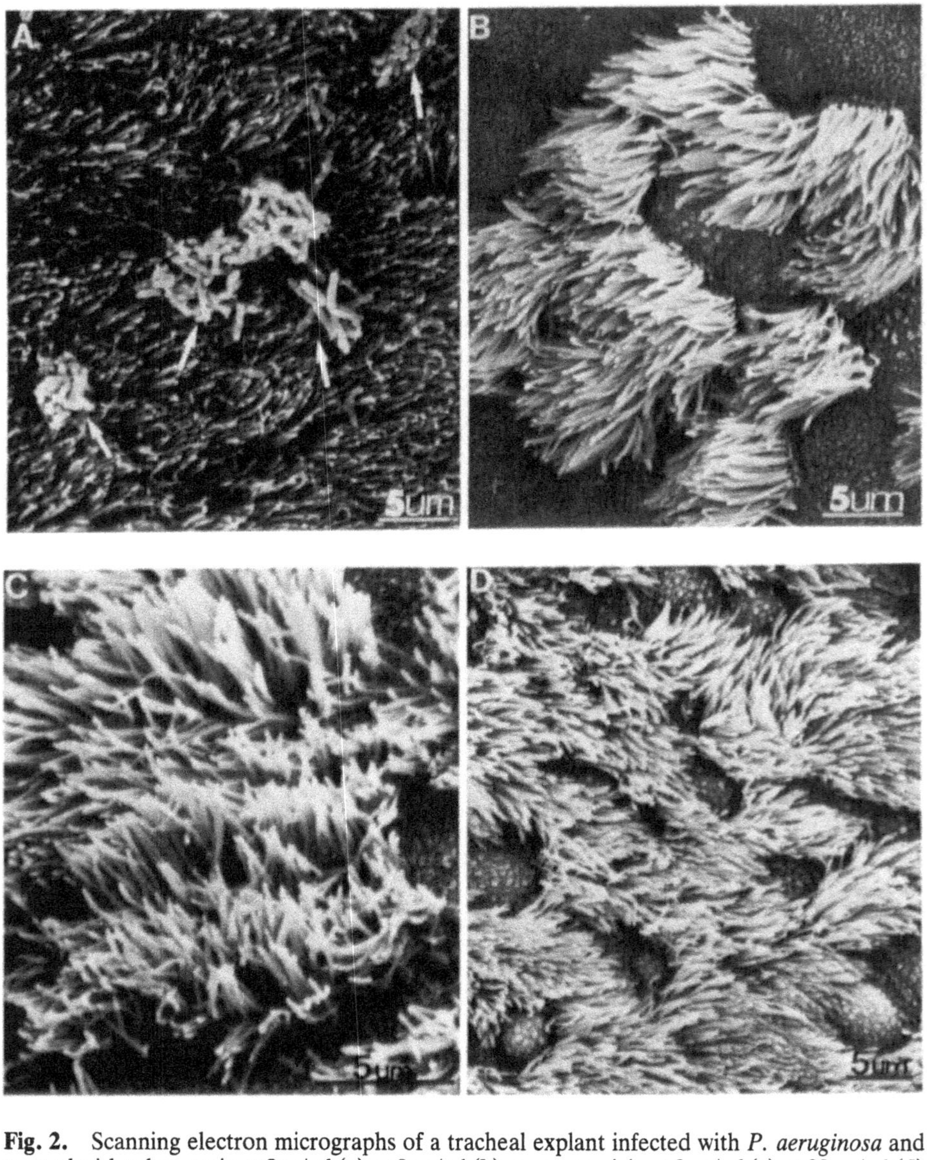

Fig. 2. Scanning electron micrographs of a tracheal explant infected with *P. aeruginosa* and treated with tobramycin at 2μg/ml (**a**) or 8 μg/ml (**b**), or gentamicin at 8 μg/ml (**c**) or 32 μg/ml (**d**) for 18 h. The MIC for tobramycin and gentamicin in the infected tracheal explants was 8 to 16 μg/ml. Occasionally clusters of bacteria (arrow) could be found on the epithelial surface with no evidence of tissue destruction (**a**)

Fig. 3. Scanning electron micrographs of a tracheal explant infected with *P. aeruginosa* for 4 h and treated with carbenicillin at 160 μg/ml (**a**) or 320 μg/ml (**b**), or ceftazidime at 16 μg/ml (**c**) or 32 μg/ml (**d**) for 18 h. The MIC's for carbenicillin and ceftazidime were 320 μg/ml and 32 μg/ml respectively. Filamentous forms of bacteria (arrows) were frequently observed on the epithelium

Synergism between the beta-lactam and aminoglycoside antibiotics

The only antibiotic combination that gave a clear indication of synergism when tested in a standard assay was carbenicillin and gentamicin. The other combinations were borderline by the criteria used in this study. All combinations of antibiotics tested, however, were synergistic in the infected tracheal explant model (Table 3).

Table 3. Synergy between β-lactam and aminoglycoside antibiotics for *P. aeruginosa* tested by standard methods and for *P. aeruginosa* infecting tracheal explants[a]

		Synergism[b]	
Strain	Antibiotics	Standard	Organ culture
381	cef/tob	−	+
	cef/gm	−	+
	cb/tob	−	+
	cb/gm	+	+
579	cef/tob	−	+
	cef/gm	−	+
	cb/tob	−	+
	cb/gm	+	+

[a] In the standard MIC determination 10^5 CFU of *P. aeruginosa* were mixed with serial twofold dilutions of the antibiotics in a checkerboard titration. In the infected organ culture assay, tracheal organ cultures were infected with *P. aeruginosa* for 4 h, washed to remove nonadherent bacteria, and exposed to serial twofold dilutions of the antibiotics. The cultures were observed for turbity after incubation for 18 h at 37 °C

[b] Synergism was defined as a fractional inhibitory concentration (FIC) of <0.5

The MIC's for tobramycin and gentamicin in the infected explant model were 1 and 4 μg/ml respectively when ceftazidime was present at 4 μg/ml. The MIC's for tobramycin, gentamicin, and ceftazidime alone were 8, 8, and 32 μg/ml respectively (Table 1). The MIC for both tobramycin and gentamicin in this model was 1 μg/ml when carbenicillin was present at 40 ug/ml; the MIC for carbenicillin alone was 320 ug/ml. The combination of carbenicillin at 40 μg/ml with an aminoglycoside at 1 μg/ml provided complete protection of the tracheal epithelium from destruction by *P. aeruginosa*. The damage observed when infected tracheas were treated with inhibitory concentrations of beta-lactam antibiotics alone was not seen (Fig. 4). Protection by subinhibitory concentrations of antibiotic combinations was not observed.

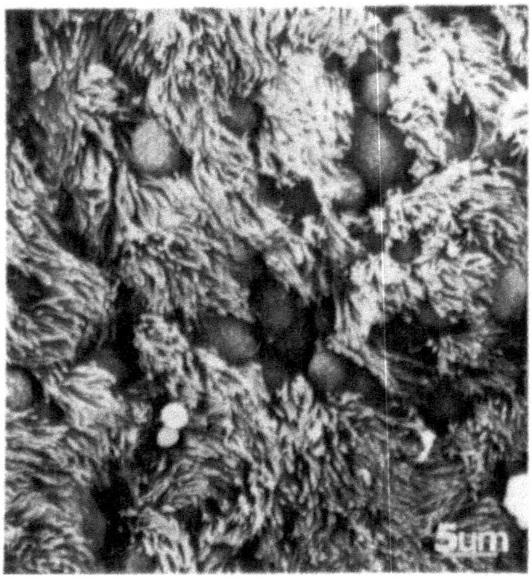

Fig. 4. Scanning electron micrograph of a tracheal explant infected with *P. aeruginosa* for 4 h and treated with 4 μg of ceftazidime per ml and 1 μg of tobramycin per ml for 18 h. The epithelium was relatively normal indicating synergy between the antibiotics. The combination of carbenicillin at 40 μg/ml with either aminoglycoside at 1 μg/ml provided similar protection

Discussion

Subinhibitory concentrations of antibiotics have been shown to cause a variety of effects on bacteria, including alteration of adherence characteristics [2, 14], bacterial cell structure [10], and export of bacterial products [4, 5, 15]. Warren et al have shown recently that aminoglycoside antibiotics inhibited the release of proteases from *P. aeruginosa* at concentrations that did not inhibit bacterial growth or protein synthesis [15]. We have shown that subinhibitory concentrations of tobramycin or gentamicin can inhibit damage to tracheal epithelium caused by infection with *P. aeruginosa*. A previous study showed that the extensive damage caused by *P. aeruginosa* in the tracheal culture model was primarily due to bacterial proteases [1]. The kinetics of the toxicity associated with exotoxin A indicated that the toxin is not responsible for the rapid destruction observed during the first 24 h following infection. Thus, it is possible that the aminoglycoside antibiotics protected the tracheal epithelium by inhibiting the release of proteases from *P. aeruginosa*.

The mechanism of inhibition may involve inhibition of bacterial protein synthesis by aminoglycosides or disruption of regulatory mechanisms. Inhibition of exoproduct release from *P. aeruginosa* by treatment with subinhibitory concentrations of ciprofloxacin, which acts on bacterial DNA, has been noted [4, 5]. *In vivo* evidence for inhibition of toxin production, but not bacterial numbers, by ciprofloxacin in a granuloma pouch model of *P. aeruginosa* infection was also reported [4]. Thus it is apparent that antibiotics can cause beneficial effects which are distinct from inhibition of bacterial growth. Further clarification of the mechanism of inhibition of exoproduct release and protection of tissue from damage by subinhibitory concentrations of antibiotics is needed.

The beta-lactam antibiotics were not effective at protecting the infected epithelium from damage; in some instances extensive tissue damage was observed when the antibiotics were present at concentrations higher than the MIC for the antibiotic. Filamentous forms of the bacteria were frequently observed on the surface of the epithelium and it is possible that these forms were continuing to make proteases. Cultures of *P. aeruginosa* exposed to inhibitory concentrations of carbenicillin continue to excrete proteases (unpublished data). Detection of the bacterial proteases in infected explants treated with beta-lactam antibiotics is currently under investigation.

Several studies have reported synergistic activity of beta-lactam and aminoglycoside antibiotics used in combination *in vitro* [6] and *in vivo* [13]. The antibiotic combinations used to treat infected tracheal explants clearly demonstrated synergistic protection of the tracheal epithelium from damage due to infection by *P. aeruginosa*. Damage to the tracheal epithelium was seen when infected explants were treated with beta-lactam antibiotics at or above the MIC, but if an aminoglycoside was added the destruction was prevented and the concentrations of antibiotic needed for protection were reduced significantly. These results support the continued combined use of an aminoglycoside and beta-lactam antibiotics, particularly for infections of the respiratory tract where it is difficult to obtain high levels of antibiotic. The antibiotic combinations used in the infected explant model were not protective at concentrations below the MIC for the combined antibiotics.

In summary, tracheal organ cultures infected with *P. aeruginosa* were used to evaluate the ability of two beta-lactam and two aminoglycoside antibiotics to protect the explants from the destructive activity of this pathogen. The MIC's for all the antibiotics were higher in the infected explants than the standard MIC's. Tobramycin and gentamycin at concentrations below their MIC protected the explants from damage, whereas carbenicillin and ceftazidime were not protective at concentrations equal to or greater than their MIC values. Synergy between the aminoglycoside and beta-lactam antibiotics was demonstrated in the infected tracheal explant model. Since the aminoglycoside antibiotics inhibit protein synthesis, they may be more effective than the beta-lactam antibiotics at inhibiting the synthesis or release of extracellular toxins by *P. aeruginosa*.

Acknowledgment. This work was supported in part by a research grant from Eli Lilly and Company.

Summary

The hamster tracheal organ explant model of respiratory tract infection by *Pseudomonas aeruginosa* was used to evaluate the ability of antibiotics to protect the epithelium from destruction by bacterial products. Hamster tracheal explants were infected with strains of *P. aeruginosa* for 4 h and washed to remove free organisms. The infected explants were incubated an additional 18 h in fresh minimal essential medium containing various concentrations of antibiotics. Combinations of antibiotics were also tested. The explants were processed and examined by scanning electron microscopy for extent of damage caused by the infection. Tobramycin and gentamicin protected the infected explants at subinhibitory and inhibitory concentrations of antibiotics. Carbenicillin and ceftazidime only protected at high antibiotic concentrations. Treatment with combinations of beta-lactam and aminoglycoside antibiotics revealed synergistic activity. Subinhibitory concentrations of aminoglycoside antibiotics may protect against tissue damage by inhibiting the release of toxic enzymes and toxins from *P. aeruginosa*.

References

1. Baker NR (1982) Role of exotoxin A and proteases of *Pseudomonas aeruginosa* in repiratory tract infections. Can J Microbiol 28:248–255
2. Beachey EH, Eisenstein BI, Ofek I (1981) Sublethal antibiotics and bacterial adhesion. In: Adhesion and micro-organism pathogenicity (Ciba Foundation Symposium 80), Tunbridge Wells, London pp 288–300
3. Collier AM, Baseman JB (1973) Organ culture technique with *Mycoplasmas*. Ann NY Acad Sci 255: 277–289
4. Dalhoff A, Doring G (1984) Effect of ciprofloxacin on the expression of exoenzymes and exotoxin of *P. aeruginosa*. Abstr 24th Intersci Conf Antimicro Agents Chemother abstr no 981, pg 263
5. Dalhoff A, Doring G, Goldstein W (1984) Comparative antibacterial activity of ciprofloxacin, β-lactams and aminoglycosides against *P. aeruginosa*. In: Lawson D (ed) Cystic Fibrosis: Horizons Proc 9th Internat Cystic Fibrosis Congress, p 289 John Wiley & Sons, Chichester, England

6. Glew RH, Pavuk RA (1984) Early synergistic interactions between amikacin and six β-lactam antibiotics against multiply resistant members of the family *Enterobacteriaceae*. Antimicrob Agents Chemother 26:78–381
7. Govan JRW, Fyfe JAM (1978) Mucoid *Pseudomonas aeruginosa* and cystic fibrosis: Resistance of the mucoid form to carbenicillin, fluçloxacillin, and tobramycin and the isolation of the mucoid variants *in vitro*. J Antimicrob Chemother 4:233–240
8. Gray L, Kreger A (1979) Microscopic characterization of rabbit lung damage produced by *Pseudomonas aeruginosa* proteases. Infect Immun 23:150–159
9. Hoiby N (1984) The management of pseudomonas chest infection-the way forward. In: Lawson D (ed) Cystic Fibrosis: Horizons Proc 9th Internat Cystic Fibrosis Congress, pp 87–95 John Wiley & Sons, Chichester, England
10. Lorian V (1980) Effects of subminimum inhibitory concentrations of antibiotics on bacteria, pp 342–408. In: Lorian V (ed) Antibiotics in laboratory medicine. The Williams & Wilkins Co, Baltimore
11. Pennington JE, Reynolds HY, Carbone PP (1973) Pseudomonas pneumonia: a retrospective study of 36 cases. Am J Med 55:155–160
12. Pennington JE, Stone RM (1979) Comparison of antibiotic regimens for treatment of experimental pneumonia due to *Pseudomonas*. J Infect Dis 140:881–889
13. Peterson LR, Gerding DN, Moody JA, Fasching CE (1984) Comparison of azlocillin, ceftizoxime, cefoxitin, and amikacin alone and in combination against *Pseudomonas aeruginosa* in a neutropenic-site rabbit model. Antimicrob Agents Chemother 25:545–552
14. Stephens DS, Krebs JW, McGee ZA (1984) Loss of pili and decreased attachment to human cells by *Neisseria meningitidis* and *Neisseria gonorrhoeae* exposed to subinhibitory concentrations of antibiotics. Infect Immun 46:507–513
15. Warren RL, Baker NR, Johnson J, Stapleton MH (1985) Aminoglycosides selectively inhibit the accumulation of extracellular proteases of *Pseudomonas aeruginosa*. Antimicrob Agents Chemoth (in press April 1985)
16. Woods DE, Cryz SJ, Friedman RL, Iglewski BH (1982) Contribution of toxin A and elastase to virulence of *Pseudomonas aeruginosa* in chronic lung infections of rats. Infect Immun 36:1223–1228

Discussion

Kadurugamuwa: Do you have any explanation for the difference in MICs between in vitro and organ culture model? Is it due to the surface growth of the organisms in organ tissue culture model?

Baker: The only explanation that we have so far is that the antibiotics are not as effective against adherent bacteria as they are against bacteria in suspension. The antibiotics are not inactivated by the explants since antibiotics were tested for activity after 18 h incubation with explants and were not inactivated.

Opferkuch: Your data are from normal epithelial tissue. Can also damaged tissue, like in patients of intensive care units, be protected by antibiotics?

Baker: We obviously would not be able to use cytological evidence for damage, since we would be starting with a damaged epithelium. We could, however, use some other parameter, such as protein synthesis of the damaged epithelium, and do experiments similar to these described.

Karch: Is it possible to neutralize cytotoxicity by antibody against the cytotoxin in order to determine whether the damage of the cells is a function of cytotoxin or elastase.

Baker: Yes, that is one approach. We are trying to detect toxins using immunohistochemistry and western-blotting of supernatants from the infected explants.

Warren: Have you looked at the effects of antibiotics on adherence of *Pseudomonas aeruginosa* to the epithelium?

Baker: Yes. We have observed previously that mucoid *P. aeruginosa* adhere well to normal hamster tracheal epithelium. If mucoid *P. aeruginosa* are grown in the presence of subinhibitory concentrations of gentamycin or tobramicin, they do not adhere as well. They also make less of the alginate-like polysaccharide which appears to be the adhesin in this system. The adherence of the nonmucoid strains of *P. aeruginosa* is not reduced by growth in the presence of subinhibitory concentrations of gentamicin or tobramycin.

Haag: You incubated Pseudomonas-infected tracheal explants in the presence of antibiotics for 18 hours. In the case of β-lactam antibiotics you showed more extensive destruction of the epithelium compared to the same relative concentrations of aminoglycosides.
1. Have you measured the β-lactam concentration after this incubation period? Had it been inactivated by P. aeruginosa?
2. The extensive destruction after 18 hours may be due to the program of surviving P. aeruginosa, which grew out after the β-lactam has been degraded. Have you looked at the epithelia at e.g. 4 hours and have you seen better results at this time?

Baker: 1. No, we did measure the β-lactam concentration after the 18 h incubation. It does not seem that inactivation of the antibiotics accounts for our observation, since the β-lactam did not protect even at MIC levels.
2. No, we have not looked earlier. Perhaps we should, but again MIC levels of the β-lactams do not protect well, thus I doubt that the damage is due to the presence surviving bacteria under these conditions.

Use of Lectins to Characterize Surface Alterations of β-lactam Resistant Mutants of *Pseudomonas aeruginosa*

G. STÜBNER[1] and R. MARRE[2]

[1] Krankenhaus Nordstadt, Institut für Bakteriologie und Hygiene, Hannover, FRG
[2] Med. Hochschule Lübeck, Institut für Medizinische Mikrobiologie, Lübeck, FRG

Introduction

Lectins are defined as sugar binding proteins of plant origin with no detectable enzymatic activity [5]. They react with cellular bound or soluble carbohydrates selectively likened to the antigen antibody reaction [10, 11]. They possess specifities for a wide range of carbohydrate structures [2, 4, 8, 11]. Bacteria have a lot of surface components which are capable of interacting with lectins, e.g. teichoic acid, lipopolysaccharides, peptidoglycan, group specific antigens and hydrophobic areas of proteins of the outer membrane [2, 3, 11].

It seemed to be of interest to us to investigate *Pseudomonas aeruginosa* strains-isolated from clinically proven infections which only differ in the expression of different amounts of a chromosomally mediated β-lactamase in view of changes in their outer membrane carbohydrate composition, to find out if change in genetic composition results in a change of bacterial surface architecture, e.g. lectin specific sugars, as well.

Material

Pseudomonas aeruginosa strain pairs isolated from the same patients before and after antibiotic treatment respectively were used. The main characteristics of the strains are outlined in Table 1.

Table 1. Characteristics of the *Pseudomonas aeruginosa* strains used. Serotyping and pyocinetyping was kindly carried out by Prof. Bauernfeind, Munich

Strain	MIC (mg/l)	hydrolytic activity[a]	Serotype	Pyocin type
S 430	8	6.3	5	10 c
S 431	512	133.2	5	10 c
391504 A	8	5.9	3	5 y
391504 B	256	102,9	3	5 y
290453 A	64	9.6	3	5 y
290453 B	512	185.0	3	5 y
319014 S	16	n.d.[b]	3	5 y
319014 R	512	n.d.	3	5 y

[a] 1 U/mg: 1nmol of nitrocefin hydrolised per min and mg of protein
[b] n.d.: not yet done

The Influence of Antibiotics on the Host-Parasite Relationship II
Edited by D. Adam, H. Hahn, W. Opferkuch

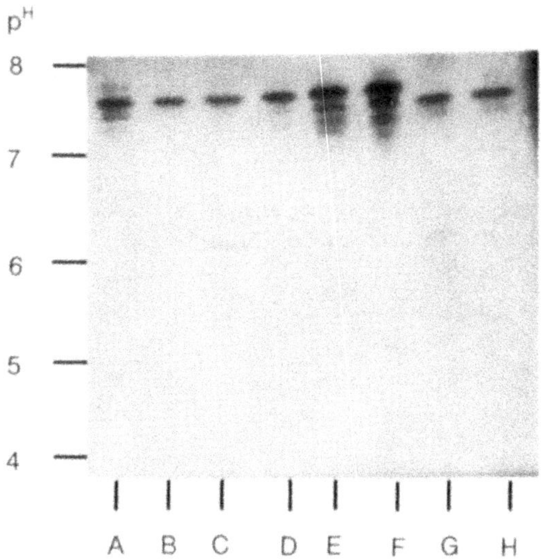

Fig. 1. Identification of sugar residues on bacteria with the Taxonolectin™-system. A lectin corresponding to an sugar localized on bacterial outer membrane is added to the bacterial suspension. The sugar residues are linked by the lectin and a agglutination reaction occurs

Analytical isoelectric focusing for detection and identification of β-lactamases was assayed with the method described by Matthew et al [6]. The existence of plasmids was excluded*.

Methods

1. *Semiquantitative determination of lectin receptors:* Lectin receptors on the bacteria were determined semi-quantitatively by use of the Taxonolectin™System (E.-Y.-Lab., San Mateo, USA). The principle of the reaction is outlined in Fig. 1.
   100 µl of bacterial suspension containing 10^9 cfu/ml respectively are filled into the wells precoated with 100 µg lectin with different carbohydrate specifities. After 20 min incubation at 37 °C the panels are rotated slightly by hand and the agglutination reactions are noted on a score card where the type of sugar responsible for the binding between lectin and glycoprotein can be elucidated. In case of agglutination quantitative determination of the amount of lectins bound to the bacteria is subsequently done.
2. *Quantitative determination of lectin receptors* was carried out according to the method described by Connelly et al [1], based on the fact that lectins can be inhibited by their corresponding sugars. Chromatographically pure preparations of rhodamin conjugated lectins (TKA Lot.No. 022 601, WGA Lot. No. 020 304, ConA Lot. No. 1 012F, DBA Lot. No. 011 011.

* We thank Prof. Laufs, Hamburg, for carrying out these investigations

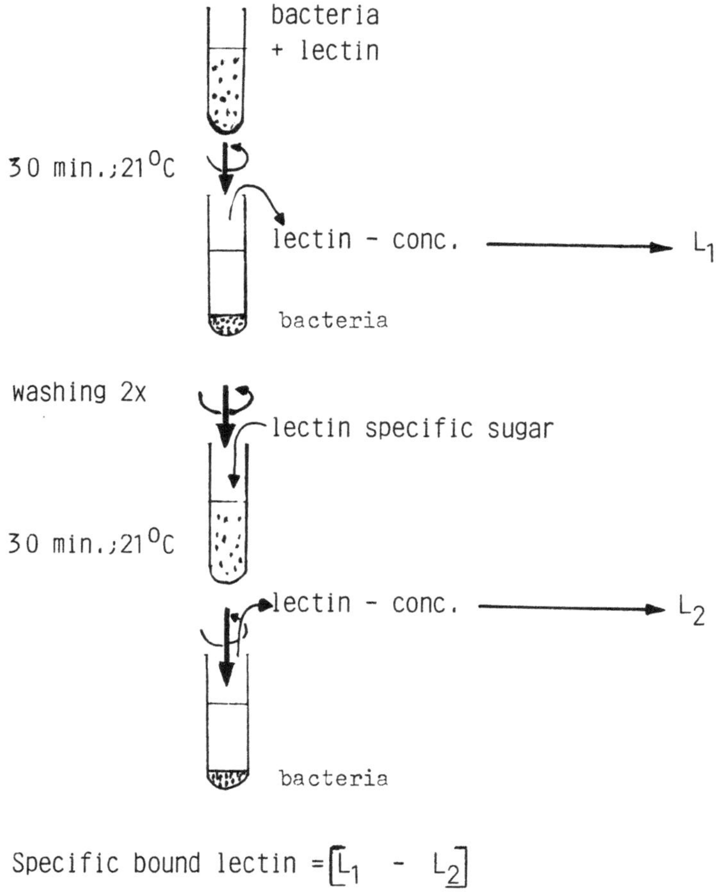

Fig. 2. Principle of quantitative determination of lectin receptors with rhodamin conjugated lectins [1]. Bacteria are incubated with lectins in the absence and presence of the specific sugars respectively. The difference between the lectin concentrations in the supernatant represents the amount of lectin bound specifically (see text)

PNA Lot. No. 021 207, SBA Lot. No. 08 190, STA Lot. No. 012 311, UEA Lot. No. 0 824, GS II Lot. No. 020 606; all preparations from E.-Y.-Lab., San Mateo, USA) are dissolved in PBS p_H 7.4 and calibration curves covering concentrations from 25 μg/ml—250 μg/ml are recorded with a Perkin Elmer Fluorescence Spectrophotometer at excitation wavelenght 555 nm. Bacteria from logarithmic the growth phase are photometrically adjusted to 2×10^8 cfu/ml and incubated with 100 μg/ml lectin in the dark at 21°C for 30 min, then centrifuged at 800 × g for 20 min following spectrophotometrical measurement of the concentration of unbound lectin from the supernatant (L_1). The pellet is washed twice with 0.1 m Tris buffer and 1.0 ml of a 0.5 m solution of the lectin specific sugar is added. After 30 min. incubation in the dark at 21°C following centrifugation at 800 × g concentration of free lectin in the supernatant is determined. (L_2). The difference of lectin concentrations before and after adding the corresponding sugar represents the amount of lectins bound specifically (Fig.2).

The quantity of lectin molecules bound to one bacterium can be calculated by the equation:

$$L_B = \frac{(L_1 - L_2) \times 6.02 \times 10^{23} \times 10^{-3}}{\text{cfu} \times \text{mw}_L}, \text{ whereby}$$

L_B = concentration of specifically bound lectin (μg/ml), (L_1–L_2) = difference of lectin concentrations in the supernatant, mw_L = molecular weight of the lectin.

Results

1. Comparison of the β-lactamase production of the strains investigated shows that sensible and resistant strains produce identical β-lactamases, but there are significant differences in the amount of production. Bearing in mind that the strains do not possess plasmids it can be concluded that a chromosomally mediated alteration of the regulation of β-lactamase activity took place (Fig. 3).
2. The semiquantitative estimation of lectin-reacting sites on the bacteria with the Taxonolectin™-system showed agglutination patterns documented in Table 2.

Although there are quite different agglutination reactions for all *Pseudomonas* strains tested, it became obvious that in general the strongest reaction occurs in case of the lectins PNA and DBA. This indicates that D-galactose and its derivatives are the most widespread sugars on the bacteria. For quantitative lectin-binding studies the parent strains 319014 S and 319014 R were selected. As the results show (Table 3) D-galactose specific lectins are bound at the highest rate to the bacterial surfaces.

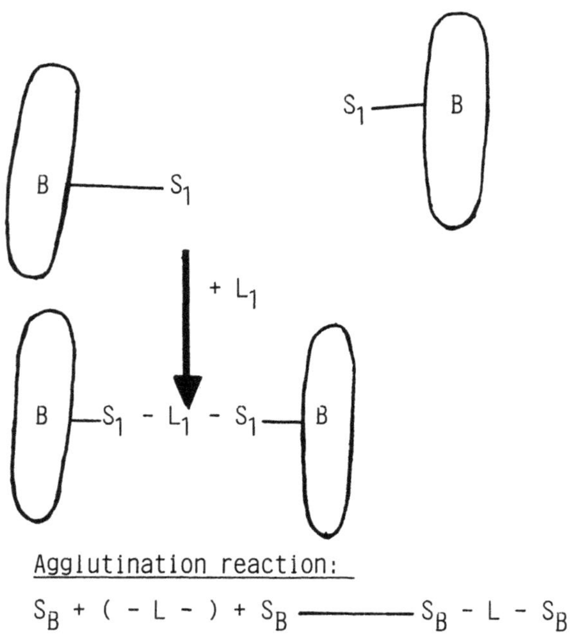

Fig. 3. Analytical isoelectrical focusing of the β-lactamases. β-lactamase bands are visible by the nitrocefin reaction. A+B = strains 390453 A/B; C+D = strains 391504 A/B; E+F = strains S 430/S 431; G+H = strains 319014 S/R

Table 2. Agglutination pattern of the *Pseudomonas* strains obtained with the Taxonolectin™-system. ConA = Concanavalin A; TKA = Trichosanthes kinlowii; PNA = Arachis hypogaea; DBA = Dolichos biflorus; STA = Solanum tuberosum; LFA = Limax flavus; UEA I = Ulex europaeus; LCH = Lens culinaris; GS I = Griffonia simplicifolia, SBA = Glycine max.; GS II = Griffonia simplicifolia; WGA = Tritium vulgaris; PHA–L = Phaseolus vulgaris; Lotus = Lotus tetragonolobus (E.-Y.-Lab. Commun., 4, 8)

	ConA	TKA	PNA	DBA	STA	LFA	UEA I	LCH	GSI	SBA	GS II	WGA	PHA	Lotus
S 430	–	+	++	+++	++	–	++	+	–	–	++	–	(+)	–
S 431	–	++	+	+++	–	–	–	–	–	–	–	–	(+)	–
391504 A	–	++	++	+++	–	–	+	(+)	++	+	++	–	(+)	–
391504 B	–	++	++	+++	–	–	+	(+)	–	–	–	+	–	–
290543 A	–	++	++	+++	–	–	–	–	–	–	–	–	–	–
290543 B	–	+	++	+++	(+)	–	–	–	+	–	+	–	–	–
319014 S	–	(+)	(+)	+++	–	–	–	–	–	–	–	–	–	–
319014 R	–	+	(+)	++	+	–	+	–	(+)	–	+++	–	–	–

Table 3. Quantity of lectins (μg/ml) bound to the bacteria (2×10^8 cfu/ml)

	PNA	DBA	WGA	STA	GS II	SBA	UEA	ConA	TK
319014 S	7.2	47	0	1.1	0	0	0	0	0
319014 R	15.5	34	0	28	43	0	33	1.2	27

Table 4. Amount of lectins expressed as mole lectins/bacterial cell (see text)

	PNA	DBA	WGA	STA	GS II	SBA	UEA	ConA	TKA
319014 S	1.9×10^9	1.2×10^{10}	0	3×10^7	0	0.3×10^7	0	0	0
319014 R	0.5×10^9	0.9×10^{10}	0	1.3×10^9	1.1×10^9	0	2.2×10^9	0.3×10^8	2.3×10^9

The amount of lectin molecules bound to one bacterium calculated as mole lectins/cell is given in Table 4.

It becomes obvious that there is a significant difference among the lectin receptors of both strains. In case of strain 319014 S DBA is the lectin preferably bound indicating that D-galactose is the most common sugar on its surface; whereas in case of strain 319014 R this is βD-glucose corresponding to lectin GS II.

Discussion

Lectins interact with carbohydrates selectively [2,3,4,11]. This makes them apt tools to characterize carbohydrate structures of bacterial surface components. To determine lectin-reactive carbohydrates on bacteria one must consider three facts of importance:

1. Lectins generally have a higher affinity to cellular bound carbohydrates than to soluble ones, because they possess 2–4 binding sites regularly and therefore react with complex carbohydrates like oligo or polysaccharides in a multivalent manner [4, 8, 11];
2. Lectin-reactive sugars on bacterial surfaces must not be hidden by steric hindrance. This occurs when certain areas are masked by capsular material or by components of the growth medium.
3. Binding of lectins is affected by different variables e.g. lectin concentration, p_H, time and temperature [2, 8].

Experiments to determine association constants between lectin and receptor sites (Scatchard equation [7]) were not carried out because saturation of the binding sites was reached by preincubating the bacteria with 100 μg/ml lectin as preliminary investigations had shown. The results mentioned above indicate that determination of lectin receptors on bacteria enables to detect subtle changes in carbohydrate composition of surface components inaccessible with biochemical or serological methods until now. Furthermore they show that genetic changes leading to chromosomally mediated resistance induce changes in the carbohydrate pattern of outer surface components of the bacteria. The next step would be to localize the different carbohydrate residues morphologically. Their significance for phagocytosis by human PMN is still under investigation [9].

Summary

Isogenic *Pseudomonas aeruginosa* strains isolated from clinically proven infections which only differ in one criterion production of different amounts of chromosomally mediated β-lactamase were investigated in order to find out whether this change in genetic composition results in a change of bacterial surface components, e.g. lectin specific sugars as well. To verify this assumption lectin-receptors on bacteria were determined quantitatively by measuring the amount of rhodamin conjugated lectins bound to the surface. The results indicate that changes in the chromosomally coded β-lactamase production effect changes in the distribution of surface carbohydrates as well.

Acknowledgement: This work was partially supported by the DFG (MA 864/1–2). The authors thank B. Graff for her skiful technical assistance.

References

1. Connelly MC, Stein DC, Young FE, Morse SE, Allan PZ (1981) Interaction with lectins and differential wheat germ agglutinin binding of pyocin 103 sensitive—and resistant *Neisseria gonorrhoeae*. J Bacteriol 48: 796–803
2. Doyle R, Keller K (1984) Lectins in diagnostic microbiology. Eur J Clin Microbiol 3:4–9

3. Firon N, Ofek I, Sharon N (1983) Carbohydrate specifity of the surface lectins of *Escherichia coli, Klebsiella pneumoniae*, and *Salmonella typhimurium*. Carbohydrate Research 120:235–250
4. Goldstein IJ, Hayes CE (1978) The lectins: carbohydrate-binding proteins of plants and animals. In: Tipson RS, Horton D (ed) Advances in Carbohydrate Chemistry and Biochemistry, Academic Press, New York–San Francisco–London, p 127
5. Hart DA (1980) Lectins in biological systems: applications to microbiology. Am J Clin Nutr 33:2416–2421
6. Matthew M, Harris AM, Marshall MJ, Ross GW (1975) The use of analytical isoelectric focusing for detection and identification of β-lactamases. J Gen Microbiol 88:169–178
7. Scatchard G (1949) The attraction of proteins for small molecules and ions. Ann NY Acad Sci 51:660–672
8. Sharon N, Lis H (1972) Lectins: cell agglutinating and sugar—specific proteins. Science 177:949–959
9. Stübner G (1984) Significance of lectins and lectin-receptors of bacteria and PMN for phagocytosis and intracellular killing. 18th Congress of the International Society of Blood Transfusion, Munich, July 22–27, 1984, Abstract S 37—0 3
10. Uhlenbruck G, van Mil A, Dreesbach K, Kich O (1982) Makrophagen und Phagocytose—Funtionsteste. Immun Infekt 10:122–129
11. Uhlenbruck G, Groß R, Koch OM, Chun-Kyung L (1983) Die Bedeutung von Lektinen für den Adhäsionsmechanismus von Bakterien. Deutsches Ärzteblatt 80:27–32

Index

GPSR Compliance
The European Union's (EU) General Product Safety Regulation (GPSR) is a set of rules that requires consumer products to be safe and our obligations to ensure this.

If you have any concerns about our products, you can contact us on

ProductSafety@springernature.com

In case Publisher is established outside the EU, the EU authorized representative is:

Springer Nature Customer Service Center GmbH
Europaplatz 3
69115 Heidelberg, Germany

www.ingramcontent.com/pod-product-compliance
Lightning Source LLC
Chambersburg PA
CBHW061200240926
51846CB00006B/1074
* 9 7 8 3 6 4 2 7 0 7 5 0 6 *